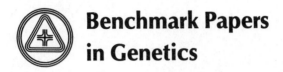

Benchmark Papers
in Genetics

Series Editor: David L. Jameson
University of Houston

Volume

RELATED TITLES IN OTHER BENCHMARK SERIES

**Benchmark Papers
in Genetics / 12**

A BENCHMARK® Books Series

POLYPLOIDY

Edited by

R. C. JACKSON
Texas Tech University

and

DONALD P. HAUBER
Texas Tech University

Hutchinson Ross Publishing Company

Stroudsburg, Pennsylvania

LIBRARY OF CONGRESS CATALOGING IN PUBLICATION DATA
Main entry under title:
Polyploidy.
   (Benchmark papers in genetics; 12)
   Includes index.
   1. Polyploidy—Addresses, essays, lectures.   2. Plant
genetics—Addresses, essays, lectures.   3. Plants,
Cultivated—Addresses, essays, lectures.   I. Jackson,
R. C.   II. Hauber, Donald P.   III. Series.
QH461.P64   1983   581.1'5   82-21388
ISBN 0-87933-088-0

Distributed worldwide by Van Nostrand Reinhold Company Inc.,
135 W. 50th Street, New York, NY 10020

CONTENTS

PART I: THEORETICAL FOUNDATIONS

PART II: SYNTHETIC AND NATURAL POLYPLOIDS

Contents

PART VI: CYTOGENETICS AND CONTROL OF CHROMOSOME PAIRING

PART VII: QUANTITATIVE METHODS FOR PREDICTING AUTOPLOID MEIOTIC CONFIGURATIONS

SERIES EDITOR'S FOREWORD

The study of any discipline assumes the mastery of the literature of the subject. In many branches of science, even one as new as genetics, the expansion of knowledge has been so rapid that there is little hope of learning of the development of all phases of the subject. The student has difficulty mastering the textbook, the young scholar must tend to the literature near his own research, the young instructor barely finds time to expand his horizons to meet his class-preparation requirements, the monographer copes with wider literature but usually from a specialized viewpoint, and the textbook author is forced to cover much the same material as previous and competing texts to respond to the user's needs and abilities.

Few publishers have the dedication to scholarship to serve primarily the limited market of advanced studies. The opportunity to assist professionals at all stages of their careers has been recognized by Hutchinson Ross and by a distinguished group of editors knowledgeable in specific portions of the genetic literature. These editors have selected papers and portions of papers that demonstrate both the development of knowledge and the atmosphere in which that knowledge was developed. There is no substitute for reading great papers. Here you can learn how questions are asked, how they are approached, and how difficult and essential it is to obtain definitive answers and clear writing.

R. C. Jackson and D. P. Hauber have selected papers from an important and complex literature that has a growing importance as genetics turn more and more to molecular approaches. The volume will provide an important resource to those outside the field as well as a significant synthesis to those working with polyploids.

DAVID L. JAMESON

PREFACE

The recognition and study of polyploidy began soon after it was clear that species usually had a constant chromosome number. Many of the early papers were concerned primarily with reporting only chromosome numbers, and this practice persists today, particularly among less-sophisticated practioners of cytotaxonomy. However, even among the earlier contributions there were papers that attempted theoretical and experimental approaches in the study of polyploidy. Two such early works stand out: the general treatise on chromosomes and the theory of arithmetical increase proposed by Winge (Paper 1), and the direct application of these ideas in genome analyses of a variety of grain crops by Kihara (Paper 2), his associates, and students soon thereafter. We have no idea of the total number of articles involving some aspects of polyploidy because they are distributed widely among journals in zoology, botany, genetics, and other more local sources for reporting such information. However, in preparation for a review several years ago, Jackson (Evolution and Systematic Significance of Polyploidy, *Annu. Rev. Ecol. Syst.* **7:**209-234, 1976) perused over 800 articles. Unfortunately, the majority of these did little or nothing to advance our knowledge of theory and mechanisms involved in the origin and maintenance of polyploidy. This is not to say that most of the papers were without merit; many of them were concerned with problems and relationships within particular taxonomic groups.

We have tried to select articles for this volume that have been instrumental in the synthesis and production of theoretical knowledge of polyploidy. Such articles can be considered useful and valuable without regard to taxonomic groups. While several of the selected papers were not the first to present a particular idea or data set, in our opinion they expressed new ideas and theory more clearly than some others did. Inevitably, a selection of articles for a book such as this shows some personal bias, but we believe the included papers fairly reflect the development of the theory and methods most useful for the analysis of polyploidy.

We have included twenty-nine papers in our selection. Twenty-three of the articles are reproduced *in toto;* the remaining five are portions of much longer works. We believe that including the entire article when possible is very important because it gives the reader a broader understanding of peripheral issues as well as the central issue of the article and a better insight into the author's thinking. The Editors' Comments are as brief as possible because most of the papers make the desired points clearly and concisely. In the Epilogue we have tried to bring the reader up to date on current concepts as we see them and to indicate the direction that future research should take.

Of necessity most of the papers deal with plant systems, for two reasons. First, polyploidy is not at all well represented among animal phyla, and it is found in sexually reproducing groups only where there is lack of differentiated sex-chromosomes. In other groups (both plants and animals) it is present in parthenogenetic forms, but we have not included this situation or apomixes among our plant papers even though it is well known. Second, quantitative data on meiotic configurations in reported animal species is almost totally lacking, so the polyploids could not even be classified according to the methods employed in the past. Almost half the articles have used a cultivated plant as the object of study. These articles were chosen because of merit, their quantification of data in most cases, and their economic importance. Generally, a great deal of necessary genetic information is available for many economically important plant crops.

The included papers encompass about sixty-six years of work and represent the contributions of scientists from eight different countries. We regret that space limitations preclude the selection here of many more of the very fine contributions produced internationally.

We would like to acknowledge the helpful comments by our visiting British colleague, Brian G. Murray, on the initial selection of articles to be included.

R. C. JACKSON
DONALD P. HAUBER

CONTENTS BY AUTHOR

POLYPLOIDY

INTRODUCTION

Cytological studies of both plant and animal chromosomes in the late 1800s demonstrated conclusively the continuity of chromosome number from generation to generation, and it was understood that this continuity of number was made possible by the process of meiosis, which reduced the somatic chromosome number by half in the gametes. The stage was thus set for recognizing deviations from normality in any part of the life cycle.

In the early 1900s chromosome numbers were increased in various ways by several experimentalists. A number of methods were used, such as wounding in various plants, temperature shock, mechanically treating fertilized animal eggs, and employing various concentrations of such chemicals as potassium chloride, chloral hydrate, and quinine sulphate. The most productive and detailed studies were those involving various plant groups.

Experimental tetraploidy, or higher ploidy levels, was induced by wounding the sporophyte of various mosses. The wound response was the production of protonemal growths (gametophyte phenotype) from the wounded regions. Although the gametophytes were heteroploid diploids, the production of gametes by mitosis was normal, and the resulting sporophytes were tetraploid. Meiosis in sporophytes yielded diploid spores that again gave diploid gametophytes, so that relatively constant tetraploid races could be maintained.

Winkler (1916), who coined the term *polyploidy,* utilized a wounding method to produce polyploids in tomato and other solanaceous species. The technique involved grafting *Solanum nigrum* and tomato interchangeably. After the graft had taken the stem was cut across the grafted surface. New shoots that grew from this region were sometimes periclinal polyploid chimeras containing a mixture of cells from the two species. Later cuttings from such tissue produced tetraploid tomatoes.

Throughout this period of cytological investigation, there were numerous measurements of somatic cells and spore sizes. In many cases an approximate doubling of cell size accompanied a change from diploidy to tetraploidy, but this relationship was not absolute, and variations in greater and lesser proportions were found. Later work in the 1930s and 1940s showed that much of the effect of polyploidy per se on cell size depended on the genotypes that were doubled. Polyploidy was found to decrease growth rate in many plants and to cause an increase in leaf thickness.

Keeping pace with the experimentalists, descriptive cytologists began to determine chromosome numbers in many varied and distantly related taxonomic groups of plants and animals. It was soon apparent that significant differences existed in the frequency of "didiploidy" or tetraploidy. While examples of polyploidy in animals derived from a small number of invertebrates, illustrations from the plant kingdom ran the gamut from the lower to the higher taxonomic groups.

Winge (Paper 1) published the first theoretical and analytical treatment of polyploidy and suggested explanations for arithmetical increases in chromosome number. Experimental verification of Winge's hypothesis on polyploidy was provided by the production of synthetic polyploids in tobacco by Clausen and Goodspeed (Paper 4). They called attention to the case of *Primula kewensis* described by Digby (1912), who apparently did not recognize its evolutionary significance. Thus, the implication of species hybridization followed by chromosome doubling was understood and appreciated during the 1920s, and some species of *Chrysanthemum*, *Nicotiana*, *Rosa*, and *Triticum* were considered good examples.

During the 1920s and 1930s much experimental work was carried out with both natural and cultivated plant species. Several reviews were published, and attempts were made to classify polyploids, largely on the basis of chromosome pairing.

As a result of many careful and detailed studies of natural and synthetic polyploids and related diploids, Clausen and colleagues (Papers 5 and 9) proposed a classification based on what they referred to as biosystematic principles. They were particularly interested in understanding the factors involved in successful polyploidy. Their work included an extensive and critical review of the literature and was followed two years later by the genetically more sophisticated analysis of Stebbins (Paper 10), which has since received wide acceptance. Two more recent reviews of polyploidy were concerned primarily with mechanisms of origin (Harlan and deWet, 1975) and evolutionary and systematic significance with emphasis on cytogenetics (Jackson, 1976). Finally, the publication of *Polyploidy: Biological Relevance* (Lewis, ed., 1980) was instrumental in bringing together more recent concepts.

Beginning with a series of papers in 1957 to 1958, the genetic control of chromosome pairing in cultivated wheats and their relatives has been studied intensively. This subject was reviewed in detail by Sears (1976). Discovery of pairing control genes in wheat represented a major breakthrough toward understanding alloploid-like behavior in polyploids.

Perhaps the most analytical and exciting work on chromosome pairing in polyploids is just emerging. Methods and models that allow quantitative predictions of various meiotic configuration frequencies were proposed for alloploids by Driscoll and coworkers (1979). Models and methods for autoploids were devised by Jackson and Casey (1980; Paper 28), and correction coefficients for these binomial methods were given by Jackson and Hauber (Paper 29). Kimber and his colleagues recently published a series of papers that describe methods of quantifying genome relationships (Alonso and Kimber, 1981; Kimber and Alonso, 1981; Kimber et al., 1981).

Biochemical analyses of polyploids have demonstrated that polyploidy per se may cause modifications in gene expression in artificial polyploids so that a dosage compensation-like effect may be operating to adjust certain enzyme levels (Albuzio et al., 1978; Dunbier et al., 1975; Nakai, 1977; Tal, 1977). Moreover, novel products apparently result from the inactivation of some loci or as a result of gene interactions. In several polyploid animals studied, the claim has been made that various numbers of gene sites have been inactivated, presumably as a result of mutation (Li, 1980). However, experimental data that distinguish inactivation by mutation from a dosage compensation-like effect are lacking.

It is apparent that polyploidy is of little importance as a significant evolutionary force in the majority of animal taxa. This is not to say that it may not have been important in the remote past in some phyletic lines, as suggested by Ohno (1970). However, successful polyploidy is unknown in higher vertebrates, and it is a lethal condition in humans (Niebuhr, 1974). The primary failure of polyploids to survive in higher vertebrates appears to be the result of genetic imbalance, which is reflected in development and physiology. Apparently, most deviations from the normal balance of diploidy will have an untoward effect. This effect is easily seen when trisomics are present in humans, and duplications or deficiences for even small chromosome arms have deleterious and debilitating effects, especially when such aberrations occur among autosomes. The x-chromosome inactivation in females offers a measure of protection when multiples are present, but males with extra x-chromosomes are sterile.

Where polyploidy is found in less-specialized animal groups, it is usually associated with parthenogenesis, hermaphroditism, gynogenesis, or what Schultz (1969) has termed hybridogenesis.

3

Polyploidy may, nonetheless, occur in sexually reproducing less-specialized vertebrates in which well-developed sex chromosomes are unknown. Bogart (1980) has indicated that there are at least five families of Anura with one or more species that are polyploid or have diploid and polyploid populations. In most cases, the reports are based on somatic or meiotic chromosome counts. The published meiotic studies usually give insufficient quantitative data. It appears, however, that some of the tetraploids are true autoploids. Our unpublished analysis of *Odontophrynus americanus* shows the tetraploid race is an autoploid, if the authors' statements on chiasma frequencies are reasonably accurate.

The role of polyploidy in the evolution of fish has been thoroughly reviewed by Shultz (1980), who compiled data suggesting polyploidy for at least six families. However, alternative explanations exist for some of the literature reports, which may consist only of chromosome counts and alpha karyotypes in the sense of White (1978). Polyploidy undoubtedly occurs in some groups, but many of the examples represent gynogenetic or hybridogenetic types. As with the data on amphibians, the necessary cytogenetic analyses are incomplete or absent, and estimates are based on cell size, DNA amounts, duplicate or multiple gene loci, or chromosome numbers, all of which can be manipulated in the absence of polyploidy. Nevertheless, the bulk of the data indicate natural polyploidy is present in some sexually reproducing taxa, and there is no doubt of its occurrence in some gynogenetic and hybridogenetic species.

Predicting the future is always precarious. However, more papers that give quantified data on chiasma frequencies and meiotic configurations will undoubtedly be published. A greater number of gene loci will be analyzed by electrophoretic methods so that it will be possible to determine to some extent the ancestry of certain polyploid species that arose from parents divergent enough to have suitable genetic markers. Searches will be made for pairing control genes in natural and cultivated plant populations because of their intrinsic interest and their economic importance in producing stable polyploid food plants. There will be an increasing number of autoploids produced from natural species to test their pairing and biochemical relationships with natural polyploids. We believe such studies will demonstrate that autopolyploidy can be successful in both plant and animal evolution in some taxonomic groups.

Certainly a major unsolved problem in polyploid genetics concerns the extent, rate, and nature of diploidization. Li (1980) has proposed methods of estimating the rate, but he has ignored the more fundamental question of whether "silenced" genes in polyploids are

4

the results of mutations. We suggest that some of the so-called silenced genes may be simply turned off in certain loci by a dosage compensation-like mechanism. This problem is deserving of careful and detailed studies in synthetic polyploids, preferably those derived vegetatively from progenitor diploids so as to allow an exact comparison of genotypes. Already there are many reports in the literature of diploidization in polyploids, but in such examples the alternative proposed here generally has not been considered.

REFERENCES

Albuzio, A., P. Spettoli, and G. Cacco, 1978, Changes in Gene Expression from Diploid to Autotetraploid Status of *Lycopersicon esculentum, Plant Physiol.* **44:**77–80.

Alonso, L. C., and G. Kimber, 1981, The Analysis of Meiosis in Hybrids. II. Triploid Hybrids, *Can. J. Genet. Cytol.* **23:**221–234.

Bogart, J. P., 1980, Evolutionary Significance of Polyploidy in Amphibians and Reptiles, in *Polyploidy: Biological Relevance,* W. Lewis, ed., Plenum Press, New York, pp. 341–378.

Digby, L., 1912, The Cytology of *Primula kewensis* and of Other Related *Primula* Hybrids, *Ann. Bot.* **36:**357–388.

Driscoll, C. J., L. M. Bielig, and N. L. Darvey, 1979, An Analysis of Frequencies of Chromosome Configurations in Wheat and Wheat Hybrids, *Genetics* **91:** 755- 767.

Dunbier, M. W., D. L. Eskew, E. T. Bingham, and L. E. Schrader, 1975, Performance of Genetically Comparable Diploid and Tetraploid Alfalfa: Agronomic and Physiological Parameters, *Crop Sci.* **15:**211–214.

Harlan, J. R., and J. M. J. deWet, 1975, On Ö. Winge and a Prayer: The Origins of Polyploidy, *Bot. Rev.* **41:**361–390.

Jackson, R. C., 1976, Evolution and Systematic Significance of Polyploidy, *Annu. Rev. Ecol. Syst.* **7:**209–234.

Jackson, R. C., and J. Casey, 1980, Cytogenetics of Polyploids, in *Polyploidy: Biological Relevance,* W. Lewis, ed., Plenum Press, New York, pp. 17–44.

Kimber, G., and L. C. Alonso, 1981, The Analysis of Meiosis in Hybrids. III. Tetraploid Hybrids, *Can. J. Genet. Cytol.* **23:**235–254.

Kimber, G., L. C. Alonso, and P. J. Sallee, 1981, The Analysis of Meiosis in Hybrids. I. Aneuploid Hybrids, *Can. J. Genet. Cytol.* **23:**209–219.

Lewis, W., ed., 1980, *Polyploidy: Biological Relevance,* Plenum Press, New York, 583p.

Li, W. H., 1980, Rate of Gene Silencing at Duplicate Loci: A Theoretical Study and Interpretation of Data from Tetraploid Fishes, *Genetics* **95:**237–258.

Nakai, Y., 1977, Variation of Esterases and Some Soluble Proteins in Diploids and Their Induced Autotetraploids in Plants, *Jpn. J. Genet.* **52:**171–181.

Niebuhr, E., 1974, Triploidy in Man. Cytogenetical and Clinical Aspects, *Humangenetik* **21:**103–125.

Ohno, S., 1970, *Evolution by Gene Duplication,* Springer-Verlag, New York, 160p.

Schultz, R. J., 1969, Hybridization, Unisexuality, and Polyploidy in the Teleost *Poeciliopsis* (Poeciliidae) and Other Vertebrates, *Am. Nat.* **108:**605–619.

Schultz, R. J., 1980, Role of Polyploidy in the Evolution of Fishes, in *Polyploidy: Biological Relevance,* W. Lewis, ed., Plenum Press, New York, pp. 313–378.

Sears, E. R., 1976, Genetic Control of Chromosome Pairing in Wheat, *Annu. Rev. Genet.* **10:**31–51.

Tal, M., 1977, Physiology of Polyploid Plants: DNA, RNA, Protein and Abscissic Acid in Autotetraploid Tomato Under Low and High Salinity, *Bot. Gaz.* **138:**119–122.

White, M. J. D., 1978, *Modes of Speciation,* Freeman, San Francisco, 455p.

Winkler, H., 1916, Uber die experimentelle Erzeugung von Pflanzen mit abweichenden Chromosomen-Zahlen, *Z. Bot.* **8:**417.

Part I

THEORETICAL FOUNDATIONS

Editors' Comments
on Papers 1, 2, and 3

1   **WINGE**
Excerpts from *The Chromosomes. Their Numbers and General Importance.*

2   **KIHARA**
Excerpts from *Cytological and Genetical Studies of Important Cereal Species with Special Consideration of the Behavior of the Chromosomes and Sterility in the Hybrids*

3   **HALDANE**
Theoretical Genetics of Autopolyploids

Paper 1 by Winge is part of a much longer treatise on chromosomes and cytology in general. Winge attempted to review the literature to that time and was probably the first to look at possible basic chromosome numbers in many taxa in a statistical fashion. He suggested that polyploids arose when the chromosomes paired incompletely or not at all in somatic tissues of hybrids between races or species, so that each parental set of chromosomes doubled. The chromosomes in each genome would then have a complete homologue with which to pair. Although somatic association of homologous chromosomes and somatic pairing are known for many different organisms, there is no evidence in the literature today indicating that lack of somatic pairing or association leads to chromosome doubling. Winge correctly suggested that many natural species may have arisen by polyploidy via hybridization between different species. This mechanism was clearly stated and diagrammed by using letters to represent genomes. Another important contribution of this work is the explanation of arithmetical increases in chromosome number.

Winge reiterated these ideas on chromosome doubling in an introduction to a later paper (Winge, 1925). He provided a further explanation of chromosome doubling in another article (Winge, 1932), in which he stated that in 1917 he had believed that the doubling of the chromosome number occurred in the zygote, but he now acknowledged that doubling could occur also by the production of unreduced gametes.

Paper 2 was translated and abstracted by Phillips and Burnham from a much longer paper by Kihara. (See also *Cytogenetics*, R. L. Phillips and C. R. Burnham, eds., Benchmark Papers in Genetics, vol. 6, Dowden, Hutchinson & Ross, Stroudsburg, Pa., 1977.) This paper is often cited as the first treatment of genome evolution as deduced from polyploid analyses, but the article by Winge (Paper 1) appeared earlier. Kihara presented very little that was theoretically new in this paper and even acknowledged and reproduced Winge's diagram of methods of polyploid analysis and origin, changing the scheme only slightly for the cereal species he was studying. He continued Winge's use of alphabetical letters to denote genomes and certainly popularized the idea of genome evolution. Kihara deserves the title "Father of Genome Analysis" given him by some workers. He did much of the early classic work on wheat cytogenetics, and later his work was used as a basis for the expanded and more sophisticated studies carried out by Sears and Riley and associates in the 1950s and 1960s.

Although there were earlier applied approaches to the study of polyploid genetics, Haldane's article (Paper 3) was the first purely theoretical mathematical analysis of chromatid segregation. An earlier analysis by Muller (1914) was based on chromosomal units of segregation, but it was later found to be inadequate for experimental data obtained in *Datura* by Blakeslee and colleagues (1923). Other segregation models were produced by Mather (1935, 1936), Sansome and Philip (1939), Fisher and Mather (1943), and Little (1945, 1958). More recent studies have examined the physical and theoretical aspects and mechanisms of converting autoploids to alloploids (Doyle, 1973, 1979*a*, 1979*b*; Jackson, 1982).

REFERENCES

Blakeslee, A. F., J. Belling, and M. E. Farnham, 1923, Inheritance in Tetraploid Daturas, *Bot. Gaz.* **76:**319-373.

Doyle, G. G., 1973, Autotetraploid Gene Segregation, *Theor. Appl. Genet.* **43:**139-146.

Doyle, G. G., 1979*a*, The Allotetraploidization of Maize. Part 1: The Physical Basis—Preferential Pairing, *Theor. Appl. Genet.* **54:**103-112.

Doyle, G. G., 1979*b*, The Allotetraploidization of Maize. Part 2: The Theoretical Basis—Cytogenetics of Segmental Allotetraploids, *Theor. Appl. Genet.* **54:**161-168.

Fisher, R. A., and K. Mather, 1943, The Inheritance of Style Length in *Lythrum Salicaria*, *Ann. Eugenics* **12:**1-23.

Jackson, R. C., 1982, Polyploidy and Diploidy: New Perspectives on Chromosome Pairing and Its Evolutionary Implications, *Am. J. Bot.* **69:**1512-1523.

Little, T. M., 1945, Gene Segregation in Autotetraploids, *Bot. Rev.* **11:**60-85.

Little, T. M., 1958, Gene Segregation in Autotetraploids. II, *Bot. Rev.* **24:**318–339.

Mather, K., 1935, Reductional and Equational Separation of Chromosomes in Bivalents and Multivalents, *J. Genet.* **30:**53–78.

Mather, K., 1936, Segregation and Linkage in Autotetraploids, *J. Genet.* **32:**287–314.

Muller, H. J., 1914, A New Mode of Segregation in Gregory's Tetraploid Primulas, *Am. Nat.* **48:**508–512.

Sansome, F. W., and J. Philip, 1939, *Recent Advances in Plant Genetics,* 2nd ed., Churchill, London, 412p.

Winge, Ö., 1925, Contributions to the Knowledge of Chromosome Numbers in Plants, *Cellule* **35:**305–324.

Winge, Ö., 1932, On the Origin of Constant Species-Hybrids, *Sven. Bot. Tidskr.* **26:**107–122.

1

Reprinted from pages 131–134 and 192–206 of *Carlsberg Lab., Copenhagen C. R. Trav.*
13:131–206 (1917)

THE CHROMOSOMES.

THEIR NUMBERS AND GENERAL IMPORTANCE.

BY

Ö. WINGE.

> "L'un des faits les plus importants, mis
> en évidence chez les plantes, consiste dans
> la fixité du nombre des éléments chro-
> matiques des noyaux sexuels."
> *L. Guignard,* 1891.

THE greatest morphological resemblance between plants and animals is that found within the walls of the cell, and more especially in the qualities of the nucleus. In the cell — and here alone — do we encounter elements which are in the strictest sense of the word homologous for the two great groups of living matter. And the occurrence of the chromosomes: the small colourable bodies which continually make their appearance during the process of nuclear division, is in a remarkable degree common to most plants and animals.

An element of so ubiquitous a character, that it may almost be said to be characteristic of living matter generally, must naturally be presumed to be of fundamental importance for the existence of the living organism. And we find also, that the chromosomes have gradually come to be regarded as of quite extraordinary importance, the more so since the technical improvements in microscopy have rendered it possible to penetrate further and further into their nature.

Students of heredity also, have shown considerable interest in these bodies, the peculiar behaviour of which during reduction division, fertilization, etc. — in a word, during sexual propagation — offers them, as it were, tangible evidence as to the correctness of the principles experimentally arrived at. On the other hand, however, it must be noted that certain authorities on heredity themselves oppose the tendency to allow the chromosomes a too direct importance in this respect: i. e. as the seat of genotypic disposition.

The fact that each species of plant or animal has as a rule a certain and constant number of chromosomes in the nuclei, more

particularly those of the sexual cells. has, from the very commencement of chromosome research in the 70's, incited cytologists
to ascertain, if possible, the precise number of chromosomes for
each separate species of plant and animal. The minuteness of
the chromosomes themselves, their variability. and frequently in-
convenient position, render the work far from easy, but in the
course of years one investigator after another has continually
furnished contributions, with the result that we know now the
numerical value for the chromosomes in several hundred species
of both kingdoms.

As long as the nature of the chromosomes is still comparatively little understood, comprehensive investigations as to their
character and occurrence generally will necessarily be a matter
of difficulty; one line of approach, however, yet seems to be
clear: to wit, by the study of this very feature: the number in
which they are found in the various organisms. In epitomizing
what is known as to chromosomes generally, it would be natural
to commence with the simplest factor, i. e. their numerical value.
If we can then succeed in discovering any laws or principles
governing the numbers characteristic of each organism, such
knowledge will in itself afford a better foundation for further study.

In seeking to form some preliminary idea as to what features
in plants might be imagined to determine the numerical value
of the chromosomes, an embarrassing number of possibilities
suggest themselves. The value in question might for instance
be supposed to stand in close relation to the phylogeny of the
particular plant — or to the volume of the cells, the size of the
plant itself, or possibly to the numerical proportions in the external
organs, or physiological conditions, or the number of genes etc.
There is thus no lack of plausible hypotheses; only by critical
investigation of the numerical values actually found for the chromosomes can we determine which theory is — or which theories
are — correct, or perhaps, per exclusionem, which are not.

When I commenced this work, in 1914, my intention was
to collect, as far as possible, from the literature already extant,
all the known values for number of chromosomes in the vegetable
kingdom, in order thus to form some view as to the importance
of their numerical value. No such work had up to that time
appeared, and there seemed good reason to believe that one
might, by statistical methods, arrive at results which the invest-

igation of single plants had not sufficed to afford. Here, however, apart from the ordinary difficulties inseparable from such compilatory work, such as that of procuring, and reading through within a reasonable time, the necessary publications, I encountered further hindrance in the fact that the various authorities consulted frequently disagreed. There are numerous plants whose number of chromosomes is still not known with certainty, despite considerable research, the different results arrived at by several investigators in each case leaving us in doubt as to which of them is correct. Statistics of this nature would soon become worthless if incorrect statements were allowed to creep in, and even though the bulk of them were indubitably correct, it would obviously be impossible to rely completely upon conclusions drawn from such material. I have therefore restricted myself to the procuring of information as to the numerical value for chromosomes in a considerable number of plants, and have more particularly endeavoured, by personal investigation, to elucidate the importance of the chromosomes, and of their numerical value, especially within a certain class of related forms. A further reason for thus departing from my original plan was the fact that I had already, from the statistical results obtained through my study of the literature, secured confirmation of the most important points which I had expected to arrive at by a thorough statistical investigation.

As is frequently the case when one is occupied with a task of any general interest, a kindred work appeared while I was thus engaged, to wit, Tischler's excellent treatise entitled "Chromosomenzahl, -Form und -Individualität im Pflanzenreiche" (Progr. Rei Botanicae, Vol. 5, 1915) which deals with the question of chromosome numbers in plants more thoroughly than has hitherto been done. Tischler's work, is, according to its purpose, more particularly a survey of our present knowledge on the subject, and the writer does not therefore go so closely into the theory of chromosomes and their numerical value as is necessary for a thorough comprehension of the question. My present work, therefore, which is especially directed towards the theoretical aspect of these chromosome numbers, will, I trust, hardly be superfluous.

The science of cytology has great tasks before it, and not least important is the function it has acquired as a most neces-

13

sary discipline for various branches of hereditary research. Nevertheless, the chromosomes and their importance, — central features, so to speak, in cytology itself — are still frequently disregarded in natural history research, though this is mostly found to be the case among non-cytologists.

It is still, of course, a question, how great importance we should attach to these peculiar elements; yet I venture to assert that if the chromosomes and their behaviour were easier to observe, and thus more generally known, then those who believed in their greater significance — which need not here be more precisely defined — would form a large majority.

The present cytological investigations were for the most part carried out in the Physiological Department of the Carlsberg Laboratory, to the Head of which department, Dr. Johs. Schmidt, I am indebted for the facilities accorded me in carrying out the work. I also wish to thank my very respected tutor, Prof., Dr. Eug. Warming, who has throughout taken an interest in my researches, and Dr. Ove Paulsen, Amanuensis at the Botanical Gardens in Copenhagen, who kindly undertook the fixation of some plant material at a time when I was prevented by absence from doing so myself. Finally, I beg to express my thanks to the Trustees of the Carlsberg Fund, who in 1911 furnished me with funds for studies at the universities of Stockholm, Paris and Chicago, and i 1914-1915 with the means for carrying out the present work.

[*Editors' Note:* Material has been omitted at this point.]

CHAPTER 4.

THEORETICAL STUDIES ON THE ORIGIN
OF THE SYSTEM OF CHROMOSOME NUMBERS.

With the empirical basis here given, it will now be natural to consider the chromosome numbers of plants from a theoretical point of view. Several investigators have devoted their attention to this question, and most of the theories previously advanced have been more or less intimately connected with the idea of "reduplication of chromosomes" — which as a matter of fact is hardly surprising, since in all the cells we find the chromosomes splitting up, and doubling their number. It has, moreover, long since been demonstrated that apogamous species have as a rule a chromosome number twice that of the normally sexed in the same genus, and in consequence, we find it frequently asserted that reduplication of the chromosomes often involves loss of the power of sexual propagation.

The considerations advanced in connection with these points are, as it seems to me, not altogether satisfactory, and I therefore propose in the following to go further into the question, the correct solution of which is of great importance for the whole study of chromosomes generally.

Strasburger's contributions in particular have served to popularize the theory that the higher chromosome numbers in certain "tetraploid" species should be due to reduplication of previously existing lower values. Other writers have, however, also supported this view. There is a considerable mass of literature dealing with the subject of chromosome reduplication, either in the form of works specially devoted to this particular point, or more incidentally touching upon the problem (Strasburger 1907, 1910, Rosenberg 1909, Tischler 1910, Němec 1910, Kuwada 1911, Ishikawa 1911, Bally 1912, Pace 1913, Gates 1914, a, b, and others).

In many genera such tetraploid species are known; in some few again, we find a longer series, i. e. species with 2x, 4x, 6x, 8x, etc. The investigations of Tischler (1910) with various races of *Musa sapientum* gave us the first known examples of the fact that races of a plant species may exist with three different chromosome numbers, the "*Dole*" race having 8, whereas "*Radjah*

Siam" has 16 and *"Kladi"* 24. As mentioned, Tischler considers it likely that one or two extra chromosome divisions or
monaster respectively may have given rise to the increased number of chromosomes.

Here, I must confess, I am unable altogether to agree with
Tischler. A doubling of the chromosome number 8 certainly
gives $x = 16$, a new extra division, however, must necessarily
give $x = 32$. As var. *Kladi* has but 24 chromosomes, it would
then be necessary to assume that the chromosomes had redivided
only in one of the daughter nuclear plates, an explanation which
seems to me highly improbable.

An extra tripartition of the chromosomes in the original form
with 8 would give $x = 24$; such a theory, however, is likewise
altogether inacceptable, and in fact does not seem to have any
supporters. No one has ever observed a threefold division of
chromosomes, and the power of repeated bipartition is so characteristic for these bodies that it must be regarded as a fundamental and unalterable feature most intimately connected with
the entire nature of the chromosomes themselves.

In his latest work (1915) Tischler seems more inclined to
explain the origin of the mentioned *Musa* races by the theory
that dispermy may have taken place, — as observed by Němec
(1912) in *Gagea lutea*. The penetration of two male nuclei into
the egg cell, and fertilization of the same would necessarily, provided the germ were capable of developement, give rise to a
plant with $3/2$ the original number of chromosomes. Tischler
thus considers it possible to explain the origin of a race with
$x = 24$ by taking $x = 16$ as starting point. The chromosome
number 8, however, cannot be brought into line with the two
other figures by this method. In other words, it would be necessary to regard the higher chromosome numbers in *Musa* as arrived at by different means.

As regards the possibility of normal developement in poly-
spermatically fertilized organisms, it must undoubtedly be admitted that there is nothing, theoretically speaking, to prevent
this from taking place under certain circumstances, though most
of the experiments made in this direction, including those with
animal material, do not exactly support the theory. That plants
may exist with doubled number of chromosomes has been fully
demonstrated by the experimental researches of Marchal and

16

Marchal (1907, 1909), by which, moreover, it has been further shown that the power of sexual propagation is not lost, even though the number of chromosomes be doubled. — In numerous other plant genera we may, as already mentioned, also find normally sexed species with "double" chromosome number, e. g. among the species of *Chenopodium* and *Atriplex* investigated by the present writer, with 9 and 18 chromosomes respectively. Since therefore plants with "augmented" number of chromosomes may very well exist, and even be normally sexed, there is, as mentioned, nothing theoretically to prevent the same applying to individuals produced by polyspermatic fertilization. Dispermatic fertilization must, however, often prevent subsequent normal reduction division.

For all this, however, it must be said, that the idea of the chromosome number of a species arising from doubling or trebling that of another species is in itself unreasonable. How could it be possible for forms essentially new to be produced by the mere occurrence of chromosomes — and possibly therewith genes — in twice the normal number? It might no doubt be imagined, that a reduplication of chromosomes could take place in some plant, and that this would result in the production of two races differing particularly in regard to size, and possibly incapable of interfertilization[1]). Such races might then be supposed to develope in slightly different directions, but it would seem more than doubtful whether this could be of any considerable phylogenetic importance. From the numerous investigations with *Oenothera*, and those of Marchal with *Amblystegium*, also on purely theoretical grounds, there seems no reason to suppose that the reduplication of chromosomes should in itself be of supreme importance.

How then, can these chromosome numbers have arisen? And how can it be, that a double chromosome number at times involves apogamy, at others not?

---

[1]) Since the completion of this work, I have received Winkler's highly interesting paper "Ueber die experimentelle Erzeugung von Pflanzen mit abweichenden Chromosomenzahlen" (Zeitschr. f. Bot., Jahrg. 8, 1916, p. 417-531). — Winkler has succeeded in raising tetraploid individuals of both *Solanum lycopersicum* and *S. nigrum* by grafting, and has also been able to show that tetraploidy is on the whole transmissible, though irregularities may also occur. The fact that the tetraploid individuals now and again locally revert to the diploid type is, Winkler considers, due to reduction division in the somatic cells. The tetraploidy makes itself apparent in the increased size of the plants, and other peculiarities, which also distinguish the tetraploid *Oenothera* forms.

Strasburger (1904) writes in his summary: "Die Annahme liegt nah, dass übermässige Mutation die Schwächung der geschlechtlichen Potenz der Eualchimillen veranlasste und durch den Ausfall der Befruchtung die Anregung zur apogame Fortpflanzung gab" . . . and further: "Auch Diöcie hat in manchen Fällen den Anstoss zur Ausbildung apogamer Fortpflanzung gegeben, weil durch Trennung männlicher und weiblicher Individuen Befruchtungsmangel sich einstellte".

That fertilization is rendered difficult owing to marked mutation or to the separation of the sexes may be reasonable enough, but that this should be the cause of apogamous propagation in plants I cannot admit, and must reject the idea as a teleological explanation. "Befruchtungsmangel" is in itself an entirely negative principle. Overton (1904) follows the same line, with regard to *Thalictrum purpurascens,* as Strasburger; with this writer also I am unable to agree.

Moreover, we know of not a few monoecists which are apogamous, and dioecious plants which are not; for this reason also the theory that dioecism should be the cause of apogamy appears hardly reasonable.

It has occurred to me that there is one likely explanation of the occurrence both of apogamy and of the multiplication of chromosome numbers, which deserves consideration, but which does not appear ever to have been discussed (see, however, Ostenfeld's statements (1910) as to apogamous *Hieracia*) — to wit, that occasional hybridization might be the cause.

It will here be necessary to go further into the peculiarities of the chromosomes themselves. The most characteristic feature in these is, as we have already noted, their tendency to occur in pairs. Chromosomes are frequently found paired in the somatic cells, and continually in the gonokonts, the parent chromosomes exhibiting a mutual affinity. They are nowhere found in threes; neither in the nuclei of the endosperm, where 3x is nevertheless found, nor in somatic, hyperchromatic cells. Fours are likewise unknown, save where the chromosomes are in repeated division, as in the gonokont nuclei. Chromosomes once separated unite in pairs.

This evidently reveals some fundamental principle in the nature of chromosomes, and must be connected with the dualism to which every sporophyte owes its origin.

18

Generally speaking, only nearly related plants are capable of interfertilization and the production of progeny able to live on; it is therefore certain that only such gametes as harmonize physiologically — or better, perhaps, physiogenetically — can enter into the formation of a duality such as the sporophytic organism.

The gametophytic cell can rightly be considered as one harmonic whole, and we may further presume that only physiogenetically uniform nuclei — and especially their chromosomes — are able to affect each other. It will indeed doubtless be only reasonable to suppose that the greater or lesser degree of physiogenetic similarity or harmony will determine what common result is to arise from the union of two gametes.

I will now draw up a scheme showing the different degrees of physiogenetic likeness between gametes — and thus also between their chromosomes — endeavouring at the same time to ascertain what result we can expect in each case from the fusion of such gametes, basing my conclusions upon our knowledge as to the chromosomes in general, and of their occurrence in pairs in particular.

I. **Philozygoty**[1]). Two gametophyte organisms very markedly resembling each other are evidently found on comparing for instance two gametes from the same species of plant. Such homogamous cells may unite, their nuclei also unite, and the chromosomes appear in pairs, corresponding to the number of chromosomes — the pairing, however, being often noticeable first in the sporogenous cells, when a new gametophyte generation is in preparation. By philozygoty I mean the relation distinguishing such units as tend, by virtue of genetic, physiological or other similarity, to form, through fertilization, a common normal offspring, the sporophyte. Philozygoty must thus be said normally to characterize gametes belonging to one and the same systematic species.

The fact that self fertilization or inbreeding in many cases gives bad offspring or fails altogether must, I consider, be due to lack of mutual complementation, as a result of the over close relationship — a point to which I shal refer later on.

We may, however, also find that gametes derived from different species or forms have a harmonizing or corresponding

---

[1]) From φιλο in compounds = loving, and ζῦγόν a yoke.

"inner physiology" i. e. possess the qualities requisite for a pairing between their chromosomes, and thus the formation of a harmonic common product, a zygote with unimpaired vitality. Heterogamous organism may thus also be philozygotic. As in the case of the homogamous organisms, so also here the parent chromosomes are united in pairs — at any rate in the gonotokont nuclei, after which they again separate, the two sets of parent chromosomes being distributed on the daughter cell during reduction division.

Philozygoty thus denotes the highest degree of harmony between the sexual cells of two organisms.

2. **Pathozygoty**[1]). By this I mean to indicate the fact that two gametes may enter into the formation of a common zygote, which, however, owing to the less marked harmony between the constitution of the gametes, may often be only partly capable of developement, or not capable of developement in the normal manner.

This category will only include gametes derived from different species or races, i. e. heterogamous gametes.

When such heterogamous gametes unite, one of the following alternatives must take place:

A. Direct union of the chromosomes. If the parent chromosomes in accordance with the nature of chromosomes in affinity pair directly, we must presume that this will as a rule give rise to a zygote which may either be apogamous or normally sexed.

a. Apogamous forms. On the fusion of heterogametes, a zygote will be formed having the sum of the chromosome numbers in the gametes. When the chromosome numbers of the parent are different, the pairing will naturally be incomplete, but where the chromosome numbers are alike, a complete union of homologous chromosomes will take place. In both cases then, we have an apogamous zygote, and as the related species in a genus usually have the same chromosome number, x, the apogamous zygote will in most cases have 2x chromosomes. Apo-

---

[1]) From παθεῖν in the sense of endure or submit to. The same word being frequently used in the sense of suffering pain or damage, a better expression might possibly be found (Dyszygoti strikes me as an awkward combination). The Latin derivative pati is also used in the special sense for instance "concubitus pati" (Ovid, Ars Amandi 3, 666) and like expressions.

gamous species with normal chromosome number (compared with the remaining species of the genus) are also as a matter of fact found in nature — as for instance in the *Marsilia Drummondii* investigated by Strasburger (1907) which, like the remaining species examined, has 2x = 32. From the synapsis stage, where the separation of the parent chromosomes is prepared, irregularities occur, so that reduction cannot take place, and the macrospore is formed by ordinary mitotic division. Some of the dioecious *Rumex* species examined by Roth (1907) are also believed to be partly apogamous, despite the fact that the chromosome number is not "doubled", and the species in question should thus agree with *Thalictrum purpurascens*, which, as Overton (1904) has shown, is capable both of normally sexed and apogamous reproduction. Should *Thalictrum purpurascens* really be produced by hybridization between two species, then Overton's results in this instance must probably be explained as due to a so complete union between the chromosomes that normal reduction may at times take place, at others not. The progeny of the *Thalictrum* in question should then be genotypically heterogeneous — a point which it would be most interesting to have investigated.

When difference in the chromosome numbers of the parent gametes renders complete pairing impossible, only a certain number of chromosomes uniting, the fertilization would probably also be imperfect, so that the propagating cells of the hybrid individual produced (somatic sexual products) would be rather of a gametic nature, and capable of further fertilization by a gamete with constitution able to give a more complete union of the chromosomes. There will thus be no question of apogamy — unless indeed the diploid sexual products of the individual were capable of independent developement despite the incomplete union of the chromosomes. This point will be further discussed later on.

b. Normally sexed forms. There is also the further possibility that the chromosomes meeting in the zygote may all, or some of them only, constitute a new chromosome set, when the zygote will thereafter represent something entirely new; a new species with a set of chromosomes rendered stable for the future, and normally sexed.

It seems to me as a matter of fact but natural to suppose, as I have also noticed in the foregoing, that the homologous or most closely harmonizing elements in the nuclei of the gametes

21

continually seek out each other in order to form harmonious unions, just as — to use a purely illustrative comparison — in chemistry, atoms of greatest mutual affinity constantly tend to unite.

B. Indirect chromosome union. Where a less marked harmony exists, we must then suppose that this will be visibly expressed by the fact that the chromosomes derived from the two gametes will not unite in pairs at all, but distribute themselves throughout the primary cell of the zygote, as if no dualistic relation of any kind existed. If the chromosomes are to find a partner, then each of the chromosomes in the zygote must divide, for thus indirectly to produce a union of chromosomes, and we must assume that this is realized in the hybrid zygotes which have any possibility at all of propagating — in accordance with what we know from experience as to the behaviour of pairs of chromosomes. The hybrid sporophyte thus produced will then have 4x chromosomes, taking the number for each of the parent gametes as x.

After this, either the chromosome pairs will have the power of further separating by reduction division, transmitting one set of the chromosomes from either parent to each of the gametes — in which case we have a new hybrid organism with the qualities of a pure species and "double" chromosome number; i. e. containing the sum of the chromosome numbers in the parent species. Or, if the power of reduction has been lost, but the power of continued existence otherwise retained, the result will be an apogamous species with 4x chromosomes.

Instances of species thus arising are, I believe fairly numerous in nature; we may doubtless assume that most of the species exhibiting "double" chromosome numbers are hybrids formed in this manner. It is perhaps not so easy to point out, in concrete instances, the parent species whence such "double" species are derived; the latter may even have altered since the time of their formation. Nevertheless, it is possible that we may succeed in doing so. By way of example, we might imagine *Atriplex patulum*, with its 18 chromosomes, as possibly derived from hybridization between *A. hastatum* and *A. littorale*, each of which has 9. The process of such formation would in such case be as schematically shown in Fig. 35.

With regard to the apogamous species, the chro-

22

mosome number here is, as we know, generally just twice that of the related normally sexed species, and as sexual sterility is itself characteristic of many hybrids, I consider it higly probable that apogamous species are derived from crossings between those normally sexed.

It is difficult to say whether the fact that the occurrence of apogamous species is relatively often associated with dioecism can also be explained by this means. On the one hand it may be urged that homozygotism is more easily maintained in

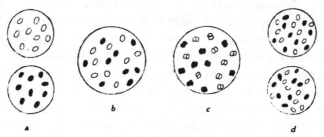

Fig. 35. Schematic view of occurrence of doubled chromosome numbers through hybridization. *a*, two heterogenous gametes each with 9 chromosomes. *b*, the hybrid zygote. *c*, indirect chromosome binding in the same. *d*, the hybrid gametes produced on reduction division, with 18 chromosomes.

monoecious than in dioecious species, where the possibility of absolute homozygoty is probably out of the question. The genotypic differences, and thus also the possibility of hybridization in species or varieties would naturally seem to be greatest in dioecious species. On the other hand it must be admitted that the dioecious species does demand a continual mingling of genotypes, whereas this is not necessarily the case in monoecists.

C. No chromosome union. Where the mutual harmony between two united gametes of heterogamous origin is so slight that their relation becomes almost disharmonious, the interaction between the two organisms will in all probability be of briefer duration and of a less intimate character. This might naturally result in the chromosomes of two gametes failing to unite either directly or indirectly. I believe I am justified in assuming that the chromosomes in sporophytes capable of sexual developement must continually act in pairs, and I must accordingly conclude that when neither direct nor indirect union of chromosomes takes place, then the organism is doomed, and can only exist as an

23

embryo — possibly with the power of cell division, as long as it is nourished by the mother tissues.

Experiments have convinced me that the pollination of a plant with pollen from another not of the same genus actually can result in developement of an embryo, and even in the formation of an apparently normal fruit, but that the embryo in such fruit is altogether incapable of independent existence. These experiments will be referred to later on.

3. **Misozygoty**[1]) Gametes of systematically widely differing organisms will as a rule altogether lack the power of uniting in a zygote. Not only is there too little similarity between the constitutions of such gametes; there is moreover no harmony between them, and in consequence, nothing calculated to produce the mutual affinity which is inseparable from all fertilization. The term misozygoty then, I use to designate this lack of mutual harmony: a reciprocal disinclination rendering fertilization impossible.

I must here repeat, that we cannot regard it as impossible for two distant organisms at times to be capable, by virtue of mutually harmonizing constitution, of uniting by cellular conjugation and forming something entirely new. We know of cases, where widely differing forms have united in symbiotic doubles. The symbiotic relation is marked by a more or less intimate union, in several instances of such a nature that the symbionts grow round or through each other, presenting the external appearance of one harmonious whole. It will suffice merely to mention the lichens and mycorrhiza as excellent examples of this. The latter moreover, further afford us an instance of one symbiont (the fungus) even growing intracellularly with the other (e. g. an orchid). I do not mean to suggest that the interval between this relation and a complete union should always be inconsiderable. It is often, in all probability, the very contrast which gives the symbiotic relation its value, each of the symbionts in question furnishing the other with the forms of nourishment which the latter is unable to obtain for itself. In my opinion, however, we can properly speaking find no just cause why a true cellular union between two organisms should not under certain circumstances take place, whereby a new organism could arise. Winkler's *Solanum* chimæras are of considerable interest, inasmuch as they

---

[1]) From μισο — in compounds indicating hate, aversion.

show that two organisms forming a common product can at times develope as a harmonious whole, despite the possession by each of the component of latent qualities which do not attain their developement. All epidermis formation, for instance, might be due to the one partner, albeit the other possessed a like power. This will serve to illustrate the conditions which we can imagine as existing in a zygote derived from two heterogamous gametes. By suitable plasmatic and nuclear combinations and amputations, a harmonious zygote might be supposed to arise.

In the supposition as to the importance of hybridization, more especially as to a more or less intimate union between the chromosomes of the fused gametes — according as these are of more or less mutually harmonious constitution, — it will be understood that the very chromosome numbers which we find in nature might arise.

We may further consider the possibility that longer series of chromosome numbers within a single genus may be explained by the same means, taking, merely by way of example, *Chry-santhemum* as type. In *Chrysanthemum* we find, according to the investigations hitherto made, the chromosome numbers 9, 18, 27, 36 and 45, as we have already seen — and, as I have further shown, just such differential series are met with also in other spheres of relation.

As starting point we may take *Chrysanthemum* species with $x = 9$, this being here the cardinal number. Such species — it will here suffice to consider three of them — we may designate A, B and C; their chromosomes 9a, 9b and 9c respectively. On crossing these, we obtain results as shown schematically on page 203.

It will here be seen, that hybridization in the manner indicated results in the production of species with 9×1, 9×2, 9×3 chromosomes in the gametophyte, the chromosome numbers of the species paired being in each case added together in the offspring.

Naturally, again, other conditions will arise when the species thus produced are either intercrossed or paired with one of the parent species, since homologous chromosomes will then meet in the zygote, and direct union may take place.

In the case mentioned, we must suppose that there would

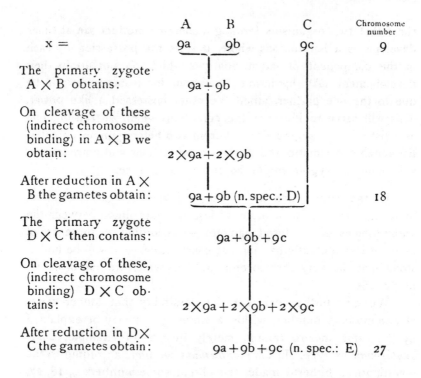

	A	B	C	Chromosome number
x =	9a	9b	9c	9

The primary zygote
A × B obtains:
9a + 9b

On cleavage of these
(indirect chromosome
binding) in A × B we
obtain:
2 × 9a + 2 × 9b

After reduction in A ×
B the gametes obtain:
9a + 9b (n. spec.: D)   18

The primary zygote
D × C then contains:
9a + 9b + 9c

On cleavage of these,
(indirect chromosome
binding) D × C ob·
tains:
2 × 9a + 2 × 9b + 2 × 9c

After reduction in D×
C the gametes obtain:
9a + 9b + 9c (n. spec.: E)   27

be formed a species, D, with 18 chromosomes (9a + 9b). On
crossing these with A, we get the following:

D   A
9a + 9b   9a

The primary zygote
D × A obtains:
2 × 9a + 9b

How this zygote will behave, must depend upon circumstances; the constitution of the sporophyte may be more or less
harmonious. It would be natural, however, that the 9a derived
from D at any rate should unite directly with the homologous 9a
from A. The 9b must either unite indirectly — i. e. if the sporo·
phyte is to be normally capable of developement — or remain
unpaired, when the sporophyte will be abnormal. Rosenberg's
investigations (1909) with *Drosera* hybrids might be an instance of
such a case. *Drosera rotundifolia* has x = 10, *D. longifolia* x = 20,
and the hybrid 2x = 2 × 10 + 10, the 10 chromosomes from
D. rotundifolia uniting with the 10 from *D. longifolia*, while the
remaining 10 continue unpaired, so that a natural sexual further

26

developement cannot take place. We must from this suppose that 10 of *D. longifolia*'s chromosomes will be homologous with *D. rotundifolia*'s.

If my theory be correct, then we should occasionally find differences in the size and shape of the chromosomes, i. e. when a species had arisen by hybridization of two others with chromosomes differing in appearance. Also, chromosomes of one and the other type respectively should then be found in a proportion according with the characteristic cardinal number for the field of relation in question. A species A, with 6 long chromosomes, and a species B, with 6 short ones should on their union give rise to another with 6 long and 6 short.

Both shape and size of the chromosomes are, however, as a rule of considerably variable nature, and it is well known that absolute dimensions are difficult to apply. Some parts of the tissue have longer, others shorter chromosomes in the spindle, and probably also conditions of nourishment etc. affect their appearance. In the gonokonts, however, the chromosomes do appear in such a form that they may, when not too numerous, be studied and characterized as to shape. It would be most interesting to investigate the varying appearance of the chromosomes there, from the point of view already indicated. I will here cite a couple of examples which serve to bear out the correctness of my summation theory.

In *Spinacia oleracea* (Stomps 1910) the chromosomes are of different size, three long and three short. Stomps shows, for instance, in his Fig. A (p. 70) a fine nuclear plate from a syndiploid root cell, in which six pair of long and six pair of short may be observed. The sexual cells thus contain 3 long and 3 short chromosomes. Now 3, as I have shown, is the cardinal number for *Chenopodiaceæ*, and we have thus here an instance of chromosomes differing in appearance occurring in a proportion corresponding to the cardinal number, whereas a reduplication of an original three would necessarily give a different result.

In the zoological literature we find an instance eminently suited for comparison with the above.

Chambers (1912) when investigating the American varieties of *Cyclops viridis* found the following chromosome. figures, which are alike for male and female individuals:

Cyclops viridis		12
»	var. Americanus	10
»	var. parcus	6
»	var. brevispinosus	4

The series of chromosome numbers, which thus has 2 as cardinal number, again affords an example of differential structure — albeit 8 is lacking here — and it is most interesting to note Chambers' statement that the 6 chromosomes in var. *parcus* were pairs of three different sizes.

Comparing this with the case of *Spinacia,* where the six chromosomes were 3 and 3 in two different sizes, we may doubtless take it that results stand in some relation to the respective cardinal numbers: in *Cyclops* 2, in *Spinacia* 3.

We have seen that series of chromosomes numbers such as we find in nature may be supposed to arise by hybridization, the chromosome numbers of the gametes thus united being generally added together in the gametes of the offspring.

But how, we naturally ask, could the chromosome numbers of the vegetable kingdom be as low as they are on the whole if they really were largely formed by addition? Evidently, it must be presumed that nature has some auto-regulative means of keeping the chromosome number down within suitable limits. There is of course some such limit for the number of chromosomes which a nucleus can, generally speaking, contain. Continued addition by repeated hybridization would soon give a theoretic chromosome number higher than any actually existing in nature. We must suppose then, that the supply of chromosomes in the non-dividing hybrid will under certain circumstances be simplified, but as to this we know nothing as yet.

It must also be noted, that the transition from misozygoty to 'philozygoty will naturally be gradual, since the categories given are based upon the supposition of more or less harmony or similarity between the gametes. And some pairs of organisms may therefore come to lie on the boundary line between two categories.

Finally, I would once more refer to the quotation from Strasburger (1910 p. 425) already given on p. 156 the last part of which is associated with his idea as to reduplication of chromosome numbers in apogamous species. In the main, I cannot but agree with him here, albeit I cannot regard the "Vervielfältigung" of a previously existent number of chromosomes as phylogenetically important in itself, nor can I accept it as being the cause of apogamy; the supposition as to hybridization, on the other hand, appears to me to afford a natural explanation.

I must in this connection maintain, that the various haploid chromosome numbers occurring in the species of a genus are not correlated on the principle of a, 2a, 4a, 8a, etc., as chromosome reduplication would involve, but on that of a, 2a, 3a, 4a, etc.; i. e. in arithmetical, not geometrical progression. That powers of 2 are particularly frequent in chromosome numbers is, moreover, but natural, since presumably original species with the same chromosome number are especially adapted to the production, by hybridization, of new species, which would thus acquire a "double" number.

In estimating the degree of relationship between the chromosome numbers of higher plants, we should accordingly attach chief importance to the ascertaining of the cardinal number. The presence of the cardinal number 2, on the other hand, can be due to the "reduplication" mentioned, and thus tells us nothing as to the possible relation between the two chromosome numbers in question. When, however, we find powers of 2 characterizing several chromosome numbers the relationship of which it is desired to estimate, so that 2n is a common factor, there is more reason to regard the 2-factor as indicative of such relationship — and higher the power of 2, the more likely will it be.

[*Editors' Note:* Only those references cited in the preceding excerpts have been included here.]

REFERENCES

Bally, W. 1912: Chromosomenzahlen bei *Triticum*- und *Aegilops*-arten. Ein cytologischer Beitrag zum Weizenproblem. — Ber. d. deutsch. Bot. Ges., V. 30, P. 163.

Chambers, R. 1912: Egg Maturation, Chromosomes and Spermatogenesis in *Cyclops*. — Univ. of Toronto Studies, Biol. Ser., Nr. 14.

Gates, R. R., 1914: Recent Aspects of Mutation. — Nature, V. 94, P. 296.

Guignard, L. 1891: Nouvelles études sur la fécondation. — Ann. de Sc. nat. Bot., Sér. 7, V. 14, P. 163.

Ishikawa, M. 1911: Cytologische Studien von Dahlien. — The Bot. Magazine, V. 25, P. 1.

Kuwada, Y. 1911: Maiosis in the Pollen Mother Cells of *Zea Mays* L. — The Bot. Magaz., V. 25, P. 163.

Marchal, Él. et Ém. 1907: Aposporie et sexualité chez les Mousses. — Bull. Acad. r. de Belgique.

Marchal, Él. et Ém. 1909: Aposporie et sexualité chez les Mousses. II. — Bull. Acad. r. de Belgique.

Mc. Clung, C. E. 1902: The Accessory Chromosomes Determinant. — Biol. Bull. Woods Holl, V. 3.

Merrell, W. D. 1900: A Contribution to the Life-History of *Silphium*. — Bot. Gaz., V. 29, P. 99.

Němec, B. 1910: Das Problem der Befruchtungsvorgänge. — Berlin.

Němec, B. 1912: Ueber die Befruchtung bei *Gagea*. — Bull. intern. Acad. Sc. Bohême, V. 17, P. 1.

Ostenfeld, C. H. 1910: Further Studies on the Apogamy and Hybridization of the Hieracia. — Zeitschr. f. ind. Abst.- und Vererb.- lehre, V. 3, P. 241.

Overton, J. B. 1904: Ueber Parthenogenesis bei *Thalictrum purpurascens*. — Ber. d. deutsch. Bot. Ges., V. 22, P. 274.

Pace, L. 1913: Apogamy in *Atamosco*. — Bot. Gaz., V. 56, P. 389.

Rosenberg, O. 1909 a: Ueber den Bau des Ruhekerns. — Svensk Bot. Tidsskr., V. 3, P. 163.

Rosenberg, O. 1909 b: Ueber die Chromosomenzahlen bei *Taraxacum* und *Rosa*. — Svensk Bot. Tidsskr., V. 3, P. 150.

Rosenberg, O. 1909 c: Cytologische und morphologische Studien an *Drosera longifolia x rotundifolia*. — K. Sv. Vet. -Akad. Handl., V. 43, Nr. 11.

Roth, Fr. 1907: Die Fortpflanzungsverhältnisse bei der Gattung *Rumex*. — Diss., Bonn.

Stomps, T. J. 1910: Kerndeeling en synapsis bij *Spinacia oleracea* L. —Diss., Amsterdam.

Strasburger, E. 1904: Die Apogamie der Eualchimillen. — Jahrb. f. wiss. Bot., V. 41, P. 88.

Strasburger, E. 1907: Apogamie bei *Marsilia*. — Flora, V. 97, P. 123.

Strasburger, E. 1910 a: Chromosomenzahl. — Flora, V. 100, P. 398.

Strasburger, E. 1910 b: Ueber geschlechtsbestimmende Ursachen. — Jahrb. f. wiss. Bot., V. 48, P. 427.

Tischler, G. 1910: Untersuchungen über die Entwicklung des Bananen-Pollens. I. — Arch. f. Zellforsch., V. 5, P. 622.

Tischler, G. 1915: Chromosomenzahl, -Form·und -Individualität im Pflanzenreiche. — Progr. Rei Bot., V. 5, P. 164.

Winge, Ö. 1914: The Pollination and Fertilization Processes in *Humulus lupulus* L. and *H. Faponicus* Sieb. et Zucc. — Compt. rend. des trav. d. Lab. d. Carlsberg, V. 11, P. 1.

Winkler, H. 1916: Ueber die experimentelle Erzeugung von Pflanzen mit abweichenden Chromosomenzahlen. — Zeitschr. f. Botanik, Jahrg. 8, P. 417.

2

Cytological and Genetical Studies of Important Cereal Species with Special Consideration of the Behavior of the Chromosomes and Sterility in the Hybrids

Hitoshi Kihara

These excerpts were translated by C. R. Burnham, R. L. Phillips, and Patrick Buescher, University of Minnesota, from pp. 3-5, 11, 29, 30, 33, 144-148, 181, 182, 183, 186-187 of "Cytologische und genetische Studien bei wichtigen Getreidearten mit besonderer Rücksicht auf das Verhalten der Chromosomen und die Sterilität in der Bastarden," Kyoto Univ. Coll. Sci. Mem., ser. B, 1(1):1-200 (1924) with the permission of the Faculty of Science, Kyoto University.

INTRODUCTION

Since the cereal species are the most important cultivated plants, people have been interested in their origin and mode of inheritance for a long time. The exact understanding of the inheritance phenomena, especially of species or genus hybrids (although it in part has already been considerably advanced through genetic studies), can only be satisfactorily solved when one does the cytology of the chromosomes.

In the *Triticum* species, the correct chromosome numbers were first found in 1918 by Sakamura. Until then an incorrect number had been generally accepted as correct, which meant no small hindrance for the progress of cytological studies of this plant. Through the establishment of the correct chromosome number of *Triticum* species, which I soon could confirm as correct, it became clear that the wheats fell cytologically into three different chromosome groups which agrees well with the phylogenetic classification established by Schulz (1913). Further, it is well known that the hybrids between these three groups were more or less fertile. This circumstance caused me to thoroughly investigate cytologically the species hybrids and their progeny, bearing in mind their genetic characteristics as well as sterility.

Moreover, I have worked with investigations concerning the chromosome number relationships in *Avena* species and the abnormal behavior of chromosomes in wheat-rye hybrids through which I have extended my cytological studies considerably.

The previous experimental results have been communicated since 1919 under the title "Cytological studies in some cereal species." (Kihara 1919a, b; 1921). The present work contains a summary of my earlier cytological and genetical investigations of the cereal species mentioned above. The first part is first of all devoted to the establishment of the correct chromosome numbers of important cereal species in order to make possible further studies concerning the behavior of chromosomes in the hybrid progeny, especially of wheat.

By drawing from previously published work of other authors and also from the results of my own experiments, I have established the degree of affinity as well as the behavior of paternal and maternal chromosomes in the reduction-division of hybrid plants and arranged them into groups.

For the X and non-X ploid relationship of chromosomes in the closely related species as well as their changes in number in the course of phylogenetic development, I have also drawn upon a large amount of literature.

In the progeny of 35 chromosome pentaploid wheat hybrids, one must pay special attention to the variation of chromosome number as well as sterility which is brought about by the recombination of the different chromosomes. The inheritance of these progeny is naturally so complicated that the law of simple Mendelian segregation is not applicable. At any rate, the distribution of chromosomes to the progeny is closely correlated with the morphological characteristics of the plants.

Therefore one is not justified to explain the sterility, manner of inheritance, etc., of the progeny of wheat hybrids without considering the variation in the number of chromosomes in the progeny. My morphological studies are communicated in the second part of this work.

*   *   *   *

[Based on the chromosome numbers found by Sakamura (1918) and confirmed by myself (1919a, 1921) and also by Sax (1921)], we are now in a position with these counts to reexamine karyologically the Schulz summary of the hereditary relationships of the Eutriticum species. In Table 2, I give only the haploid numbers.

Table 2. Chromosome numbers and hereditary relationships of Triticum species.

Groups of cultivated forms			Einkorn	Emmer	Spelt (Dinkel)
	Naked types	malformed		14 *T. polonicum* ↑	
		normal		14 *T. durum* 14 *T. turgidum*	21 *T. compactum* 21 *T. vulgare*
		Spelt types	7 *T. monococcum* ↑	14 *T. dicoccum* ↑	21 *T. spelta* ↑
		original species	7 *T. aegilopoides*	14 *T. dicoccoides*	unknown
		Series	Einkorn	Emmer	Spelt (Dinkel)

THE REDUCTION DIVISION IN F_1 HYBRIDS BETWEEN PARENTS WITH DIFFERENT CHROMOSOME NUMBERS.

Pentaploid hybrids between Emmer and Dinkel series.

[*Editors' Note:* This section is not translated. See paragraphs 2, 3, 4, and 18 in the summary.]

Triploid hybrids

T. dicoccum × *T. monococcum*
T. aegilopoides × *T. dicoccum*

Both of these hybrids were always completely sterile in my experiments, but according to Tschermak (1914), they should have been almost but not completely sterile. The behavior of the chromosomes in the reduction-division of pollen mother cells in these two hybrids is more or less identical; therefore I would like to concern myself here only with meiosis in *T. dicoccum* × *monococcum* hybrids.

The distribution of chromosomes to the daughter cells in the heterotypic nuclear division is irregular. It differs considerably from that of the above mentioned pentaploid hybrids. The bivalent pairing is loose, and their number is also variable. Already in the diakinesis stage the affinity of the parental chromosomes is somewhat weak. They often associate in the heterotypic metaphase only at their ends. The number of bivalents and univalents varies between 4 and 7, 13 and 7, respectively [i.e., there were cells with 7 bivalents and 7 univalents, 6 and 9, 5 and 11, or 4 and 13, a total of 21 chromosomes in each case]. Sax (1922) had determined in the triploid hybrid (*T. turgidum* × *monococcum*) 7 bivalent and 7 univalent chromosomes.

The division of the 4–7 bivalents follows in a normal manner as in the pentaploid hybrids. The equational division of univalents does not always occur after the halves of the bivalents have arrived at both poles. Some univalents reach the pole unsplit; others split lengthwise, and the longitudinal halves diverge from each other toward the poles as in the pentaploid hybrids. In rare cases I have seen 7 chromosomes in anaphase, all of which remained between the two groups of chromosomes that had arrived at the poles.

In telophase, however, some of the lagging chromosomes after the equational division go toward both poles to participate there in the formation of daughter nuclei. One can, however, often observe chromosomes not reaching either pole that remain isolated in the cytoplasm as a small dark stained clump, exactly as in the pentaploid hybrids.

The homeotypic nuclear plate usually shows 11–12 chromosomes, among which there are dyad and monad chromosomes. With regard to the dyad chromosomes, the second division occurs in the normal manner. The monad chromosomes are delayed naturally once again. Their number amounts usually to 1–4. [*Ed. note:* As in the pentaploid hy-

33

brid, the monad chromosomes are distributed unsplit among the four microspores in the tetrad.] Tetrad formation is usually regular. From one pollen mother cell, usually 4 microspores are formed. However, there are frequently also micronuclei that contain the lagging chromosomes in the homeotypic nuclear division.

The heterotypic nuclear division of these hybrids is characterized by the homologous parental chromosomes showing weak affinity and some of the univalent chromosomes going unsplit to one pole while the others are at the equator and split into two longitudinal halves. According to the literature, the *Pilosella* hybrids (for example, *Hieracium auricula* × *H. aurantiacum*, Rosenberg 1917) and *Papaver atlanticum* × *dubium* (Ljungdahl 1922) should behave similarly. However, the chromosome affinity in our hybrids seems to me to be stronger than in those just mentioned.

[MODE OF ORIGIN OF X-PLOID NUMBERS IN RELATED SPECIES]

Winge (1917) has in his work "The chromosomes, their numbers and general importance" considered this question also of the numbers in related species and how they might have originated. I will give a short account of his view.

If we assume that there are 3 different species (A, B, and C) with the same chromosome number ($x = 9$), then the hybrids among them would change the chromosome number to tetraploidy, hexaploidy in the following manner:

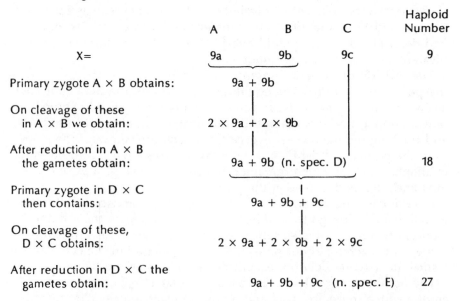

	A	B	C	Haploid Number
X=	9a	9b	9c	9
Primary zygote A × B obtains:		9a + 9b		
On cleavage of these in A × B we obtain:		2 × 9a + 2 × 9b		
After reduction in A × B the gametes obtain:		9a + 9b (n. spec. D)		18
Primary zygote in D × C then contains:		9a + 9b + 9c		
On cleavage of these, D × C obtains:		2 × 9a + 2 × 9b + 2 × 9c		
After reduction in D × C the gametes obtain:		9a + 9b + 9c (n. spec. E)	27	

About the behavior of the chromosomes in the zygote of the F_1-hybrid, he said:

How this zygote will behave, must depend upon circumstances; the constitution of the sporophyte may be more or less harmonious. It would be natural, however, that the 9a derived from D at any rate should unite directly with the homologous 9a from A. The 9b must either unite indirectly, i.e., if the sporophyte is to be normally capable of development or remain unpaired, when the sporophyte will be normal. Rosenberg's investigations (1909) with *Drosera* hybrids might be an instance of such a case. *Drosera rotundifolia* has x = 10, *D. longifolia* x = 20, and the hybrid 2x = 2 × 10 + 10, the 10 chromosomes from *D. rotundifolia* uniting with the 10 from *D. longifolia* while the remaining 10 continue unpaired, so that a natural sexual further development cannot take place.

According to this assumption, all hybrids with the *Drosera*-scheme should be sterile. There are, however, many fertile hybrids in this scheme (for example, triploid *Oenothera*, pentaploid *Triticum* hybrids). Although Winge's opinion cited above has a certain significance, it must still not be accepted incontestably.

I will therefore now give the possibilities for the change in number from diploid to tetraploid to hexaploid in *Triticum*. Possibility I is the following:

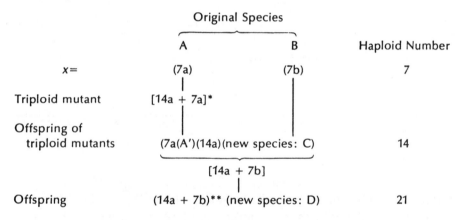

	Original Species		Haploid Number
	A	B	
x=	(7a)	(7b)	7
Triploid mutant	[14a + 7a]*		
Offspring of triploid mutants	(7a(A')(14a)(new species: C)		14
	[14a + 7b]		
Offspring	(14a + 7b)** (new species: D)		21

* This doubling of the chromosome number occurs before fertilization.

** If in these cases the chromosome numbers of the offspring of C × B are not (14a + 7b) + (14a + 7b), namely 3x + 3x = 6x, but 5x + n (where n < x), then their further offspring will ultimately reach an even 6x.

The opportunity, whereby the tetraploid plant can change into hexaploid, must be very rare because here high sterility and mortality among the zygotes prevail.

Percival (1921) is of the opinion that the plants of the Spelt (Dinkel) series may be none other than the offspring of the hybrid *Aegilops* × Emmer. One can therefore also assume that the hexaploidy might be engendered by hybridization between the 28-chromosome tetraploid plants. If we, for example, represent the chromosome makeup of two tetraploid 28-chromosome species A and B, with 7a + 7c—7b + 7c— then both C-chromosome constellations could form pairs (with each other), whereas the affinity for pair formation is lacking between the a- and b-chromosome constellations. The change of the chromosome number through their hybridization will then be as follows for Possibility II:

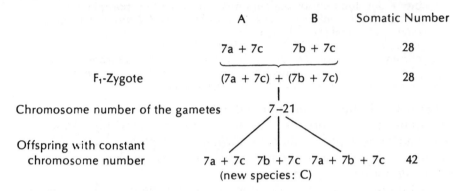

	A	B	Somatic Number
	7a + 7c	7b + 7c	28
F$_1$-Zygote	(7a + 7c) + (7b + 7c)		28
Chromosome number of the gametes	7–21		
Offspring with constant chromosome number	7a + 7c 7b + 7c 7a + 7b + 7c		42
	(new species: C)		

The third possibility derived from the above cited possibilities is the following:

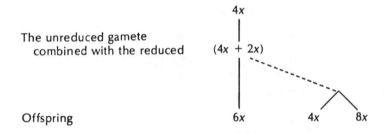

	4x		
The unreduced gamete combined with the reduced	(4x + 2x)		
Offspring	6x	4x	8x

As this sequence shows, the zygote obtained through the union of reduced and unreduced gametes from tetraploid plants is therefore hexaploid. Their chromosome number is, however, generally not constant because again univalent chromosomes segregate. This is shown on the right side of the above diagram.

The possibility is not excluded that the 2–4–8 series might be obtained in this manner inasmuch as the extra chromosomes form no bivalents. It is not impossible that these mutant plants at a certain time in the course of the phylogenetic development had shown, as in diploid plants, pairing of the 2x extra chromosomes and that this affinity of the chromosomes was lost later.

It is also thinkable that soon after the origin of the new tetraploid species from the diploid, the affinity [for pair formation as in the diploid] of the homologous chromosomes was not lost. Then through the union of these gametes (4x + 2x) a constant new hexaploid species with the reduced chromosome number 3x would be formed immediately.

Through possibility III we can better understand the different degrees of relationship between the three *Triticum* series, the 14II and 21II species are more closely related than are the 7II and 14II species.

* * * *

The plants of the Spelt (Dinkel) series are hexaploid. They have 21 haploid chromosomes (3 × 7), which had been obtained through trebling of the original chromosome composition with 7 chromosomes. If we next assume that a gene for a character (for example, seed color) is located in any one chromosome of every chromosome set, then one can easily understand that the hexaploid wheats possess 3 synonymous Mendelian factors for this character. From that, one also can conclude that tetraploid wheats possess 2 because here one chromosome set (a–g) is missing. We name, for example, the three genes A, B, and D. Then emmer wheats have the formula A_E A_E B_E B_E and Spelt (Dinkel) wheats the A_D A_D B_D B_D D D. The mode of inheritance of the pentaploid hybrids with reference to these characters can be shown in the following:

$$A_E \ A_E \ B_E \ B_E \qquad A_D \ A_D \ B_D \ B_D \ D \ D \qquad P$$
$$(28) \qquad\qquad (42)$$

$$A_E \ A_D \ B_E \ B_D \ D \qquad\qquad F_1$$
$$(35)$$

$A_E \ A_E \ B_E \ B_E$	$A_E \ A_E \ B_E \ B_E \ D \ D$	Homozygous offspring with 28 and 42 chromosomes.
$A_E \ A_E \ B_D \ B_D$	$A_E \ A_E \ B_D \ B_D \ D \ D$	
$A_D \ A_D \ B_E \ B_E$	$A_D \ A_D \ B_E \ B_E \ D \ D$	
$A_D \ A_D \ B_D \ B_D$	$A_D \ A_D \ B_D \ B_D \ D \ D$	
(28)	(42)	

This conception of the inheritance is indeed also valid with reference to different characters. Therefore very many new combinations of different genes appear in the offspring of this hybrid.

Because D represents all Dinkel genes (D_1, D_2, D_3, etc.), which the chromosome set (a–g) possesses, it is not incorrect to suppose that they all together give the corresponding characters of *T. vulgare*, *T. spelta*, and *T. compactum*. The manner of inheritance of spike form of *T. durum* × *vulgare* and *T. polonicum* × *Spelta* is a good example. *D* for spike form is in this case epistatic to *A* and *B* for this character. That some 42-chromosome offspring of *T. turgidum* × *compactum* possess *speltoides*-spikes shows a complicated mutual relation which thereby takes place among the genes (A, B, and D). Therefore it is understandable that they [may] retain the characteristics of the Dinkel plants (for example, hollow stems, not keeled glumes). Yet they may not show the typical Dinkel form because they do not have all the D-genes.

SUMMARY

1. The chromosome number of the *Triticum-*, *Aegilops-*, *Secale-*, *Hordeum-*, and *Avena-*species has been established (for certain) by the present investigations. Their haploid numbers are the following:

*Triticum-*species	7	14	21
*Aegilops-*species		14	
Secale cereale	7 or 8		
*Hordeum-*species	7		
*Avena-*species	7	14	21

2. The pentaploid hybrids between the 14-chromosome Emmer and the 21-chromosome Spelt (Dinkel) series have 35 chromosomes, corresponding to the sum of the haploid chromosomes of the parental plants. In the heterotypic nuclear division of the pollen mother cells, 14 Spelt (Dinkel) chromosomes form 14 bivalents with just as many Emmer chromosomes, the 7 surplus Spelt (Dinkel) chromosomes remain as univalents.

3. The heterotypic nuclear division of the pentaploid hybrids is, in relation to the 14 bivalents, a normal reduction division and in relation to the 7 univalents, a longitudinal splitting. The homeotypic nuclear division of these hybrids is in relation to the 14 dyad chromosomes, which come from the 14 bivalents in the heterotypic nuclear division, an equational division. The 7 surplus chromosomes distribute themselves unsplit to the 4 microspores and indeed according to the law of probability. Therefore the resulting pollen grains receive 14 + i chromosomes, where $i = 0$–7.

4. Often are seen chromosomes delayed and not arrived at the pole. They do not participate in the future division process (chromatin-diminution). At times they form small pollen.

5. The triploid hybrids between the Einkorn- and the Emmer-series have 21 somatic chromosomes. The number of bivalents of these wheat hybrids varies between 4–7, the number of univalents correspondingly between 13–7. The bivalents show a regular behavior during the entire meiotic nuclear division. A part of the univalents divide lengthwise in the heterotypic nuclear division, while the others go unsplit to any one pole. In the homeotypic nuclear division, the dyad chromosomes divide normally lengthwise, and then the delayed monad chromosomes go to the poles.

18. Constant 40-chromosome [nullisomic] plants were found among the progeny of pentaploid wheat hybrids (21 II × 14 II). The first plant and its progeny were dwarf in growth habit. They lacked one pair of chromosomes. Another constant 40-chromosome plant and its progeny were semidwarf.

REFERENCES

Kihara, H. 1919a. Über cytologische Studien bei einigen Getreidearten. Mit. I. Spezies-Bastard des Weizens und Weizenroggen-Bastard. *Bot. Mag.* **33**:17–38.

———. 1919b. Mit. II. Chromosomenzahl und Verwandtschaftsverhältnisse unter *Avena* Arten. *Bot. Mag.* **33**:94–97.

———. 1921. Mit. III. Über die Schwankungen der Chromosomenzahlen bei den Spezies-bastarden der *Triticum*-Arten. *Bot. Mag.* **35**:19–44.

Ljungdahl, H. 1922. Zur Zytologie der Gattung *Papaver. Svensk. Bot. Tidskr.* **16**:103–114.

Percival, J. 1921. *The wheat plant.* A monograph. London.

Rosenberg, O. 1909. Cytologische und morphologische Studien an *Drosera longifolia* × *rotundifolia. K. Sven. Vetenskapsakad Handl.* **43**(11):1–65.

———. 1917. Die Reduktionsteilung und ihre Degeneration in *Hieracium. Svensk. Bot. Tidskr.* **11**:145–206.

Sakamura, T. 1918. Kurze Mitteilung über die Chromosomenzahlen und die Verwandtschaftsverhältnisse der *Triticum* Arten. *Bot. Mag.* **32**:150–153.

Sax, K. 1921. Sterility in wheat hybrids. 1. Sterility relationships and endosperm development. *Genetics* **6**:399–416.

———. 1922a. II. Chromosome behavior in partially sterile hybrids. *Genetics* **7**:513–558.

Schulz, A. 1913. *Die Geschichte der kultivierten Getreide I.* Halle.

Tschermak, E. v. 1914. Die Verwertung der Bastardierung für phylogenetische Fragen in der Getreidegruppe. *Z. Pflanzenzucht.* **2**:291–312.

Winge, Ø. 1917. The chromosomes. Their number and general importance. *C. R. Trav. Labor. Carlsberg* **13**:131–275.

[*Editors' Note:* This translation also appears as Paper 42 in *Cytogenetics*, R. L. Phillips and C. R. Burnham, eds., Benchmark Papers in Genetics, vol. 6, Dowden, Hutchinson & Ross, Stroudsburg, Pa., 1977, pp. 379-387.]

3

Reprinted from *J. Genet.* **22**:359-372 (1930)

THEORETICAL GENETICS OF AUTOPOLYPLOIDS.

By J. B. S. HALDANE, M.A.

(The John Innes Horticultural Institution.)

An increasing amount of practical genetics is now being done on polyploid plants. These may be divided into allopolyploids, in which each chromosome has only one normal mate, although occasionally an abnormal pairing occurs, and autopolyploids in which a chromosome is equally likely to pair with any of two or more others. There appear to be intermediate types as well. At this institution researches are at present being carried out on three autopolyploids, namely the tetraploid forms of *Primula sinensis*, *Campanula persicifolia*, and *Lycopersicum esculentum*. The genetics of *Dahlia variabilis*, which behaves in some respects as an autopolyploid, are also being studied. The present paper deals with the genetical behaviour to be expected from such plants. Linkage is not considered, as this is being dealt with in a forthcoming paper by de Winton and Haldane on *Primula sinensis*. A gamete or zygote containing x of the factor A and y of the factor a is called $A^x a^y$. Thus $A^3 a$ denotes $AAAa$, and so on. We shall only consider orthoploids, *i.e.* polyploids containing an even number of sets of chromosomes.

GAMETIC SERIES.

The gametic series to be expected from each class of zygote are given in Table I. The results for tetraploids and octaploids have already been given by Muller (1914) and Lawrence (1929). To save space the gametic output of homozygotes is omitted. Thus an A^{10} decaploid produces only A^5 gametes, an a^{10} only a^5, and so on. After the hexaploid some of the gametic series are omitted. Thus a decaploid of constitution $A^2 a^8$ would give a gametic series $2A^2 a^3 : 5Aa^4 : 2a^5$, the same as that of $A^8 a^2$, with A and a interchanged.

The method of calculation is simple, as the following example shows. Consider a decaploid $A^7 a^3$. Five out of ten factors can be chosen in $\frac{10.9.8.7.6}{5.4.3.2.1}$ or 252 ways. Of these either an A^5 or $A^2 a^3$ gamete can be chosen in $\frac{7.6}{2.1}$ or 21 ways, an $A^4 a$ or $A^3 a^2$ in $\frac{7.6.5}{3.2.1} \times 3$, or 105 ways.

TABLE I.

$2m$	Zygote	Gametes
4	$A^3 a$	$1A^2$: $1A a$
	$A^2 a^2$	$1A^2$: $4A a$: $1\ a^2$
	$A a^3$	$1A a$: $1\ a^2$
6	$A^5 a$	$1A^3$: $1A^2a$
	$A^4 a^2$	$1A^3$: $3A^2a$: $1A a^2$
	$A^3 a^3$	$1A^3$: $9A^2a$: $9A a^2$: $1a^3$
	$A^2 a^4$	$1A^2a$: $3A a^2$: $1a^3$
	$A a^5$	$1A a^2$: $1a^3$
8	$A^7 a$	$1A^4$: $1A^3a$
	$A^6 a^2$	$3A^4$: $8A^3a$: $3A^2a^2$
	$A^5 a^3$	$1A^4$: $6A^3a$: $6A^2a^2$: $1A a^3$
	$A^4 a^4$	$1A^4$: $16A^3a$: $36A^2a^2$: $16A a^3$: $1a^4$
	etc.	
10	$A^9 a$	$1A^5$: $1A^4a$
	$A^8 a^2$	$2A^5$: $5A^4a$: $2A^3a^2$
	$A^7 a^3$	$1A^5$: $5A^4a$: $5A^3a^2$: $1A^2a^3$
	$A^6 a^4$	$1A^5$: $10A^4a$: $20A^3a^2$: $10A^2a^3$: $1A a^4$
	$A^5 a^5$	$1A^5$: $25A^4a$: $100A^3a^2$: $100A^2a^3$: $25A a^4$: $1a^5$
	etc.	
12	$A^{11}a$	$1A^6$: $1A^5a$
	$A^{10}a^2$	$5A^6$: $12A^5a$: $5A^4a^2$
	$A^9 a^3$	$2A^6$: $9A^5a$: $9A^4a^2$: $2A^3a^3$
	$A^8 a^4$	$1A^6$: $8A^5a$: $15A^4a^2$: $8A^3a^3$: $1A^2a^4$
	$A^7 a^5$	$1A^6$: $15A^5a$: $50A^4a^2$: $50A^3a^3$: $15A^2a^4$: $1A a^5$
	$A^6 a^6$	$1A^6$: $36A^5a$: $225A^4a^2$: $400A^3a^3$: $225A^2a^4$: $36A a^5$: $1a^6$
	etc.	
16	$A^{15}a$	$1A^8$: $1A^7a$
	$A^{14}a^2$	$7A^8$: $16A^7a$: $7A^6a^2$
	$A^{13}a^3$	$1A^8$: $4A^7a$: $4A^6a^2$: $1A^5a^3$
	$A^{12}a^4$	$5A^8$: $32A^7a$: $56A^6a^2$: $32A^5a^3$: $5A^4a^4$
	$A^{11}a^5$	$1A^8$: $10A^7a$: $28A^6a^2$: $28A^5a^3$: $10A^4a^4$: $1A^3a^5$
	$A^{10}a^6$	$1A^8$: $16A^7a$: $70A^6a^2$: $112A^5a^3$: $70A^4a^4$: $16A^3a^5$: $1A^2a^6$
	$A^9 a^7$	$1A^8$: $28A^7a$: $196A^6a^2$: $490A^5a^3$: $490A^4a^4$: $196A^3a^5$: $28A^2a^6$: $1A a^7$
	$A^8 a^8$	$1A^8$: $64A^7a$: $784A^6a^2$: $3136A^5a^3$: $4900A^4a^4$: $3136A^3a^5$: $784A^2a^6$: $64A a^7$: $1a^8$
	etc.	

Ratios of dominants to recessives when a given heterozygous type is crossed with a recessive (nulliplex):

Tetraploid	A^2a^2,	$5:1$; $A a^3$, $1:1$
Hexaploid	A^3a^3,	$19:1$; A^2a^4, $4:1$; $A a^5$, $1:1$
Octaploid	A^4a^4,	$69:1$; A^3a^5, $13:1$; A^2a^6, $11:3$; $A a^7$, $1:1$
Decaploid	A^5a^5,	$251:1$; A^4a^6, $41:1$; A^3a^7, $11:1$; A^2a^8, $7:2$; $A a^9$, $1:1$
Dodecaploid	A^6a^6,	$923:1$; A^5a^7, $131:1$; A^4a^8, $32:1$; A^3a^9, $10:1$; A^2a^{10}, $17:5$; $A a^{11}$, $1:1$

Heccaidecaploid A^8a^8, $12,869:1$; A^7a^9, $1429:1$; A^6a^{10}, $285:1$; A^5a^{11}, $77:1$; A^4a^{12}, $25:1$; A^3a^{13}, $9:1$; A^2a^{14}, $23:7$; $A\ a^{15}$, $1:1$

The gametic series is therefore

$$21A^5 : 105A^4a : 105A^3a^2 : 21A^2a^3,$$

or

$$1A^5 : 5A^4a : 5A^3a^2 : 1A^2a^3.$$

In general from a zygote $A^r a^{2m-r}$, m factors can be chosen in $\dfrac{(2m)!}{(m!)^2}$

ways. A gamete $A^s a^{m-s}$ can be chosen in $\dfrac{r!}{(r-s)!\,s!} \cdot \dfrac{(2m-r)!}{(m-r+s)!\,(m-s)!}$ ways. Hence the probability of a zygote $A^r a^{2m-r}$ producing a gamete $A^s a^{m-s}$, is

$$\frac{(m!)^2\,(2m-r)!\,r!}{(2m)!\,(m-s)!\,(m-r+s)!\,s!\,(r-s)!}.$$

When $r = m$, i.e. the zygote is $A^m a^m$, this probability reduces to

$$\frac{(m!)^4}{(2m)!\,(m-s!)^2\,(s!)^2}.$$

RESULTS OF SELFING AND CROSSING.

The results of selfing any type can easily be calculated. For example, the result of selfing an octaploid of composition $A^3 a^5$ is given by the algebraic process of formally squaring the gametic series

$$1A^3 a : 6A^2 a^2 : 6Aa^3 : 1a^4,$$

i.e. by expanding $(A^3 a + 6A^2 a^2 + 6Aa^3 + a^4)^2$. The expected series is therefore

$$1A^6 a^2 : 12A^5 a^3 : 48A^4 a^4 : 74A^3 a^5 : 48A^2 a^6 : 12Aa^7 : 1a^8,$$

i.e. one complete recessive, or nulliplex, in 196.

When two allelomorphic pure lines of a $2m$-ploid are crossed, the result is a zygote $A^m a^m$. On selfing this we may expect one recessive in a number of zygotes equal to the square of the gametic series, i.e. one in $\dfrac{(2m!)^2}{(m!)^4}$, e.g. 1 in 4 when $m = 1$. The values of this number are given in Table II, up to $m = 16$.

TABLE II.

Diploid number $2m$	2	4	6	8	10	12	14	16
Number needed in F_2	4	36	400	4900	63,504	853,776	11,778,624	165,636,900

It is clear that in practical breeding work it would rarely be worth attempting to obtain pure recessive (nulliplex) plants in F_2 from anything higher than a hexaploid.

The results of crossing can easily be calculated by symbolic multiplication. Thus from $A^5 a^3 \times A^3 a^5$ we should expect a zygotic series corresponding to

$$(1A^4 + 6A^3 a + 6A^2 a^2 + 1Aa^3)(1A^3 a + 6A^2 a^2 + 6Aa^3 + 1a^4),$$

i.e. $1A^7 a : 12A^6 a^2 : 48A^5 a^3 : 74A^4 a^4 : 48A^3 a^5 : 12A^2 a^6 : 1Aa^7.$

If selfing is carried on for several generations, homozygotes will appear in reasonable numbers, and in the long run the whole population will become homozygous. The final ratio of $A^{2m} : a^{2m}$ is the same as that of $A : a$ in the original population. We shall consider two cases where this ratio is unity.

When, in a tetraploid, A^4 and a^4 are crossed, F_1 is A^2a^2. Let the zygotic series in F_n be:

$$p_n A^4 : q_n A^3 a : r_n A^2 a^2 : q_n A a^3 : p_n a^4,$$

where $$2p_n + 2q_n + r_n = 1.$$

Now the result of selfing $A_3 a$ is $1A^4 : 2A^3 a : 1A^2 a^2$, that of selfing $A^2 a^2$ is $1A^4 : 8A^3 a : 18A^2 a^2 : 8A a^3 : 1a^4$.

$$\therefore \quad p_{n+1} = p_n + \tfrac{1}{4} q_n + \tfrac{1}{36} r_n,$$
$$q_{n+1} = \tfrac{1}{2} q_n + \tfrac{2}{9} r_n,$$
$$r_{n+1} = \tfrac{1}{2} q_n + \tfrac{1}{2} r_n.$$

Also $p_1 = q_1 = 0$, $r_1 = 1$.

$$\therefore \quad 3q_{n+1} + 2r_{n+1} = \tfrac{5}{6} (3q_n + 2r_n).$$
$$\therefore \quad 3q_n + 2r_n = 2 \left(\tfrac{5}{6}\right)^{n-1},$$
$$3q_{n+1} - 2r_{n+1} = 6 (3q_n - 2r_n).$$
$$\therefore \quad 3q_n - 2r_n = -2 \left(\tfrac{1}{6}\right)^{n-1}.$$
$$\therefore \quad 2q_n + r_n = \tfrac{7}{12} (3q_n + 2r_n) + \tfrac{1}{12} (3q_n - 2r_n)$$
$$= \tfrac{7}{5} \cdot \left(\tfrac{5}{6}\right)^n - \left(\tfrac{1}{6}\right)^n.$$

This is the proportion of heterozygotes of various kinds expected in F_n, *i.e.* after n generations of selfing. It is tabulated in Table III.

TABLE III.

n	1	2	3	4	5	6	∞
$2q_n + r_n$	1·0	·94	·805	·6744	·5625	·4685	0

In the corresponding case in the hexaploid the F_1 is $A_3 a_3$, the zygotic series in F_n:

$$p_n A^6 : q_n A^5 a : r_n A^4 a^2 : s_n A^3 a^3 : r_n A^2 a^4 : q_n A a^5 : p_n a^6,$$

where $2p_n + 2q_n + 2r_n + s_n = 1$ and $p_1 = q_1 = r_1 = 0$, $s_1 = 1$.

$$\therefore \quad p_{n+1} = p_n + \tfrac{1}{4} q_n + \tfrac{1}{25} r_n + \tfrac{1}{400} s_n,$$
$$q_{n+1} = \tfrac{1}{2} q_n + \tfrac{6}{25} r_n + \tfrac{9}{200} s_n,$$
$$r_{n+1} = \tfrac{1}{4} q_n + \tfrac{12}{25} r_n + \tfrac{99}{400} s_n,$$
$$s_{n+1} = \tfrac{12}{25} r_n + \tfrac{41}{100} s_n.$$

Solving the identity

$$q_{n+1} + \lambda r_{n+1} + \mu s_{n+1} \equiv k\,(q_n + \lambda r_n + \mu s_n),$$

we find $k = \frac{1}{20}, \frac{11}{25}$, or $\frac{9}{10}$.

$$\therefore\ 80 q_{n+1} - 144 r_{n+1} + 89 s_{n+1} = \tfrac{1}{20}\,(80 q_n - 144 r_n + 89 s_n) = \tfrac{1}{20} x_n,$$
$$25 q_{n+1} - 6 r_{n+1} - 12 s_{n+1} = \tfrac{11}{25}\,(25 q_n - 6 r_n - 12 s_n) = \tfrac{11}{25} y_n,$$
$$10 q_{n+1} + 16 r_{n+1} + 9 s_{n+1} = \tfrac{9}{10}\,(10 q_n + 16 r_n + 9 s_n) = \tfrac{9}{10} z_n.$$

$$\therefore\ x_n = 89\,(\tfrac{1}{20})^{n-1};\ y_n = -12\,(\tfrac{11}{25})^{n-1};\ z_n = 9\,(\tfrac{9}{10})^{n-1},$$

$$1 - 2 p_n = 2 q_n + 2 r_n + s_n$$

$$= \frac{23 x_n + 1190 y_n + 7007 z_n}{50{,}830}$$

$$= \frac{4851}{3910} \left(\frac{9}{10}\right)^{n-1} - \frac{84}{299} \left(\frac{11}{25}\right)^{n-1} + \frac{89}{2210} \left(\frac{1}{20}\right)^{n-1}.$$

This expression gives the proportion of heterozygotes in F_n. After the first few generations it approximates to a geometric series whose successive terms are in the ratio $9 : 10$, so the heterozygotes disappear quite slowly. Similar but more complicated expressions give the results of selfing in higher polyploids.

RANDOM MATING.

Some autopolyploids are self-fertile, even when the corresponding diploid is self-sterile. This is not, however, always the case. Thus Lawrence (1929) has shown that the self-sterile *Dahlia variabilis* behaves in some respects at least as an autotetraploid. The question therefore arises as to the results of random mating in a large population. Under any system of mating the ratio of dominant to recessive allelomorphs remains constant. This ratio will be called u. It is legitimate to regard the gametes of all plants as pooled in each generation, as this will obviously not affect the numbers of pairings of each type of gamete.

The following theorem holds good for all autopolyploids: "When equilibrium is reached under random mating of a $2m$-ploid the gametes are produced in proportions given by the expansion of $(uA + 1a)^m$, the proportions of zygotic types being given by the expansion of $(uA + 1a)^{2m}$."

Thus in the case of a tetraploid the gametes will be in the ratios:

$$u^2 AA : 2uAa : 1aa;$$

the zygotes $\quad u^4 A^4 : 4 u^3 A^3 a : 6 u^2 A^2 a^2 : 4 u A a^3 : 1 a^4.$

In the case of a tetraploid it can readily be shown that such a population is in equilibrium. The general proof follows.

If the proportion of $A^s a^{m-s}$ gametes produced by one generation is

$$\frac{m!\, u^s}{(m-s)!\, s!\, (u+1!)^m},$$

as follows from the above expansion, the proportion of $A^r a^{2m-r}$ zygotes is clearly

$$\frac{(2m)!\, u^r}{(2m-r)!\, r!\, (u+1!)^{2m}}.$$

The probability of such a zygote producing an $A^s a^{m-s}$ gamete is

$$\frac{(m!)^2\, (2m-r)!\, r!}{(2m)!\, (m-r+s)!\, (m-s)!\, (r-s)!\, s!},$$

as pointed out above. Hence the total proportion of $A^s a^{m-s}$ gametes produced by the next generation is

$$\sum_{r=0}^{2m} \frac{(2m)!\, u^r}{(2m-r)!\, r!\, (u+1)^{2m}} \cdot \frac{(m!)^2\, (2m-r)!\, r!}{(2m)!\, (m-r+s)!\, (m-s)!\, (r-s)!\, s!}$$

$$= \frac{(m)!\, u^s}{(m-s)!\, s!\, (u+1)^m} \sum_{r=0}^{2m} \frac{(m)!\, u^{r-s}}{(m-r+s)!\, (r-s)!\, (u+1)^m}$$

$$= \frac{(m)!\, u^s}{(m-s)!\, s!\, (u+1)^m}.$$

Hence the population is in equilibrium.

The following example shows how the above theory might be applied: "Three-quarters of a population of *Dahlia variabilis* have yellow flavone, *i.e.* possess the factor Y (Lawrence, 1929). If mating is at random, what proportions of the different genotypes may be expected?"

In a diploid we should expect $1YY : 2Yy : 1yy$. As Y shows tetraploid inheritance, $(u+1)^{-4} = \frac{1}{4}$, $\therefore u = \sqrt{2} - 1$. Hence we should expect to find:

$$\frac{(\sqrt{2}-1)^4}{4}\, Y^4 : (\sqrt{2}-1)^3 Y^3 y : \tfrac{3}{2}(\sqrt{2}-1)^2\, Y^2 y^2 : (\sqrt{2}-1)\, Yy^3 : \tfrac{1}{4}y^4,$$

or $0.74\%\ Y^4,\ 7.1\%\ Y^3 y,\ 25.74\%\ Y^2 y^2,\ 41.42\%\ Yy^3,\ 25.0\%\ y^4.$

It is significant that Lawrence, who analysed a number of individuals from a population where y^4 is a fairly common type, found no Y^4, and only one $Y^3 y$.

THE RATE OF APPROACH TO EQUILIBRIUM.

Whereas a diploid population reaches equilibrium for an autosomal factor after one generation, this is not of course the case for a sex-linked factor, nor yet for an autosomal factor in a polyploid.

Consider a tetraploid population in which the nth generation produces (pooled) gametes in the proportions $x_n AA : 2y_n Aa : z_n aa$, where $x_n + 2y_n + z_n = 1$. Hence the $(n + 1)$th generation consists of:

$$x_n^2 A^4 : 4x_n y_n A^3 a : (4y_n^2 + 2x_n z_n) A^2 a^2 : 4y_n z_n Aa^3 : z_n^2 a^4.$$

$$\therefore \; x_{n+1} = x_n^2 + 2x_n y_n + \tfrac{2}{3}y_n^2 + \tfrac{1}{3}x_n z_n = x_n + \tfrac{2}{3}(y_n^2 - x_n z_n),$$
$$y_{n+1} = x_n y_n + \tfrac{4}{3}y_n^2 + \tfrac{2}{3}x_n z_n + y_n z_n = y_n - \tfrac{2}{3}(y_n^2 - x_n z_n),$$
$$z_{n+1} = \tfrac{2}{3}y_n^2 + \tfrac{1}{3}x_n z_n + 2y_n z_n + z_n^2 = z_n + \tfrac{2}{3}(y_n^2 - x_n z_n).$$

Let $t_n = y_n^2 - x_n z_n$.

$$\therefore \; t_{n+1} = (y_n - \tfrac{2}{3}t_n)^2 - (x_n + \tfrac{2}{3}t_n)(z_n + \tfrac{2}{3}t_n)$$
$$= \tfrac{1}{3}t_n.$$
$$\therefore \; t_n = 3^{-n}t_0.$$

If u be the ratio of $A : a$,

$$\therefore \; x_n + y_n = \frac{u}{u+1}, \quad y_n + z_n = \frac{1}{u+1}.$$

$$\therefore \; t_n = y_n^2 - \left(\frac{u}{u+1} - y_n\right)\left(\frac{1}{u+1} - y_n\right) = y_n - \frac{u}{(u+1)^2}.$$

$$\therefore \; x_n = \frac{u^2}{(u+1)^2} - 3^{-n}t_0,$$

$$y_n = \frac{u}{(u+1)^2} + 3^{-n}t_0,$$

$$z_n = \frac{1}{(u+1)^2} - 3^{-n}t_0.$$

Hence the approach to equilibrium is very rapid. Thus in the case of a population originally consisting only of homozygotes A^4 and a^4, $u = 1$, $t_0 = -\tfrac{1}{4}$.

Hence the percentages of homozygotes in successive generations are 100, 50, 22·$\dot{2}$, 15·43, 13·44, 12·81, 12·60, etc., the final value being 12·50.

In a hexaploid population let the gametic series produced by the nth generation be
$$p_n A^3 : 3q_n A^2 a : 3r_n Aa^2 : s_n a^3,$$

where
$$p_n + 3q_n + 3r_n + s_n = 1,$$

so that
$$p_n + 2q_n + r_n = \frac{u}{u+1}, \quad q_n + 2r_n + s_n = \frac{1}{u+1}.$$

Then the zygotic series produced is:

$$p_n^2 A^6 : 6p_n q_n A^5 a : (9q_n^2 + 6p_n r_n) A^4 a^2 : (2p_n s_n + 18q_n r_n) A^3 a^3, \text{ etc.}$$
$$\therefore \; p_{n+1} = p_n^2 + 3p_n q_n + \tfrac{1}{5}(9q_n^2 + 6p_n r_n) + \tfrac{1}{10}(p_n s_n + 9q_n r_n)$$
$$= p_n + \tfrac{1}{10}(2q_n^2 - p_n r_n + q_n r_n - p_n s_n).$$
$$q_{n+1} = q_n + \tfrac{3}{10}(4p_n r_n - 4q_n s_n + p_n s_n - q_n r_n + 2r_n^2 + 2q_n s_n), \text{ etc.}$$

On putting
$$x_n = q_n{}^2 - p_n r_n,$$
$$y_n = p_n s_n - q_n r_n,$$
$$z_n = r_n{}^2 - q_n s_n.$$
$$\therefore\ p_{n+1} = p_n + \tfrac{9}{10}(2x_n - y_n),$$
$$q_{n+1} = q_n + \tfrac{3}{10}(-4x_n + y_n + 2z_n),$$
$$r_{n+1} = r_n + \tfrac{3}{10}(2x_n + y_n - 4z_n),$$
$$s_{n+1} = s_n + \tfrac{9}{10}(-y_n + 2z_n).$$

Substituting
$$p_n = \frac{u}{u+1} - 2q_n - r_n, \quad s_n = \frac{1}{u+1} - q_n - 2r_n.$$

$$\therefore\ x_n = (q_n + r_n)^2 - \frac{ur_n}{u+1},$$

$$y_n = 2(q_n + r_n)^2 - \frac{2q_n + uq_n + r_n + 2ur_n}{u+1} + \frac{u}{(u+1)^2},$$

$$z_n = (q_n + r_n)^2 - \frac{q_n}{u+1}.$$

$$\therefore\ q_{n+1} = q_n + \tfrac{3}{10}\left[\frac{-4q_n - uq_n - r_n + 2ur_n}{u+1} + \frac{u}{(u+1)^2}\right],$$

$$r_{n+1} = r_n + \tfrac{3}{10}\left[\frac{2q_n - uq_n - r_n - 4ur_n}{u+1} + \frac{u}{(u+1)^2}\right].$$

Let
$$v_n = q_n + r_n, \quad w_n = q_n - r_n.$$

$$\therefore\ v_{n+1} = \tfrac{2}{5}v_n + \tfrac{3}{5}\frac{u}{(u+1)^2}.$$

$$\therefore\ v_n = \frac{u}{(u+1)^2} + \left(\frac{2}{5}\right)^n\left[v_0 - \frac{u}{(u+1)^2}\right],$$

$$w_{n+1} = \tfrac{1}{10}w_n + \frac{9(u-1)}{10(u+1)}v_n.$$

$$\therefore\ w_n = \frac{(1 - 10^{-n})\,u\,(u-1)}{(u+1)^3}$$
$$\div 10^{-n}w_0 + 3\left[\left(\frac{2}{5}\right)^n - 10^{-n}\right]\frac{(u-1)}{(u+1)}\left[v_0 - \frac{u}{u+1}\right].$$

$$\therefore\ q_n = \frac{u^2}{(u+1)^3} \div \frac{\left(\frac{2}{5}\right)^n(1-2u)}{u+1}\left[\frac{u}{(u+1)^2} - q_0 - r_0\right] + \frac{10^{-n}u\,(u-1)}{(u+1)^3}$$
$$- \frac{10^{-n}\left[(u-2)q_0 + (2u-1)r_0\right]}{u+1}.$$

$$r_n = \frac{u}{(u+1)^3} + \frac{\left(\frac{2}{5}\right)^n(u-2)}{u+1}\left[\frac{u}{(u+1)^2} - q_0 - r_0\right] - \frac{10^{-n}u\,(u-1)}{(u+1)^3}$$
$$+ \frac{10^{-n}\left[(u-2)q_0 + (2u-1)r_0\right]}{u+1}.$$

The corresponding expressions for p_n and s_n can readily be calculated. These numbers settle down to equilibrium only a little more slowly than in the tetraploid case, the ratio of successive differences from the final value being approximately $\frac{2}{5}$ instead of $\frac{1}{3}$. The higher polyploids approach equilibrium more slowly still.

EQUILIBRIUM IN A TETRAPLOID POPULATION WHICH IS PARTLY SELF-FERTILISED.

Suppose that a proportion λ of each generation is formed by random mating, the remainder by self-fertilisation. Let u have the same meaning as before, and let the population in equilibrium consist of:

$$pA^4 : 4qA^3a : 6rA^2a^2 : 4sAa^3 : ta^4,$$

the pooled gametic series being $xA^2 : 2yAa : za^2$, where

$$x = p + 2q + r, \quad y = q + 2r + s, \quad z = r + 2s + t.$$

Let $$p + 4q + 6r + 4s + t = x + 2y + z = 1,$$

$$\therefore \ x + y = \frac{u}{1+u}, \quad y + z = \frac{1}{1+u}.$$

Then, since all these quantities are unchanged from one generation to another:

$$p = (1 - \lambda)(p + q + \tfrac{1}{6}r) + \lambda x^2,$$
$$q = (1 - \lambda)(\tfrac{1}{2}q + \tfrac{1}{3}r) + \lambda xy,$$
$$r = (1 - \lambda)(\tfrac{1}{6}q + \tfrac{1}{2}r + \tfrac{1}{6}s) + \lambda(\tfrac{2}{3}y^2 + \tfrac{1}{3}xz),$$
$$s = (1 - \lambda)(\tfrac{1}{3}r + \tfrac{1}{2}s) + \lambda yz,$$
$$t = (1 - \lambda)(\tfrac{1}{6}r + s + t) + \lambda z^2.$$

$$\therefore \ y = q + 2r + s = \frac{5(1-\lambda)}{6}(q + 2r + s) + \lambda(xy + \tfrac{4}{3}y^2 + \tfrac{2}{3}xz + yz).$$

$$\therefore \ (1-\lambda)y = 4\lambda(xz - y^2) = 4\lambda\left[\frac{u}{(u+1)^2} - y\right].$$

$$\therefore \ y = \frac{4\lambda u}{(1 + 3\lambda)(u + 1)^2}.$$

$$6r = (1 - \lambda)(q + 3r + s) + 2\lambda(2y^2 + xz)$$
$$= (1 - \lambda)(r + y) + 2\lambda\left[3y^2 + \frac{(1-\lambda)y}{4\lambda}\right].$$

$$\therefore \ r = \frac{3y(1 - \lambda + 4\lambda y)}{2(5 + \lambda)}.$$

y can be eliminated from this expression, and by somewhat tedious but quite straightforward algebra we arrive at the equations:

$$p = \frac{u}{u+1} - \frac{4\lambda u}{(1+3\lambda)(u+1)^2}\left[1 + \frac{(1-\lambda)(7-\lambda)}{2(1+\lambda)(5+\lambda)}\right.$$
$$\left. + \frac{4\lambda u}{(1+\lambda)(u+1)} - \frac{24\lambda^2 u}{(1+3\lambda)(5+\lambda)(u+1)^2}\right],$$

$$q = \frac{4\lambda u}{(1+3\lambda)(u+1)^2}\left[\frac{(1-\lambda)^2}{(1+\lambda)(5+\lambda)}\right.$$
$$\left. + \frac{2\lambda u}{(1+\lambda)(u+1)} - \frac{24\lambda^2 u}{(1+3\lambda)(5+\lambda)(u+1)^2}\right],$$

$$r = \frac{6\lambda u}{(1+3\lambda)(5+\lambda)(u+1)^2}\left[1 - \lambda + \frac{16\lambda^2 u}{(1+3\lambda)(u+1)^2}\right],$$

$$s = \frac{4\lambda u}{(1+3\lambda)(u+1)^2}\left[\frac{(1-\lambda)^2}{(1+\lambda)(5+\lambda)}\right.$$
$$\left. + \frac{2\lambda}{(1+\lambda)(u+1)} - \frac{24\lambda^2 u}{(1+3\lambda)(5+\lambda)(u+1)^2}\right],$$

$$t = \frac{1}{u+1} - \frac{4\lambda u}{(1+3\lambda)(u+1)^2}\left[1 + \frac{(1-\lambda)(7-\lambda)}{2(1+\lambda)(5+\lambda)}\right.$$
$$\left. + \frac{4\lambda}{(1+\lambda)(u+1)} - \frac{24\lambda^2 u}{(1+3\lambda)(5+\lambda)(u+1)^2}\right].$$

Each of these quantities thus varies continuously between the values corresponding to inbreeding and random mating as λ varies between 0 and 1. The proportion t of recessives is in general little affected by a small amount of random mating in a self-fertilised population, but a good deal by a small amount of self-fertilisation in a random mating population. Thus putting $u = 9$, 10 per cent. of self-fertilisation increases the proportion of recessives from 0·01 to 0·12 per cent., while 10 per cent. of random mating only diminishes it from 10 to 5·5 per cent. Similar calculations could be made from higher polyploids, but would be very tedious except in the symmetrical case $u = 1$.

Cases involving several factors.

All these cases can be generalised so as to apply to organisms heterozygous for several factors. Thus the gametic series produced by a plant of composition $A^4 a^4 B^2 b^6$ is the expansion of

$$(1A^4 + 16A^3 a + 36A^2 a^2 + 16Aa^3 + 1a^4)(3B^2 b^2 + 8Bb^3 + 3b^4),$$

a series of 15 terms. If dominance were complete, the series would be

$$759AB : 207Ab : 11aB : 3ab.$$

The results in F_2 from homozygous lines differing in two unlinked dominant factors are $9 : 3 : 3 : 1$ in a diploid; $1225 : 35 : 35 : 1$ in a tetraploid; $159{,}201 : 399 : 399 : 1$ in a hexaploid, and so on. In practice one would not, even in a tetraploid, attempt to obtain a^4b^4 in the F_2 from $A^4b^4 \times a^4B^4$, but self a^4 and b^4 individuals in F_2, most of which would give double recessives in F_2.

The most striking consideration arising from an extension of the calculations on selfing to m factors is the extreme difficulty of establishing a homozygous dominant line by mere selfing. This is fully borne out by experience in *Primula sinensis*. Thus in F_6, *i.e.* after five generations of selfing an F_1, Table III shows that in a tetraploid there would be 26·6 per cent. of homozygous (quadriplex) dominants for a single factor, 46·85 per cent. of various heterozygotes, and 26·6 per cent. of recessives. Thus 36·2 per cent. of the dominants would be homozygous. But if the original population had been heterozygous for three unlinked factors only $0·362^3$, or 4·71 per cent. of the triple dominants would be homozygous for all three factors. In the corresponding case in a diploid 93·9 per cent. of single dominants would be homozygous for one factor, and 82·9 per cent. of the triple dominants homozygous for all three. Thus the probability of establishing a pure line in a self-fertile tetraploid is very small. In a self-sterile tetraploid or a higher polyploid it is negligible.

The populations reached as the result of random mating can be calculated with ease. Thus if u is the ratio of $A : a$, v that of $B : b$ in the population, the population in equilibrium is given by the expansion of $(uA + 1a)^{2m} (vB + 1b)^{2m}$. The approach to equilibrium remains fairly rapid in a tetraploid even when several factors are involved.

DOUBLE REDUCTION.

In a diploid organism equational non-disjunction, leading to the presence in one gamete of two chromosomes descended from the same single zygotic chromosome, implies an abnormal chromosome complement in the gamete. This is not so in a polyploid. Before meiosis each chromosome of a set has split into two chromatids, and if both the subsequent divisions are reductional, as opposed to equational, it is theoretically possible for both the chromosomes in a gamete to be derived from a pair of chromatids derived from one chromosome.

It is easy to calculate what should happen if the eight chromatids of a tetraploid are distributed at random into the gametes. A zygote $Aaaa$ gives eight chromatids $AAaaaaaa$. There are 28 combinations of these, two at a time, namely $1AA : 12Aa : 15aa$, so this is the gametic series to be expected. Similarly $AAaa$ should give $3AA : 8Aa : 3aa$. In general the probability of obtaining from a zygote $A_r a_{2m-r}$ a gamete $A_s a_{m-s}$ is

$$\frac{3m!\, m!\, 2r!\, (4m - 2r)!}{4m!\, S!\, (2r - s)!\, (m - s)!\, (3m - 2r + s)!}.$$

When $Aaaa$ is crossed with $aaaa$ we should obtain a zygotic ratio $13A : 15a$, as compared with $1A : 1a$ on the simple theory, while $AAaa$ would give $11A : 3a$, as compared with $5A : 1a$. The most striking differences are that 1 in 13 of the A zygotes from $Aaaa \times aaaa$ should be of composition $AAaa$, and that $AAAa \times aaaa$ should give 1 recessive to 27 dominants.

In *Primula sinensis* neither of these phenomena have been observed, and the ratios obtained agree pretty well with Muller's theory. In *Datura stramonium* Blakeslee, Belling and Farnham dealt with two unlinked factors P and A. P gave results agreeing very well with Muller's theory. A diverged in the direction here indicated, the figures obtained being intermediate between those expected on Muller's theory and on a basis of random assortment of chromatids.

On the latter theory the numbers in the second row of Table VI would be successive values of $\left(\dfrac{4m!\, m!}{3m!\, 2m!}\right)^2$, and the first five would be 4, 21·7, 121, 676, and 3785·2, so nulliplex individuals would be much commoner in F_2 than on Muller's theory.

The proportion of heterozygotes in tetraploid F_n would be

$$\tfrac{77}{65}\left(\tfrac{11}{14}\right)^{n-1} - \tfrac{12}{65}\left(\tfrac{6}{49}\right)^{n-1},$$

giving an ultimately rather faster rate of decrease than in Table III.

The stable population under random mating gives a pooled gametic series:

$$(5u^2 + u)\, AA : 8uAa : (u + 5)\, aa,$$

so that the proportion of nulliplex zygotes is $\dfrac{(1 + u/5)^2}{(1 + u)^4}$, the zygotic series is $9A^4 : 24A^3a : 34A^2a^2 : 24Aa^3 : 9a^4$.

In cases where Muller's type of segregation is not followed, the true values doubtless lie somewhere between these and the values given elsewhere in this paper. However, in most cases so far known the latter seem to be nearly correct.

SEX-LINKAGE OF THE *HUMULUS* TYPE.

Winge (1929) has reported that in *Humulus japonicus* the female is of composition XX, the male XXX. It is worth considering the probable mode of inheritance of a sex-linked factor in this case, which borders on polyploidy. There are two possible types of heterozygous male. AAa males should give $1AA : 2Aa$ male-producing, and $2A : 1a$ female producing pollen-grains. Aaa males should give $2Aa : 1aa$, and $1A : 2a$. The expected offspring from the 12 different types of mating are given in Table IV. If A is dominant the possible ratios among females are $5 : 1, 2 : 1, 1 : 1$ and $1 : 2$; among males $5 : 1, 2 : 1$ and $1 : 1$. In a population in equilibrium under random mating with a factorial ratio of $uA : 1a$ it can easily be verified that the population is in equilibrium if it consists of:

females, $u^2AA : 2uAa : 1aa$; and males, $u^3AAA : 3u^2AAa : 3uAaa : 1aaa$.

Hence the proportion of recessive females is $(u + 1)^{-2}$, of recessive males $(u + 1)^{-3}$, the latter being thus the $\frac{3}{2}$ power of the former, instead of the square, as with normal sex-linkage. The equilibrium is not reached at once. The rate of approach and the effects of inbreeding may easily be calculated as in the former cases. The departure from equilibrium is roughly reduced by $\frac{1}{3}$ in each generation.

TABLE IV.

Parents		Offspring		
♀ ♂		♀	♂	
$AA \times AAA$		AA	AAA	
$AA \times AAa$		$2AA : 1Aa$	$1AAA : 2AAa$	
$AA \times Aaa$		$1AA : 2Aa$		$2AAa : 1Aaa$
$AA \times aaa$		Aa		Aaa
$Aa \times AAA$		$1AA : 1Aa$	$1AAA : 1AAa$	
$Aa \times AAa$		$2AA : 3Aa : 1aa$	$1AAA : 3AAa : 2Aaa$	
$Aa \times Aaa$		$1AA : 3Aa : 2aa$		$2AAa : 3Aaa : 1aaa$
$Aa \times aaa$		$1Aa : 1aa$		$1Aaa : 1aaa$
$aa \times AAA$		Aa	AAa	
$aa \times AAa$		$2Aa : 1aa$	$1AAa : 2Aaa$	
$aa \times Aaa$		$1Aa : 2aa$		$2Aaa : 1aaa$
$aa \times aaa$		aa		aaa

SUMMARY.

The gametic series to be expected from various types of heterozygous autopolyploids are given, and the effects of self-fertilisation and random mating on populations are considered.

REFERENCES.

MULLER, H. J. (1914). "A new mode of segregation in Gregory's tetraploid Primulas." *Amer. Nat.* XLVIII, 508–12.

LAWRENCE, W. J. C. (1929). "The genetics and cytology of Dahlia species." *Journ. Genetics*, XXI, 125–58.

WINGE, Ö. (1929). "On the nature of the sex chromosomes in Humulus." *Hereditas*, XII, 53–63.

Part II

SYNTHETIC AND NATURAL POLYPLOIDS

Editors' Comments
on Papers 4 Through 8

4   **CLAUSEN and GOODSPEED**
Interspecific Hybridization in Nicotiana. II. A Tetraploid
Glutinosa-tabacum *Hybrid, an Experimental Verification of*
Winge's Hypothesis

5   **CLAUSEN, KECK, and HIESEY**
Excerpt from *Experimental Studies on the Nature of Species.*
II. Plant Evolution Through Amphiploidy and Autoploidy with
Examples from the Madiinae

6   **OWNBEY**
Natural Hybridization and Amphiploidy in the Genus Tragopogon

7   **SMITH-WHITE**
Polarised Segregation in the Pollen Mother Cells of a Stable
Triploid

8   **BEÇAK, BEÇAK, and RABELLO**
Further Studies on Polyploid Amphibians (Ceratophrydidae)
I. Mitotic and Meiotic Aspects

    This series of papers contains studies of both natural and synthetic polyploids and represents both normal and anomalous genetic systems. Paper 4 is important for two reasons, one of which is given in the title. Clausen and Goodspeed's example represents an experimental verification of the hypothesis proposed by Winge on the origin of polyploids. The second reason is the proof that spontaneous chromosome doubling in a highly sterile F_1 hybrid gave rise to a highly fertile polyploid. The authors also noted in this paper that the celebrated *Primula kewensis* apparently had the same cytological behavior as found in *Nicotiana*, and its fertility could be explained in the same way. We should point out that although the word *tetraploid* was used to describe the doubled *N. glutinosa* X *N. tabacum* hybrid ($2n = 72$), in actuality the authors were describing a hexaploid ($2n = 6X = 72$) because *N. tabacum* is a tetraploid ($2n = 4X = 48$) and *N. glutinosa* is a diploid ($2n = 2X = 24$).

The article by Clausen and colleagues (Paper 5) represents some of the best work of the time. It involved detailed field work, experimental hybridization and garden studies, cytological analyses, and morphological studies. It shows clearly the difference between species in a genetic sense and the kinds of morphological species that are acceptable to the plant systematist. The example is also interesting because of the high frequency of polyploids derived from selfing a rather sterile F_1 hybrid. This phenomenon demonstrates the efficacy of unreduced gametes as a mechanism for polyploid formation in some taxa, a method strongly advocated by Harlan and deWet (1975).

Ownbey's article (Paper 6) is of considerable interest for several reasons. The parental diploid taxa were introduced into the United States from Europe and are now widespread. The time of introduction of the parental species into the area of origin of the tetraploids is known probably within an accuracy of twenty years, so the time of origin of the natural tetraploids is better documented than any other known example. Although the hybrids giving rise to the newly described tetraploid species occur frequently, they apparently have evolved successful polyploids only rarely. The original populations still persist, and one of the tetraploid species has extended its range, as discussed in Paper 15.

Paper 7 by Smith-White describes an anomalous meiotic system in microsporocytes of *Leucopogon juniperinus,* a sexually reproducing triploid species. Megasporogenesis and embryology were discussed in a later paper (Smith-White, 1955) in the same journal. Disjunctions of four univalents during anaphase I of microsporogenesis differs significantly from random expectations. In megasporogenesis, there is again nonrandom disjunction of univalents, but only the n + n micropylar megaspore is functional. The presence of a diploid egg, triploid embryo, and pentaploid endosperm show that the functional pollen grains must be haploid. Because three of the four nuclei from each microsporocyte abort in this genus, there is selection for nuclei with a haploid chromosome number. This system in *Leucopogon juniperinus* is of considerable interest because in some ways it represents a transition type between completely random univalent disjunction and the highly directed type found during megasporogenesis in the well-known example of *Rosa canina* described by Täckholm (1922).

The article by Beçak and colleagues (Paper 8) is not the first to report polyploidy in a vertebrate species, but it is the first to give quantitative data on meiotic chromosome configurations that are amenable to testing by various models that predict configuration frequencies. The species discussed, as well as other frog taxa, are of interest because they represent vertebrate animals that are true breeding, sexually reproducing polyploids. The first paper in this series by the same authors (Beçak et al., 1966) merely reported the tetraploid condition. Later studies have demonstrated that the diploid and tetraploid populations are separate.

REFERENCES

Beçak, M. L., W. Beçak, and M. L. Rabello, 1966, Cytological Evidence of Constant Tetraploidy in the Bisexual South American Frog *Odontophrynus americanus, Chromosoma* **19:**188–193.

Harlan, J. R., and J. M. J. deWet, 1975, On Ö. Winge and a Prayer: The Origins of Polyploidy, *Bot. Rev.* **41:**361-390.

Smith-White, S., 1955, The Life History and Genetic System of *Leucopogon juniperinum, Heredity* **9:**79-91.

Täckholm, G., 1922, Zytologische Studien über die Gattung *Rosa, Acta Horti Bergiani* **7:**97–381.

4

Reprinted from *Genetics* **10**:278–284 (1925)

INTERSPECIFIC HYBRIDIZATION IN NICOTIANA. II. A TETRAPLOID *GLUTINOSA-TABACUM* HYBRID, AN EXPERIMENTAL VERIFICATION OF WINGE'S HYPOTHESIS

R. E. CLAUSEN AND T. H. GOODSPEED

University of California, Berkeley, California

Received January 29, 1925

Numerous investigations have shown that in plants the chromosome numbers of the species of a genus are often in arithmetical progression. Excellent examples are afforded by Triticum, $n = 7$, 14 and 21; by Rosa, $n = 7$, 14, 21, 28; and by Chrysanthemum, $n = 9$, 18, 27, 36 and 45. Appeal has often been made to tetraploidy as a method of increase in chromosome number, but WINGE (1917) has pointed out that successive doubling of chromosome number would give rise to geometrical rather than arithmetical series, and has suggested as an alternative hypothesis interspecific hybridization followed by doubling of chromosome number. The process suggested by WINGE would establish tetraploid interspecific hybrids having $2(n_1 + n_2)$ chromosomes, where n_1 and n_2 represent the haploid numbers of the parent species; and since such forms are essentially homozygous diploids, they may reasonably be expected to be fertile and constant. WINGE was unable, however, to present any experimental evidence in support of his hypothesis. In this article a preliminary account is given of a tetraploid hybrid of *Nicotiana glutinosa* and *N. tabacum*, and attention is invited to *Primula kewensis*, apparently a tetraploid *P. floribunda-verticillata* hybrid, which together offer the necessary experimental verification of the hypothesis.

The two species of Nicotiana employed in the present investigations are very distinctly different. *N. glutinosa* is particularly characterized by its villose pubescence; its distinctly petioled, cordate leaves; its bilabiate flowers; and its sparsely branched, racemous inflorescence. It is one of the most distinct species of the genus. *N. tabacum* is familiar to almost everyone as the tobacco of commerce. Descriptions and more complete illustrations of the forms employed in the experiments described below may be found in SETCHELL's (1912) account of the genus. The chromosome number of *N. glutinosa* is $n = 12$ (GOODSPEED 1923), of *N. tabacum*, $n = 24$ (WHITE 1913, GOODSPEED 1923).

The two species have frequently been crossed. Reciprocal hybrids may be obtained, although hybridization is attended with some difficulty. Usually only a few viable seeds are produced in a capsule; in our experience an average of about ten or twelve. The F_1 hybrids are weak in germination and development, but they grow on to maturity. This, in brief, is the behavior which has been noted for all varieties of *N. tabacum* which have been tested, except one, the variety "Cuba" (cf. GOODSPEED 1915). From "Cuba" ♀ × *glutinosa* ♂ full capsules of seed are obtained. These seeds are of the same order of viability as pure seed of the species; the seedlings are vigorous; the hybrid plants develop to a height approximately equal to that of "Cuba"; and they branch profusely. GÄRTNER (1849) apparently also observed marked differences in the vigor of F_1 *glutinosa-tabacum* hybrids, when different *tabacum* varieties were employed; but as FOCKE (1881) points out, it is difficult to know how to judge these results because of the numerous discrepancies in GÄRTNER'S account of his observations. Despite these differences in vigor, the F_1 hybrids are always intermediate in appearance and they are apparently completely sterile. Numerous attempts to secure seed by backcrossing to the parental species under a variety of conditions have failed, and no seed has been found in open-pollinated capsules.

In 1922 from a single capsule of *glutinosa* ♀ × *tabacum* var. *purpurea* ♂, three plants were secured, which were grown under the garden number, 22062. Two of these plants were obvious hybrids. They were both small plants, about two feet in height, with few, slender branches and small leaves. The flowers exhibited a strong tendency towards the bilabiate shape of *N. glutinosa*, but the color was carmine, like that of *purpurea*. The leaves showed distinct evidences of *glutinosa* in their cordate shape. One of these plants was partially fertile, the other completely sterile. No other differences were noticed at that time. The third plant was very strikingly different from the other two; in fact, if our notes and memory may be depended upon, it was identical with the *purpurea* haploid which was obtained later (cf. CLAUSEN and MANN 1923). Unfortunately, we lost it during the winter of 1922-1923 because of unfavorable greenhouse conditions.

A number of flowers on the single partially fertile plant were hand-pollinated and gave selfed seed without difficulty. The capsules were harvested separately. In the season of 1923, 155 plants were obtained from one of these capsules. These, however, were set out in the field late in the season, and they did not mature. In 1924 a culture of 65 plants was grown under the garden number, 24123. For purposes of comparison there was available at the same time a culture, 24192, of 15 F_1 plants of *purpurea* ♀ ×*glutinosa* ♂.   Much to our surprise, with one exception, the 65 F_2 plants of 24123 were uniform and almost identical with the F_1 plants of 24192. There were, however, important minor differences, which were found constantly to characterize the two populations. The F_1 plants were completely sterile. They set no seed on open-pollination, and twenty-five attempts at back-pollination with each of the parental species failed to give seed.   Under these circumstances, capsules were retained for as long as three weeks, but in no case did they reach maturity. The plants of the F_2 population were reasonably fertile, and uniform in this respect. Large plump capsules were obtained from open-pollinated and hand-pollinated flowers and also from crosses with the parent species. These capsules contained a fair quantity of seed, but not so much as capsules of normal species.

The plants of the two populations exhibited a close correspondence in morphological characters. Both populations were very uniform. The F_2 plants had slightly, but constantly, larger flowers than F_1 plants, and the anthers were conspicuously larger. F_2 plants produced abundant pollen, most of the grains of which were normal in appearance: F_1 plants produced scanty pollen, consisting entirely of shrivelled empty grains. F_1 plants averaged about two feet in height, F_2 plants about a foot and a half. Despite the difference in height, which was probably due to their unfavorable start in the flats, the general impression given by F_2 plants was that of a slight enlargement to scale of characters of F_1, aside from those features obviously connected with the difference in fertility.

One F_2 plant stood out from the rest by reason of its remarkable robustness. This plant eventually attained a height of six feet, and produced numerous stout branches. Despite the difference in size, however, the general morphological characters were those of the other plants of the population on an enlarged scale. Vegetative characters were proportionately enlarged, flower size only slightly. The plants of this population were all very weak as seedlings, and they grew very feebly during the time they were in flats. It is believed that the general small size of the plants in the population was due to this stunting during their early growth and that 24123P55, the robust individual, merely by some fortunate chance overcame this difficulty.

The uniformity of F_2 and its close resemblance to F_1 immediately suggested the need for cytological examination. Excellent aceto-carmine smears of pollen mother cells were easily secured; the stage of development of anthers containing them in proper condition being rather later than is usually the case. Examinations were made of material from several plants, including the robust plant described above, which gave exactly the same results as the others. The general impression of the cytological figures was one of regularity of meiotic division rather different from the irregular distribution seen in normal F_1 *glutinosa-tabacum* hybrids. Numerous counts of first-metaphase figures showed 36 bivalents. In a few instances it was possible to count both metaphase plates in the second division, and to determine that each contained 36 chromosomes. There was of course some doubt as to the exact count in a number of figures, but only to the extent of one or two chromosomes. There were minor irregularities in distribution, evidenced by precocious splitting, lagging, and microcyte formation; but these features, while noticed, were not studied in detail. There is no doubt that the chromosome number of the plants of this population was uniformly $n = 36$, $2n = 72$.

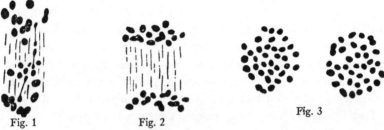

FIGURE 1.—Portion of an anaphase of the normal F_1 *glutinosa-tabacum* hybrid.
FIGURE 2.—The same of the tetraploid hybrid.
FIGURE 3.—Homotypic metaphase, polar view, of the tetraploid hybrid.

Pollen-mother-cell heterotypic anaphase conditions in the sterile F_1 *glutinosa-tabacum* hybrid and in the tetraploid hybrid are illustrated in figures 1 and 2, while figure 3 shows two homotypic metaphase plates of the tetraploid, each polar view containing 36 chromosomes. Figure 1 was drawn from fixed material, the other two figures from aceto-carmine preparations. As will be noted (figure 1), the behavior of the bivalent and univalent chromosomes closely parallels that found in the F_1 *tabacum-sylvestris* hybrid elsewhere described (GOODSPEED 1923). The bivalent partners are approaching the poles while the univalents are in the equatorial zone, either dividing or preparing to divide. In the tetraploid, on the other hand, there appear to be no univalent chromosomes and the bivalent partners move in regular fashion to the poles. No attempt is made in either figure 1 or 2 to represent the full chromosome complement.

Pollen conditions are illustrated in figures 4 and 6 for the tetraploid hybrid and in figure 5 for the normal sterile F_1. These figures are reproduced from photomicrographs of pollen preparations stained with aceto-carmine. As will be noted, the pollen of the normal hybrid consists exclusively of shrivelled grains devoid of contents. The pollen of the tetraploid hybrid consists mostly of large grains apparently normal in protoplasmic contents. Measurements were made of pollen grains of the tetraploid hybrid and of its *tabacum* parent, but unfortunately no pollen of *glutinosa* was available at the time measurements were made. The average diameter of pollen grains in the tetraploid hybrid was found to be 46.3 microns, of *tabacum*, 36.7. The volumes are therefore, in the ratio of approximately 2 : 1 (106 : 49).

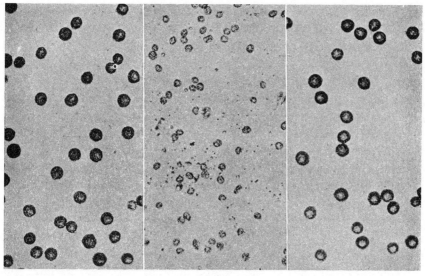

Fig. 4   Fig. 5   Fig. 6

FIGURE 4.—Portion of a photomicrograph of pollen of the tetraploid *glutinosa-tabacum* hybrid, 24123P55.

FIGURE 5.—Of the normal F₁, 24192P4.

FIGURE 6.—Of another plant of the tetraploid hybrid, 24123P58. The preparations were stained in aceto-carmine and the photomicrographs were taken at the same magnification.

The cytological findings supply an obvious explanation for the uniformity and constancy of this hybrid. Since in *glutinosa*, $n = 12$, and in *tabacum*, $n = 24$, the F₁ hybrid normally has 36 chromosomes. This was undoubtedly the case in the sterile F₁ described above. The original fertile F₁ plant, 22066P2, must have arisen from a doubling of the chromosome number immediately or soon after fertilization, by which a tetraploid hybrid with 36 pairs of chromosomes was produced. Such a plant may be represented by the chromosomal formula, 12 GG + 24 TT;

and, if *glutinosa* and *tabacum* homologues pair regularly, fertility and constancy follow as a matter of course, for every gamete would then contain 12 *glutinosa* and 24 *tabacum* chromosomes. If this explanation is correct, an interspecific hybrid may be expected to become fertile and constant by simple doubling of its chromosome number.

These observations naturally recall the case of *Primula kewensis*, the much discussed hybrid of *P. floribunda* with *P. verticillata*. According to accounts of its origin as described by Miss DIGBY (1912) and by the Misses PELLEW and DURHAM (1916), the original hybrid was sterile; but it eventually produced a fertile bud-sport which gave rise immediately to the fertile, comparatively constant form now known as *P. kewensis*. Miss DIGBY found that the chromosome numbers of *P. floribunda* and *P. verticillata* were both $n=9$ and $2n=18$, that the sterile hybrid had 18 chromosomes and the fertile *P. kewensis*, 36. *P. kewensis*, therefore, is evidently a tetraploid hybrid; and as WINKLER (1920) and RENNER (1924) suggest, this fact probably accounts for its genetic behavior. If it contains 9 pairs each of chromosomes of *P. floribunda* and *P. verticillata*, and homologues of each species pair regularly, the situation is exactly the same as that described in the tetraploid Nicotiana. It seems more reasonable to adopt this explanation of its chromosome number, since it accounts so well for the genetic results thus far obtained with it, rather than that of transverse fission suggested by FARMER and DIGBY (1914) on the basis of chromosome measurements, which has been accepted by GATES (1924) in his recent discussion of polyploidy.

The confirmation of WINGE's hypothesis afforded by the instances described above extends only to establishment of the tetraploid chromosome condition, and not to the method of origin described by him. The establishment of the condition is evidently a mutational event, analogous to that which occurs in the establishment of the tetraploid condition in pure species. The tetraploid hybrid condition may, however, arise in a variety of ways: (1) by doubling of chromosome number immediately subsequent to fertilization; (2) by bud-variation in an F_1 interspecific hybrid; (3) by crossing together tetraploid representatives of two different species; and (4) by irregular distribution of chromosomes in an interspecific hybrid in which the chromosomes do not pair in meiosis, as suggested by COLLINS and MANN (1923). It may be possible, therefore, that tetraploid hybrids have the significance in the origin of new chromosome numbers ascribed to them by WINGE.

LITERATURE CITED

CLAUSEN, R. E., and MANN, MARGARET C., 1924 Inheritance in *Nicotiana Tabacum*. V. The occurrence of haploid plants in interspecific progenies. Proc. Nation. Acad. Sci. **10**: 121-124.

COLLINS, J. L., and MANN, MARGARET C., 1923 Interspecific hybrids in Crepis. II. A preliminary report on the results of hybridizing *Crepis setosa* Hall. with *C. capillaris* (L.) Wallr. and with *C. biennis* L. Genetics **8**: 212-232.

DIGBY, L., 1912 The cytology of *Primula kewensis* and of other related Primula hybrids. Annals of Bot. **36**: 357-388.

FARMER, J. B., and DIGBY, L., 1914 On dimensions of chromosomes considered in relation to phylogeny. Phil. Trans. Roy. Soc. B **205**: 1-25.

FOCKE, W. O., 1881 Die Pflanzen-Mischlinge. 569 pp. Berlin: Borntraeger.

GATES, R. R., 1924 Polyploidy. British Jour. Exp. Biol. **1**: 153-182.

GÄRTNER, C. F., 1849 Versuche und Beobachtungen über die Bastarderzeugung im Pflanzenreich. 790 pp. Stuttgart: The Author.

GOODSPEED, T. H., 1915 Parthenogenesis, parthenocarpy and phenospermy in Nicotiana. Univ. California Publ. Bot. **5**: 249-272.

1923 A preliminary note on the cytology of Nicotiana species and hybrids. Svensk Bot. Tid. **17**: 472-478.

PELLEW, CAROLINE, and DURHAM, FLORENCE M., 1916 The genetic behavior of the hybrid *Primula kewensis*, and of its allies. Jour. Genetics **5**: 159-182.

RENNER, O., 1924 Vererbung bei Artbastarden. Zeit. indukt. Abstamm. u. Vererb. **33**: 317-347.

SETCHELL, W. A., 1912 Studies in Nicotiana. I. Univ. California Publ. Bot. **5**: 1-86.

WHITE, O. E., 1913 The bearing of teratological development in Nicotiana on theories of heredity. Amer. Nat. **47**: 206-228.

WINGE, O., 1917 The chromosomes. Their numbers and general significance. Compt. Rend. Lab. Carlsberg **13**: 131-275.

WINKLER, H., 1920 Verbreitung und Ursache der Parthenogenesis im Pflanzen und Tierreiche. 231 pp. Jena: Gustav Fischer.

5

Reprinted from *Carnegie Inst. Washington Publ. No. 564*, pp. 22–45 (1945)

EXPERIMENTAL STUDIES ON THE NATURE OF SPECIES.
II. PLANT EVOLUTION THROUGH AMPHIPLOIDY AND AUTOPLOIDY WITH EXAMPLES FROM THE MADIINAE

J. Clausen, D. D. Keck, and W. M. Hiesey

[*Editors' Note:* In the original, material precedes this excerpt.]

III

MADIA CITRIGRACILIS

The *Eumadia* section of *Madia* contains as a subgroup a highly variable species complex of several closely related units native on the west coasts of both North and South America. The most widely known member of this complex is *Madia sativa* Mol., a species often found as a weed in various parts of the world and having 16 pairs of chromosomes (Johansen, 1933). *Madia gracilis* (Sm.) Keck, a more slender and smaller type, is one of the nearest relatives of *sativa*. Indeed, certain 16-chromosome forms of *gracilis* are difficult to separate from forms of *sativa* because of their close resemblance and similar behavior. These species intercross fairly readily and produce hybrids that are partially fertile. Their chromosomes are homologous, as is indicated by regular pairing, but the partial sterility of the hybrids shows that the chromosomes are not freely interchangeable.

Madia gracilis has been known for a century under the name *M. dissitiflora* (Nutt.) T. et G., but it was originally described by Sir James Edward Smith as a species of *Sclerocarpus*, a fact which was overlooked until recently (Keck, 1940). It is a very common plant on relatively dry soils from the Upper Sonoran up to the Hudsonian life zone, and is distributed from the Canadian border southward through the western United States to central Utah, Nevada, and northern Baja California, as is shown on the map, figure 44 (p. 41). It differs from the late-flowering, larger *M. sativa* in being principally a member of the spring flora and rarely a weed.

In the first period of these investigations a number of races of *gracilis* were cytologically examined and found to have 16 pairs of chromosomes. These races were widely distributed, but all were from relatively low elevations. Then a montane race, which was originally thought to belong to this species, was discovered to have 24 pairs of chromosomes. This race was grown from seed taken from a herbarium specimen collected by Frank W. Peirson on Burney Mountain, Shasta County, California, an area from which this species was not otherwise known. During a visit to this locality in 1938 additional material was obtained, which also had 24 pairs of chromosomes.

When this difference in chromosome number was first discovered, the 24-chromosome form was inspected closely for characters whereby it might be distinguished from the 16-chromosome form. Its rosettes and

67

lower leaves were found to have a denser, softer, appressed pubescence, and the cauline leaves had a wider, somewhat more flaring base. Also, its central stem was overtopped considerably by the lateral branches, whereas in the 16-chromosome form the central stem equals or even overtops the laterals.

The discovery of a 24-chromosome *Madia* was of evolutionary interest, as this is the highest number found in any species of the Madiinae. The next highest is $n = 16$ of *Madia gracilis* and *sativa*. Other relatives in *Madia* have 8 pairs of chromosomes. Accordingly, these obviously related species are members of a polyploid series with a basic number of 8. The common form of *Madia gracilis* is therefore tetraploid, and the new form hexaploid.

It is very unlikely that the hexaploid form would arise from a tetraploid by autoploidy, for doubling of the chromosomes of the tetraploid would result in an octoploid with 32 pairs. Accordingly, this 24-chromosome form was suspected of having arisen from a cross between an 8- and a 16-chromosome species.

COMPARISON OF THE NEW SPECIES WITH ITS SUPPOSED PARENTS. On account of their close morphological resemblance, *Madia gracilis* was thought to have been the 16-chromosome parent of the 24-chromosome form. On the basis of morphology, distribution, chromosome numbers, and known genetic behavior of *Madia* species, *M. citriodora* Greene was considered the only possible 8-chromosome ancestor in the present flora. Accordingly, the 24-chromosome material from Burney Mountain was given the provisional name *Madia citrigracilis* after its supposed parents.

Until the experimental investigations were made, *Madia citriodora* had never been thought to be a close relative of the *Madia sativa* complex. Rather, it had been thought closer to *M. elegans* D. Don, $n = 8$, principally because both species have sterile disk akenes. Asa Gray had even transferred *citriodora* from *Madia* to *Hemizonia* because of its broad ray akenes that were an exception to the key difference between these genera. Unlike *elegans* but like *gracilis*, however, it has inconspicuous, pale rays and is self-incompatible, and even from its morphology, *citriodora* is now universally accepted as a species of *Madia*.

Madia citriodora is a spring-flowering species like *gracilis*. Even though it has somewhat more showy ray florets and occupies similar habitats, yet it is much less frequently collected and must be regarded as comparatively rare. Its distribution, which extends from southern Washington to northern California in relatively dry environments east of the Cascade Mountains, is shown in figure 44 (p. 41).

An analysis of the more important morphological differences between *Madia citriodora, gracilis,* and *citrigracilis* is given in table 4. It is evident from this table that in most of its vegetative characters the new species resembles the diploid *citriodora,* but in the characters of the flowers and fruit it is very similar to its tetraploid relative, *gracilis.* The differences in habit among the three species are illustrated by the upper row of plants in figure 19. Considering their chromosome numbers, it is the morphological similarities rather than the differences that are striking.

TABLE 4

MORPHOLOGICAL DIFFERENCES BETWEEN MADIA CITRIODORA,
M. GRACILIS, AND NATURAL M. CITRIGRACILIS

Character	M. citriodora	M. citrigracilis	M. gracilis
Side branches	surpassing main axis	surpassing main axis	shorter than or equaling main axis
Pubescence	densely villous, appressed	± densely villous, ± appressed	coarsely pilose, spreading
Herbage color	gray-green	gray-green	light green
Odor	lemon-scented	not lemon-scented	not lemon-scented
Stem leaves	broad at base	broad at base	narrow at base
Involucre	turbinate to urceolate	ovoid to globose	ovoid to globose
Ligule length	7–10 mm.	6–8 mm.	4–7 mm.
Receptacle	densely villous	glabrous or nearly so	glabrous or nearly so
Ray akenes	triangular in cross section	intermediate	laterally flattened
Disk akenes	sterile	fertile	fertile
Chromosome number	$n = 8$	$n = 24$	$n = 16$

SYNTHESIS OF MADIA CITRIGRACILIS

The discovery of the hexaploid *Madia,* and the deduction that it possibly arose through amphiploidy from a crossing between the diploid *Madia citriodora* and the tetraploid *gracilis,* prompted an attempt to synthesize it through hybridization (Clausen, Keck, and Hiesey, 1937). A race of *citriodora* from Parker Creek, Warner Mountains, Modoc County, and one of *gracilis* from the Middle Fork of Smith River, Del Norte County, from the two northern corners of California, were selected for crossing. Both were typical of their respective species, al-

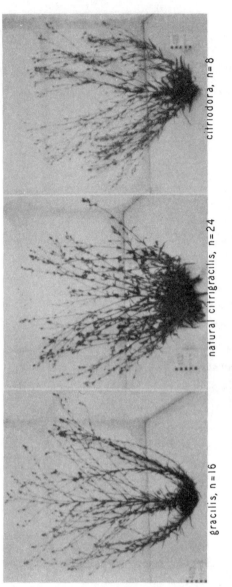

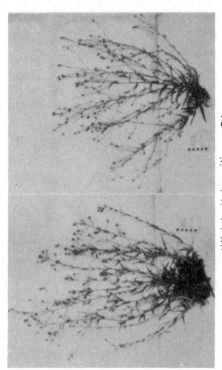

FIG. 19. *Above: Madia gracilis* from Middle Fork of Smith River, Del Norte County, California; *M. citri-gracilis* from Burney Mountain, Shasta County; and *M. citriodora* from Parker Creek, Warner Mountains, Modoc County. *Below:* Plant 4131, F₃ amphiploid derivative of he *gracilis* and *citriodora* races shown above; and plant 4134, a spontaneous F₃ amphiploid of different parentage. (All plants in a uniform garden at Stanford.)

though the progenitors of the wild hexaploid could not have come from so widely separated localities.

Mechanical difficulties offer problems in emasculation of these regularly self-fertile species, because pollen is shed very early in the tiny, very glandular buds. Since some outcrossing does take place, however, open pollination in isolation plots was resorted to in the hope of obtaining hybrids. In 1937, in a plot well away from other species of *Madia*, one plant of *gracilis* was surrounded by a circle of seven plants of *citriodora*,

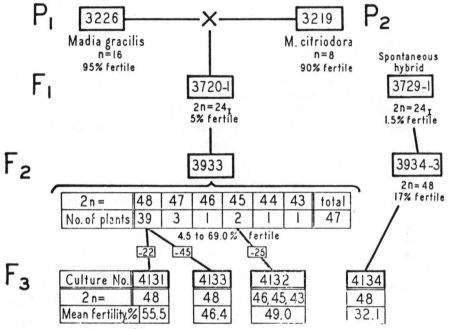

FIG. 20. Pedigrees of *Madia citrigracilis*, showing parentage, number of plants, chromosome numbers, and fertilities.

and in another, one plant of *citriodora* was surrounded by seven plants of *gracilis*. In each plot the seeds of the central plant were harvested.

From the plant of *citriodora* approximately 400 akenes were obtained, from which, in 1940, only 34 plants were grown to maturity. All these proved to be selfed *citriodora*. About 1500 apparently good akenes were harvested on the *gracilis* parent in the reciprocal cross, and part of these were sown. They produced a culture of 399 plants, one of which, 3720-1 of figure 20, proved to be the desired F_1 hybrid. The others were selfed *gracilis*.

THE F_1 GENERATION. The hybrid was recognized among the selfed plants of *gracilis* by its wider leaf bases, its denser, more appressed pubescence, its longer side branches, and its faint *citriodora* odor. This plant lacked the alternate arrangement of the branches typical of both parents, for it was repeatedly subverticillate or subumbellate, a peculiarity often accompanied by a degeneration of the terminal bud of the central leader. This character was reproduced in later generations, but was not encountered in other hybrids between *citriodora* and members of the *sativa* complex. The final proof of the hybrid origin of this plant was found in its chromosome number, $n = 24$.

The F_1 plant was self-pollinated in an insect-proof greenhouse, and although it was quite highly sterile, a few akenes developed from both ray and disk florets. The disk akenes of *citriodora* are sterile, whereas those of *gracilis* are fertile. Therefore, the partial fertility of the disk akenes in the F_1 hybrid indicates that fertile disks are at least partially dominant over sterile disks, as they were in the *Madia nutans* $\times$ *Rammii* cross. Fertility of disk akenes has been employed as a specific, sectional, and even generic character in the Madiinae. It is consequently noteworthy that characters of such taxonomic importance are inherited in a manner similar to those used for distinguishing minor subspecific categories. Such evidence points to a fundamental similarity in the nature of the differences that distinguish both the large and the small taxonomic categories in plants.

THE F_2 POPULATION. Among some 1900 akenes harvested on the single first-generation hybrid, 94 appeared to be good, indicating a fertility of approximately 5 per cent. The following year these produced a second generation of 47 plants. This F_2 was quite uniform and faithfully reproduced the characters of the F_1, even to the peculiar subumbellate branching. The variation was of a minor nature and involved such characters as the citric odor of *citriodora*, which was present in varying degree in different plants and completely undetectable in others, and the slight dwarfing of a few individuals. On the whole, however, the culture consisted of plants as tall and vigorous as those of the parental species and of the natural form of *citrigracilis*.

This F_2 population was morphologically rather similar to garden plants of natural *citrigracilis*. The latter had paniculate instead of subumbellate branching above, no citric odor, and heads less strongly glandular. In more fundamental flower and fruit characters the two forms were essentially the same.

Among the 47 F_2 plants, 39 were fully amphiploid, with $2n = 48$ chromosomes, double the number of the F_1 hybrid. The remaining 8 plants varied in chromosome number between $2n = 43$ and 47, so they also were nearly amphiploid. This is shown in the pedigree, figure 20. Almost all the functional sex cells of the F_1 hybrid, therefore, must have contained all the chromosomes of both *gracilis* and *citriodora*.

The F_2 plants are listed in table 5, on page 32, and belong to three cultures, 3933 to 3935. The last two of these were derived from spontaneous hybrids to be mentioned later. All the plants of 3934 and 3935 and the plants 3933-1 to -19 were kept in a screened greenhouse to test their self-fertility, but the others were planted in the garden. It was found that the protected plants in the greenhouse were as fertile as those pollinated in the garden.

Forty-three of the F_2 plants of 3933 matured so that their fertility could be determined. As is shown in table 5, this varied between 4.5 and 69.0 per cent, averaging 41.6 per cent. This is a substantial increase from the 5 per cent fertility of the F_1 hybrid, and also much higher than the fertility of 5.5 per cent observed in the F_2 plants of *Madia nutrammii*. Even the eight plants with less than 48 pairs of chromosomes were from 4.5 to 62.0 per cent fertile, averaging 33.6 per cent. As in *gracilis* and natural *citrigracilis*, the ray and disk akenes were equally fertile.

SPONTANEOUS AMPHIPLOIDS BETWEEN MEMBERS OF THE M. SATIVA COMPLEX AND M. CITRIODORA. Three additional hybrids arose spontaneously in the garden as a result of insect pollination between *M. citriodora* and members of the *sativa* complex. They were detected by their morphological characters, their odor, and their chromosome number (24 somatic chromosomes instead of 32). The *Madia gracilis* parents of these hybrids were themselves sister intraspecific hybrids between the rare broad-akened form that has been described as *M. anomala* Greene and the common narrow-akened form, which are completely interfertile. These crossings took place in two different years, but the three hybrids were discovered in 1940 when many cultures of *Madia* were grown. The hybrids were as follows:

1. Plant 3729-1: *Madia gracilis* × *M. citriodora* F_1 (partial pedigree in fig. 20). The *citriodora* parent could have been either the Parker Creek strain, from Modoc County, or one from near Timbered Crater, in Siskiyou County. The F_1 hybrid was 1.5 per cent fertile; it resembled very closely the F_1 hybrid 3720-1 described above, and had similar cytological characteristics. From unprotected heads of this plant, pollinated in the garden, four F_2 offspring (3934) were obtained, three having $2n = 48$ chromosomes, and one $2n = 50$ (see table 5). The fertilities of these F_2 plants were much lower than those of the amphiploids of the previous hybrid. The plant with the highest fertility, 17 per cent, was

3934-3, with 48 regularly paired chromosomes. A vigorous F_3 generation, 4134 in the pedigree (fig. 20), was obtained from this plant, which was selfed in a screened greenhouse. Only part of the akenes were sown, but approximately 250 of them germinated. One of these plants is pictured in figure 19, in the lower row to the right. Unlike the previous amphiploid, the derivatives of this lineage resembled native *citrigracilis* from Burney Mountain in their branching. Two plants that were investigated were found to have $2n = 48$ chromosomes. The fertilities of twenty plants varied between 2 and 74 per cent, with an average of 32.1 per cent, a considerable improvement over the 17 per cent of the F_2 plant.

2. Plant 3737-1: *Madia (gracilis × sativa) × citriodora*, Parker Creek, F_1. The maternal parent was an F_1 hybrid between a plant of the *gracilis* parentage mentioned above ($n = 16$) and the closely related *sativa* ($n = 16$). The chromosomes of *gracilis* and *sativa* pair regularly, but some genetic incompatibility between these species is indicated by the fact that their hybrids are only 3 to 30 per cent fertile, depending upon the races crossed. In spite of the fact that the maternal genome was a mixture from two forms of *gracilis* and one of *sativa*, this hybrid behaved like 3720-1, which contained a maternal genome of only one species. The influence of the genes of *sativa* was found in the slightly more robust and more densely glandular stems as compared with the two F_1 hybrids described above. This hybrid was completely sterile, for no good akenes were found among approximately 1500 empty and undeveloped ones.

3. Plant 3737-2: Same origin as 3737-1, described above. Although this hybrid appeared to be identical morphologically with its sister plant, it differed in being 3.1 per cent fertile, for among some 350 akenes, 11 apparently good ones were found, from which seven F_2 plants were grown, all with $2n = 48$ chromosomes. This F_2 population is culture 3935 of table 5 (p. 32). All its plants were straight amphiploid, and their fertilities varied between 12 and 49 per cent, but no F_3 was grown from it. That this amphiploid contained some *sativa* genes was evident from the inflorescences, which differed from the others in being less branched and in having larger, more congested heads. The stems were also coarser and their upper leaves had a wider base than in the preceding hybrids. This amphiploid lacked the odor of *citriodora* (although the F_1 parent had it), and, like the offspring of 3729-1, it closely approached native *Madia citrigracilis* in habit.

Considerable other experimental evidence points to the close cytogenetic relationship between *sativa*, *gracilis*, and the *anomala* form of *gracilis.* Therefore it is not so surprising that the different combinations of their chromosomes in a hybrid did not interfere with the production of an amphiploid when a genome of *citriodora* was introduced. It appears that the combinations of forms of *gracilis* or *sativa* with *citriodora* produce only amphiploids, because three F_1 plants of considerably different parentage have produced 58 F_2 offspring, of which 49 are straight amphiploids and the remainder deviate but slightly.

CYTOLOGY OF MADIA CITRIGRACILIS AND THE PARENT SPECIES. Pairing of the chromosomes is perfectly regular in both parents of the artificial amphiploid. The 16 pairs of *gracilis* are somewhat smaller, on the average, than the 8 pairs of *citriodora,* but the differences were scarcely large enough to be recognized. Somatic and meiotic chromosomes of the *gracilis* parent are shown in figures 21 and 22, and the corresponding plates of the *citriodora* parent in figures 23 and 24. There is apparently no significant difference in the size of the cells of the two species, for *citriodora* has remarkably large cells for a diploid.

Suitable material for cytological study of the first-generation hybrids between *gracilis* and *citriodora* was fairly scarce, because buds are small when meiosis takes place, and the best material was too old by the time the hybrids were mature enough to be recognized. The study of the pairing of the chromosomes in the F_1 hybrids has therefore been made on all the available bud material, including that from the artificially produced F_1 plant (3720-1), and from two of the spontaneous hybrids (3737-1 and -2). The meiotic behavior of the chromosomes was essentially the same in all three.

Very little pairing of chromosomes was observed in the F_1 hybrids. Occasionally one or two (very rarely three or four) very loose pairs could be observed in first metaphase. Very commonly, however, all chromosomes were single, although they aligned in a fairly regular equatorial plate. Obviously, the chromosomes of *citriodora* are not homologous with any of those of *gracilis.*

In the comparatively fertile artificial hybrid, 3720-1, no pairing was observed, but suitable material of this plant was very scarce. A somatic cell in the floral region had 24 chromosomes (fig. 25), which indicates that this plant was a hybrid and that it had not doubled its chromosomes somatically. Meiosis was found in only one anther lobe, which contained 9 to 10 cells. These were in metaphase, each with one chromosomal plate, and 24 chromosomes could be counted (fig. 26). Side views gave the impression that 24 pairs were present, but this interpretation is excluded because the somatic number was only 24. These cells were there-

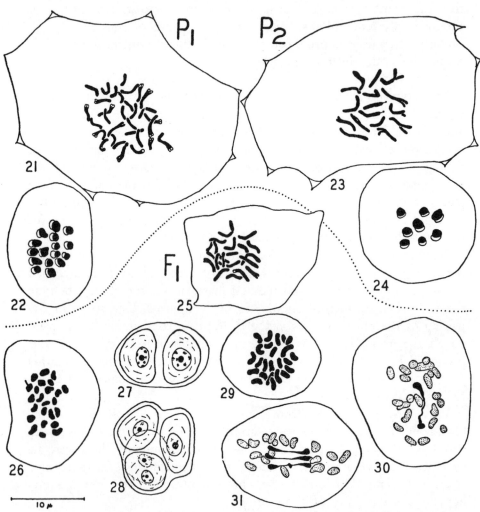

Figs. 21–31. Somatic and meiotic chromosomes of *Madia gracilis* (P$_1$), *M. citriodora* (P$_2$), and F$_1$ hybrids. Univalent chromosomes are stippled. Further explanation in text. × 2000.

fore interpreted as being in a stage corresponding to second metaphase of restitution nuclei resulting from complete nondisjunction during first metaphase. No cells in typical second metaphase were found in this lobe to support such an interpretation, but both dyads and tetrads were observed in this plant (figs. 27, 28). In a study of 46 microspore groups in anther lobes, 20 tetrads, 23 dyads, and 3 triads were found. Both random distribution and nondisjunction therefore occurred in almost equal proportions, and the high percentage of diploid sex cells explains the comparatively high fertility of this hybrid. The pollen was extremely variable in size, ranging from very large to minute grains.

No stages early enough to permit observation of meiosis were found in the spontaneous hybrid 3729-1, but pollen very variable in size was observed, and this plant's amphiploid offspring in the cultures 3934 and 4134 are sufficient evidence that its meiosis was comparable to that of 3720-1.

Good first metaphases but no second metaphases were observed in both of the spontaneous hybrids of 3737, which carried some *sativa* genes. Pairs were frequently observed in them, and even four loose pairs were seen in one cell of each. Figures 29 to 31 illustrate first metaphases of the fertile hybrid 3737-2, which produced the culture 3935 with seven amphiploids. In two of these cells the singles are scattered over the spindle surrounding the very loose pairs. Dyads and tetrads were found intermixed in the anther lobes, and great differences in size were seen in the mature pollen, ranging from dwarf to very large and well developed grains. It is therefore evident that the patterns of pollen development were quite similar in all four F_1 hybrids despite their differences in origin. Judging from the offspring obtained, only the diploid or near-diploid sex cells survived.

CYTOLOGY OF THE SECOND-GENERATION OFFSPRING. The chromosomes of both root tips and pollen mother cells were studied in all the F_2 plants so far as material was available. The resultant counts are listed in table 5 for the F_2 individuals of the synthesized hybrid, 3720-1, as well as for the eleven F_2's originating from the two partially fertile spontaneous hybrids. In addition to the somatic and meiotic numbers, the mode of pairing during meiosis and the fertility of each plant are listed. The fertility figures were determined on a sample of 200 akenes from twenty or more ripe heads.

The second generation of *Madia gracilis* $\times$ *citriodora* was cytologically much more regular than that of *nutans* $\times$ *Rammii*. Out of the total

TABLE 5

Number and pairing of chromosomes and fertility in F_2 plants of Madia gracilis × M. citriodora

PLANT NUMBER	2n	PAIRING IN FIRST METAPHASE		FERTILITY OF AKENES (PER CENT)
		Usual	Deviation	
OFFSPRING OF 3720-1				
3933-1	48	24_{II}	$23_{II} + 2_{I}$	42.5
-2	48	not seen	....	44.0
-3	48	24_{II} + fragment	$23_{II} + 2_{I}$ + fragment	died
-4	48	....	....	died
-5	47	$23_{II} + 1_{I}$	$22_{II} + 3_{I}$	27.5
-6	48	24_{II}, regular	....	56.0
-7	48	24_{II}, regular	....	13.5
-8	48	24_{II}, regular	....	51.5
-9	48	24_{II}, regular	....	43.5
-10	43	$20_{II} + 3_{I}$	$19_{II} + 5_{I}$	36.0
-11	48	24_{II}, regular	....	52.5
-12	48	not seen	....	54.0
-13	48	24_{II}	$23_{II} + 2_{I}$	47.5
-14	48	not seen	....	died
-15	48	24_{II}, regular	....	45.5
-16	48	24_{II}, regular	....	52.0
-17	48	24_{II}	$23_{II} + 2_{I}$	44.0
-18	48	24_{II}	$23_{II} + 2_{I}$	37.5
-19	48	24_{II}, regular	....	58.0
-20	48	not seen	....	38.0
-22	48	24_{II}	$23_{II} + 2_{I}$	69.0
-23	46	23_{II}	$22_{II} + 2_{I}$	52.5
-24	44	$21_{II} + 2_{I}$	....	35.0
-25	45	$22_{II} + 1_{I}$	$21_{II} + 3_{I}$	62.5
-26	45	$21_{II} + 3_{I}$	$20_{II} + 5_{I}$	10.0
-27	48	24_{II}	$23_{II} + 2_{I}$	39.5
-29	48	24_{II}, regular	....	47.0
-30	48	24_{II}	$23_{II} + 2_{I}$	26.0
-31	48	24_{II}	$23_{II} + 2_{I}$, $22_{II} + 4_{I}$	40.0
-32	48	24_{II}, regular	....	60.0
-33	47	$23_{II} + 1_{I}$	$22_{II} + 3_{I}$	47.0
-34	48	24_{II}, regular	....	26.0
-35	48	24_{II}, regular	....	32.5
-36	48	24_{II}, regular?	....	30.0
-37	48	24_{II}, regular	....	46.0
-38	48	24_{II}, regular	....	33.5
-39	48	24_{II}	....	32.5
-40	48	not seen	....	died

(*Continued on following page*)

TABLE 5—*Continued*

PLANT NUMBER	2n	PAIRING IN FIRST METAPHASE		FERTILITY OF AKENES (PER CENT)
		Usual	Deviation	
3933-41	47	$22_{II} + 3_I$	. . . .	4.5
-42	48	24_{II}	$23_{II} + 2_I$	41.0
-43	48	24_{II}	$23_{II} + 2_I$	53.5
-44	48	24_{II}	$23_{II} + 2_I$	42.5
-45	48	24_{II}	$23_{II} + 2_I$	58.0
-46	48	24_{II}	$23_{II} + 2_I$	26.5
-47	48	$23_{II} + 2_I$	$22_{II} + 4_I$	21.5
-48	48	24_{II}	$23_{II} + 2_I$	26.5
-49	48	24_{II}	$23_{II} + 2_I$	18.0

OFFSPRING OF 3729-1

3934-1	48	$23_{II} + 2_I$	24_{II}	4.0
-2	48	24_{II}, regular	. . . .	6.0
-3	48	24_{II}, regular	. . . .	17.0
-4	50	$24_{II} + 2_I$	$23_{II} + 4_I$	10.0

OFFSPRING OF 3737-2

3935-1	48	24_{II}, regular	. . . .	49.5
-2	48	24_{II}, regular	. . . .	26.0
-3	48	24_{II}, regular	. . . .	43.5
-4	48	24_{II}, regular	. . . .	33.5
-5	48	not seen	. . . .	31.5
-6	48	24_{II}	$23_{II} + 2_I$	24.0
-7	48	24_{II}	$23_{II} + 2_I$	12.0

of 58 F$_2$ individuals listed in table 5, 49 had $2n = 48$ chromosomes. Pairing was generally regular in first metaphase except for the occasional occurrence of cells with 23 pairs and 2 singles, indicating a loose attraction between the members of one pair. This tendency was observed to be fairly frequent in 20 of the 49 plants. Figures 32 to 34 show somatic and meiotic chromosomes of a plant in this class.

The chromosomes of the natural hexaploid from Burney Mountain also occasionally show two univalents during first metaphase. This natural species, however, shows this type of irregularity much less frequently than does the average F$_2$ amphiploid of artificial origin. Somatic and meiotic plates, including one with 23 pairs plus 2 univalents, are shown in figures 35 to 37 in order to facilitate comparisons with those of the artificial hexaploid just above them.

Chromosome plates are shown in figures 38 to 43 of aneuploid F_2 plants with 43, 45, 46, 47, and 50 chromosomes. The presence of these plants in the F_2 populations indicates that some sex cells with a slight

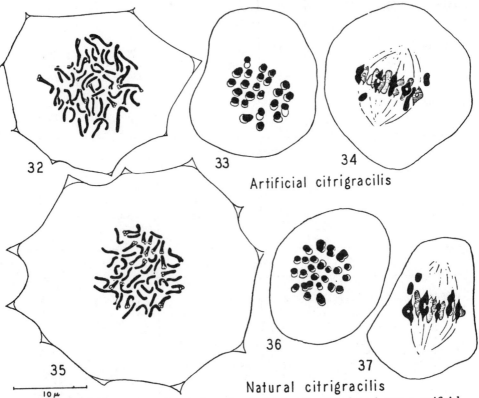

FIGS. 32–37. Comparison of somatic and meiotic chromosome plates between artificial *Madia citrigracilis* (plant 3933-22) and the natural form from Burney Mountain. × 2000.

deviation in chromosome number from the straight hexaploid were viable. Most of these showed no striking decline in either vigor or fertility as compared with some 48-chromosome plants. The individual 3933-23, which has 23 pairs of chromosomes instead of 24 (fig. 41), even suggests a way in which aneuploid numbers may arise from members of a polyploid series, for this plant was both reasonably fertile and vigorous. Although *citrigracilis* belongs to a series of species with strictly polyploid chromosome numbers, the genic balance of this group is more flexible on the hexaploid than on the lower levels, permitting survival and propagation of plants that deviate slightly from the strictly hexaploid number.

THE F₃ POPULATIONS. Akenes were harvested from each plant of the F₂, and four F₃ populations were grown the following year from selected plants. Three were derived from straight amphiploids with $2n = 48$, and one from a plant with $2n = 45$ chromosomes (fig. 20).

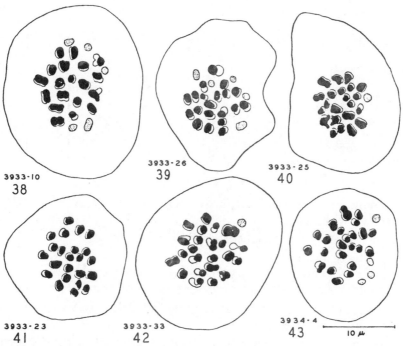

FIGS. 38–43. Polar views of first metaphase in meiotic divisions of six aneuploid *Madia citrigracilis* F₂ plants having, respectively, $2n =$ 43, 45, 45, 46, 47, and 50 chromosomes (cf. table 5). Univalents are stippled. × 2000.

From the three 48-chromosome plants approximately 650 seedlings were obtained, although only some of the akenes were sown. Part of these were transferred to the garden, where 201 plants matured. These were generally vigorous and even more uniform in appearance than the F₂. Table 6 indicates the vigor and uniformity of the combined 48-chromosome F₃ cultures and compares them with the parental species and natural *citrigracilis*. The height of the plants, when grown in a uniform garden, is found to be a satisfactory indication of vigor. Although the akenes for these cultures had been harvested from unprotected plants in the garden, no cross pollination had taken place, for the peculiarities of the strains had been faithfully reproduced. Two representative F_3 plants are shown in figure 19 in the lower row.

TABLE 6

FREQUENCY DISTRIBUTION OF HEIGHTS OF ARTIFICIAL MADIA CITRIGRACILIS, ITS PARENTS, AND NATURAL CITRIGRACILIS IN A UNIFORM GARDEN

Species	Class intervals (cm.)									Total plants	Mean height (cm.)
	25	35	45	55	65	75	85	95	105		
Madia citriodora, Parker Creek, $n = 8$......	..	1	15	12	46	21	2	..		97	68.14
Madia gracilis, Smith River, $n = 16$......	..	1	6	35	59	95	35	2		233	75.19
Artificial M. citrigracilis, F_3, $n = 24$..............	..	7	35	62	68	28	1	..		201	63.88
Natural M. citrigracilis, Burney Mountain, $n = 24$.	3	19	34	62	26	..	..	..		144	56.18

A sample of five F_3 individuals from the three 48-chromosome parents was investigated cytologically. Each had $2n = 48$ and, like the F_2 parents, showed only slight irregularities in pairing.

The fertility of the 48-chromosome F_3 was not yet as high as in natural *citrigracilis*. A sample of twenty plants from each population was tested and found to vary between 2 and 84 per cent fertile, with means of 55.5, 46.4, and 32.1 per cent, as is indicated in figure 20. This is an advance over the F_2, which indicates that further improvement would be likely in subsequent generations, and that through natural selection artificially synthesized *citrigracilis* could doubtless become more uniformly successful. The fertilities already established are better than those found in some other species of *Madia*, and high enough to enable a single plant to yield about a thousand offspring.

In contrast with the F_3 progeny of the 48-chromosome parents, those of the 45-chromosome plant were distinctly weaker, slower in development, and more variable. The fertility of the parent had been determined as 62.5 per cent, but the germination was poor; out of 77 seedlings obtained, 34 died early, and among the 43 survivors only 23 flowered. The fertilities of these, as determined by the appearance of the akenes, ranged between 7 and 70 per cent, with a mean of 49 per cent.

Six F_3 offspring of the 45-chromosome plant were studied cytologically. Four had $2n = 46$, one $2n = 45$, and one $2n = 43$ chromosomes.

The preponderance of 46-chromosome plants in this small sample may indicate a tendency for the chromosome numbers to become established at this figure, but this strain appears unable to compete successfully with the 48-chromosome strains.

COMPARISON OF ARTIFICIAL AND NATURAL MADIA CITRIGRACILIS. The cytologically balanced derivatives of the artificial hexaploid appear to constitute a stable and successful race. In the garden they are on the average slightly smaller than their parental strains, but they are larger than the natural hexaploid from Burney Mountain (table 6) and than many races of *gracilis*. The artificial race is so similar morphologically and cytologically to natural *citrigracilis* as to bring very strong support to the theory that the natural hexaploid arose through amphiploidy from an original hybrid between *gracilis* and *citriodora*. The various lines of evidence in support of this theory may be summarized as follows: (1) both *citriodora* ($n = 8$) and *gracilis* ($n = 16$) occur in the same general region as the hexaploid ($n = 24$); (2) the artificially produced amphiploid matches natural *citrigracilis* in morphology, vigor, and cytological characteristics; (3) amphiploids between *citriodora* and *gracilis* have arisen spontaneously in garden cultures on three separate occasions, under conditions of pollination such as may exist in the wild; and (4) the existence of a cytological mechanism for doubling the chromosomes in the F_1 hybrid between these species is now clearly established.

The final link in the chain of evidence that would prove the conspecific nature of the hexaploids has not been attained. This involves the production of hybrids between the two hexaploids and a test of the fertility of these hybrids. Unfortunately, these forms are not suitable for crossing with emasculation, and their extreme morphological and cytological similarity essentially precludes the possibility of distinguishing between their hybrids and selfed offspring. The strong evidence makes it reasonably certain, however, that natural *citrigracilis* is an amphiploid of the origin suggested.

MADIA GRACILIS AS A POLYPLOID COMPLEX

The repeated production of *M. citrigracilis* from different races of the parental species in the garden suggested that it should be frequently found in the wild and not merely localized as at Burney Mountain. Furthermore, the morphological similarity between natural *citrigracilis* and *gracilis* from northern California needed to be considered before the hexaploid could be proposed as a taxonomic species. These facts led to an investigation of the chromosome numbers of the races of *Madia gracilis* that were in culture.

Counts were made of 58 populations from various localities, ranging from central Washington to southern California, and from near sea level up to 2380 meters elevation. This cytological survey showed that what had been passing for *M. gracilis* (*M. dissitiflora*) is in fact a polyploid complex, for among the 58 cultures studied, 2 were diploid ($n = 8$), 45 were tetraploid ($n = 16$), and 11 were hexaploid ($n = 24$). In table 7

TABLE 7

CHROMOSOME COUNTS IN THE MADIA GRACILIS COMPLEX

Locality	Elevation (meters)
Madia subspicata, $n = 8$	
California:	
SW. of Richardson's Springs, Butte Co.	150
Wildcat Canyon, Stanislaus Co.	50
Tetraploid Madia gracilis, $n = 16$	
British Columbia:	
Thames Creek, Vancouver Island	25
Idaho:	
Thatuna Hills, Latah Co.	610
Washington:	
Bingen, Klickitat Co.	30
Oregon:	
Hood River, Hood River Co.	30
Grants Pass, Josephine Co.	290
Mount Ashland, Jackson Co.	1675
California:	
Davis Creek, Modoc Co.	1525
Castella, Shasta Co.	600
Lake Almanor, Plumas Co.	1340
Quincy, Plumas Co.	1070
N. of Nevada City, Nevada Co.	850
Grass Valley, Nevada Co.	730
Baxter's Camp, Placer Co.	1220
American River Canyon, Eldorado Co.	1160
Dry Creek, N. of Eldorado, Eldorado Co.	425
SW. of Plymouth, Amador Co.	305
N. of Sutter Creek, Amador Co.	350
*N. of Jackson, Amador Co.	425
Vallecita, Calaveras Co.	550
Angels Camp, Calaveras Co.	455
Priest, Tuolumne Co.	800
E. of Groveland, Tuolumne Co.	890

(Continued on following page)

Locality	Elevation (meters)
Smith River, Del Norte Co.	100
W. of Weaverville, Trinity Co.	425
W. of Willow Creek, Humboldt Co.	240
Buck Mountain, Humboldt Co.	760
S. of Weott, Humboldt Co.	60
N. of Willits, Mendocino Co.	430
Clear Lake Oaks, Lake Co.	365
*Napa Range, Napa Co.	335
E. of Muir Beach, Marin Co.	90
*Mill Valley, Marin Co.	30
South San Francisco, San Mateo Co.	3
W. of Belmont, San Mateo Co.	120
*La Honda, San Mateo Co.	125
*Grant's Ranch, Mount Hamilton, Santa Clara Co.	450
Hall's Valley, Mount Hamilton, Santa Clara Co.	455
Llagas Creek, Santa Clara Co.	165
N. of Ben Lomond, Santa Cruz Co.	105
Tularcitos Creek, Monterey Co.	230
San Luis Obispo, San Luis Obispo Co.	60
Nojoqui Pass, Santa Barbara Co.	275
San Marcos Pass, Santa Barbara Co.	850
Mandeville Canyon, Los Angeles Co.	535
San Dimas Canyon, Los Angeles Co.	535

Hexaploid Madia gracilis, n = 24

Washington:

N. of Ellensburg, Kittitas Co.	500

California:

Modoc Lava Beds, Modoc Co.	1430
Mount Shasta, Shasta Co.	1525
Near Lake Tahoe, Eldorado Co.	1920
Cottonwood Meadow, Yosemite National Park	1830
Dark Hole, Yosemite National Park	2375
Mariposa Grove, Yosemite National Park	1890
South Fork Mountain, Humboldt Co.	1680
Chews Ridge, Monterey Co.	1480
Chews Ridge, Monterey Co.	1540
Jolon Grade, Monterey Co.	455

Madia citrigracilis, n = 24

California:

Burney Spring, Shasta Co.	1465
Burney Spring, Shasta Co.	1430

*Chromosome counts made by Donald A. Johansen (1933).

the places of origin of these populations and of *citrigracilis* are listed, and in figure 44 their general location is indicated.

The two diploid populations had previously been recognized in these investigations as members of a morphologically distinct unit, called by us *subspicata*, but not thought to be of specific rank. These plants are very low, with very short lateral branches, if any, and the heads are arranged in a spikelike raceme; the herbage is yellow-green, and the plants bloom exceptionally early in the spring. This form is sporadic and rare, thus far having been collected only at half a dozen localities in the blue oak belt in the lowest foothills of the Sierra Nevada. The two cultures whose chromosomes were counted originated from localities 150 miles apart, but nevertheless they were morphologically indistinguishable. The sporadic occurrence of the diploid suggests that its present-day colonies may be relics of a once continuous distribution. When cultivated in a uniform garden these forms are fairly distinct in appearance as compared with other members of the *gracilis* complex.

Very few colonies of tetraploid *Madia gracilis* in the Pacific states occur above 1000 meters, and their upper limit appears to be around 1400 meters. Morphologically these plants are extremely variable, probably in part because of occasional gene exchange with the equally tetraploid *Madia sativa* of the Coast Ranges. Although tetraploid *gracilis* is morphologically fairly similar to the diploid *subspicata*, the cytological evidence suggests that it is not an autoploid, for the 16 *gracilis* chromosomes in the *gracilis* × *citriodora* hybrid do not pair inter se. In spite of the morphological resemblance between *gracilis* and *subspicata*, their chromosomal difference indicates that they are genetically distinct species.

The hexaploids occur almost exclusively at higher elevations (mainly above 1400 meters). They have been found to occur from central Washington to the central Sierra Nevada, and also in the Santa Lucia Mountains of the Coast Range. The occurrence of *gracilis* at these elevations in mountain masses as far east as central Idaho and Utah and as far south as the Tehachapi region of California suggests that hexaploid forms may extend to these areas. Two collections of the hexaploid from low elevations are listed in table 7. One of these is from Jolon Grade, where it may have come down from near-by localities in the higher Santa Lucia Mountains. The other is from a wash north of Ellensburg, Washington, but the species is not on record from the mountain masses surrounding that locality.

There are no good morphological characters by which the hexaploid forms of *Madia gracilis* can be distinguished from the tetraploids. Like other annual plants occurring at elevations above 1500 meters, they are, on the average, more slender and shorter than the tetraploids from lower elevations. Also, they tend to be slower in development and much more

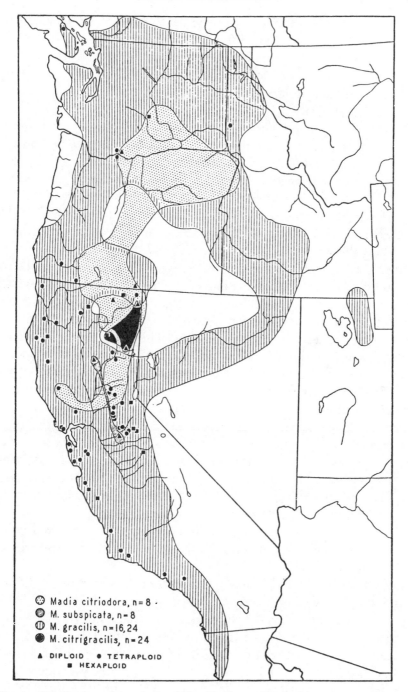

FIG. 44. Distribution of members of the *Madia gracilis* complex. The localities from which chromosome counts have been made are indicated.

difficult to germinate at Stanford. The hexaploids from the Santa Lucia Mountains, however, do not differ from the tetraploids from the same region either in size or in seasonal reactions. Although genetically determined, the characteristics of this southern material are evidently correlated with the climatic zone in which they are native rather than with chromosome number.

Hexaploid *Madia gracilis* without any doubt constitutes a genetic species distinct from tetraploid *gracilis*, but the two cannot be told apart except by chromosome number and distribution. One of the most extreme cases of a similar kind is in *Viola Kitaibeliana* Roem. et Schult., in which there are no morphological means by which forms with 7, 8, 18, and 24 pairs of chromosomes can be distinguished (J. Clausen, 1931*b*).

Difficulty is also encountered in parts of northern California in distinguishing between *Madia citrigracilis* and hexaploid forms of *M. gracilis*. It is apparent that some gene exchange has taken place between these units. A careful examination of herbarium material shows that *citrigracilis* from Burney Mountain in Shasta County is duplicated by plants of several collections from Lassen County, and probably the same form is found in eastern Modoc County. The collections of hexaploid and supposedly hexaploid forms westward toward Mount Shasta are more like *gracilis*, although they approach *citrigracilis* in certain characteristics. The probable *citrigracilis* influence is seen especially in paniculate inflorescences, larger and wider akenes, and sometimes also in the pubescence. Some populations, such as the one from Mount Shasta, are quite variable and contain, besides typical *gracilis*, individuals that approach *citrigracilis*. Still farther westward, through Siskiyou and Trinity counties to eastern Humboldt County, the material from these elevations is progressively more typical of *gracilis*. The possibility must therefore be considered that *citrigracilis* may have arisen several times in the wild, but that it has lost its identity through crossings with hexaploid *gracilis* except in Lassen and Shasta counties, where it appears to be isolated and unadulterated.

Madia citrigracilis and hexaploid *gracilis* are apparently of different origin, although both probably arose through amphiploidy. The pairing of the chromosomes is perfectly normal in hexaploid *gracilis*, with no evidence of multivalent association. Also, an autoploid *gracilis* would be expected to have 32 instead of 24 pairs of chromosomes. Possibly hexaploid *gracilis* arose through crossing of the now nearly extinct diploid *subspicata* with tetraploid *gracilis*, followed by chromosome doubling, although the possibility also exists that its diploid ancestor was some now extinct species morphologically rather similar to *subspicata*.

The situation found in this complex may be illustrated by a diagram (fig. 45). Five biological units exist at three levels of polyploidy: two diploids, one tetraploid, and two hexaploids. Unbroken lines indicate the relationship between *citrigracilis*, *gracilis*, and *citriodora* as established by experiment, and the broken lines indicate other possible relationships arrived at by circumstantial evidence.

The classification of the five biological units in the *gracilis* complex should be considered from three distinct viewpoints. From the phylogenetic point of view all five are equally important, for each has

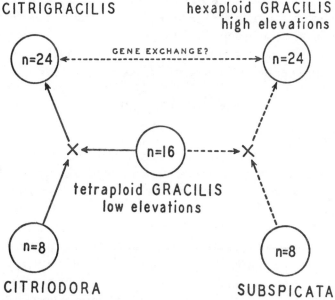

CITRIGRACILIS hexaploid GRACILIS high elevations

GENE EXCHANGE?

n=24 n=24

n=16

tetraploid GRACILIS low elevations

n=8 n=8

CITRIODORA SUBSPICATA

Fig. 45. *Madia gracilis* and allies. Relationships arrived at by experiment indicated by solid lines; those by circumstantial evidence, by broken lines.

undoubtedly had a different origin and history, but we may expect never to be able to discover more than a fraction of the history of such a group. From the genetic viewpoint, the two diploids and the tetraploid are all distinct species, but the two hexaploids might have to be considered as one species, since there is some circumstantial evidence of interbreeding. From the morphological point of view, only *citriodora* is very clearly set off from the four others, although *subspicata* and *citrigracilis* are distinguishable from the two *gracilis* forms.

In a case like this, where the three points of view do not lead to the same conclusion, the classification should preferably be practical. At present, it is impossible to distinguish tetraploid from hexaploid *gracilis* by morphological means. A practical solution for this problem in classification is therefore to include both of these in one taxonomic species, *Madia gracilis*, but to recognize two diploid species, *citriodora* and *subspicata*, and also to maintain as a species the hexaploid *citrigracilis*. In this fourth species of the complex there has been included only that material from Modoc, Lassen, and Shasta counties that closely matches the type collection and therefore also the synthesized *citrigracilis*. The hexaploids from Siskiyou County, that have some characteristics of *gracilis* and some of *citrigracilis*, are considered members of *gracilis*.

Madia citriodora, *subspicata*, tetraploid and hexaploid *gracilis*, and *citrigracilis* clearly constitute an example of a sexual polyploid complex (Babcock and Stebbins, 1938), and of the biological problems that may be found in such a group. Still other species with problems almost equally intricate belong to this same cenospecies, but their exposition is withheld for a later publication on this genus.

SYSTEMATIC NOTES

Madia citrigracilis Keck sp. nov.

Annua, 25–50 cm. alta; caule deorsum hispido-hirsuto superne villoso et viscido-puberulo cum prominenti stipitatis glandulis ornato, subsimplici vel (praecipue superne) stricte paniculato-ramoso; foliis lineari-oblongis, inferioribus 4–8 cm. longis 4–6 cm. latis, basi saepe subamplexicaulibus; capitulis stricte racemosis vel ad apices ramorum solitariis; involucro obovato 6–8 mm. alto, squamis lineari-lanceolatis aliquanto hirsutis et dense stipitato-glandulosis, marginibus villoso-ciliatis, apice acuminato; corollis radii 5–14, ligulis inconspicuis 5–8 mm. longis, tubo moderate pubescente et viscido-puberulo 1.5 mm. longo gracile; corollis disci 8–18 (–30), 2.5–3 mm. longis, tubo et lobis brevi-pubescentibus viscidisque; antheris nigris; acheniis radii ca. 4 mm. longis minute striatis muriculatisque nigris vel brunneo-maculosis late oblanceolatis (late lanceolatis in sectione) paullo arcuatis (nervo intus vix incurvo), areola terminale subsessile; acheniis disci fertilibus ad achenia radia similibus. $n = 24$.

Type, from Burney Spring, south side of Burney Mountain, Shasta County, California, at 1465 m. elevation, under *Pinus ponderosa* in volcanic ash, where not very plentiful and behaving like a ruderal, August 9, 1938, *David D. Keck 4892* (Dudley Herbarium of Stanford University). Isotypes being distributed to University of California, Carnegie Institution, Gray Herbarium, Kew, New York Botanical Garden, Pomona College, and United States National Herbarium.

First collected near the type locality, at 1430 m. elevation, by Frank

W. Peirson, no. 10297 (CI,[1] SU), in dry soil on meadow borders, July 9, 1932. The following three collections from Lassen County are morphologically indistinguishable from the type collection and so are to be referred to this species: 5 miles south of Likely, on sagebrush-covered hills with scattered junipers, elevation 1400 m., *Keck & Clausen 3729* (CI); Spalding's, Eagle Lake, elevation *ca.* 1585 m., *J. T. Howell 12494* (CAS, Ch, CI); Susanville, among yellow pine, elevation 1280 m., *Keck 949* (CAS, CI, SU). Collections from Duncan Horse Camp, 6 miles east of Perez, *Howell 12268A* (CAS, CI), and Plum Valley, above Davis Creek, Warner Mountains, *Howell 12025* (CAS, Ch, CI), both from Modoc County, appear to be the same.

Madia subspicata Keck sp. nov.

Caule 10–15 cm. alto stricto tenue praecipue simplici vel superne sparsim breve ramoso sicut foliis aliquanto flavoviride piloso viscido-puberulo cum prominenti stipitatis glandulis ornato; foliis linearibus 2–7 cm. longis usque ad 3 mm. latis; capitulis subspicatis axillaribus quam foliis brevioribus; pedunculis brevissimis; involucro ovato 6–7 mm. alto; corollis disci 5–15; antheris nigris; acheniis radii ca. 3 mm. longis minute striatis purpureo-maculosis cuneatis (lineari-oblongis in sectione); acheniis disci similibus. $n = 8$.

Type, from the mouth of Wildcat Canyon, Stanislaus River, near Knights Ferry, Stanislaus County, California, Sec. 31, T. 1 S., R. 12 E., at 50 m. elevation, June 2, 1933, in fruit, *David D. Keck 2453* (Dudley Herbarium of Stanford University); isotypes, Carnegie Institution, Pomona College.

Other collections referable to this species are all from California, as follows: 3 miles southwest of Richardson Springs, along Mud Creek, Butte County, 150 m. elevation, May 30, 1933, *Keck 2404* (CI, Po, SU); Chico Creek, 5 miles east of Chico, Butte County, May 17, 1915, *Heller 11878* (CAS, Ch, GH, M, NY, OSC, SU, US); Sec. 15, T. 2 N., R. 11 E. (east of Milton, Calaveras County), 395 m. elevation, *Roseberry 207* (V); Bear Creek, Tuolumne County, 320 m. elevation, May 14, 1919, *Williamson 78*, in part (CAS, Po, RM, SU); Mormon Bar, Mariposa County, May 28, 1895, *Congdon* (Minn).

[1] The abbreviations within parentheses on this page refer to the following United States herbaria: CAS, California Academy of Sciences; Ch, Chicago Natural History Museum; CI, Carnegie Institution; GH, Gray Herbarium, Harvard University; M, Missouri Botanical Garden; Minn, University of Minnesota; NY, New York Botanical Garden; OSC, Oregon State College; Po, Pomona College; RM, Rocky Mountain Herbarium, University of Wyoming; SU, Dudley Herbarium, Stanford University; US, United States National Herbarium; V, Vegetation Type Map Herbarium, University of California.

[*Editors' Note:* Material has been omitted at this point. Only those references cited in the preceding excerpt have been included here.]

REFERENCES

BABCOCK, E. B., and G. L. STEBBINS, JR. 1938. The American species *Crepis.* Carnegie Inst. Wash. Pub. 504. iii + 199 pp.

CLAUSEN, J. 1931. Cyto-genic and taxonomic investigations on *Melanium* violets. Hereditas **15**:219–308.

CLAUSEN, J., DAVID D. KECK, and W. M. HIESEY. 1937. Experimental taxonomy. Carnegie Inst. Wash. Year Book No. **36**:209–214.

JOHANSEN, DONALD A. 1933. Cytology of the tribe Madinae, family Compositae. Bot. Gaz. **95**:177–208.

KECK, DAVID. D. 1940. The identity of *Madia dissitiflora* (Nutt.) Torr. & Gray. Madroño **5**:169–170.

6

Copyright ©1950 by the American Journal of Botany
Reprinted from Am. J. Bot. **37**:487–499 (1950)

NATURAL HYBRIDIZATION AND AMPHIPLOIDY IN THE GENUS TRAGOPOGON[1]

Marion Ownbey

THE OLD-WORLD GENUS *TRAGOPOGON* (Compositae) is represented in North America by three introduced weedy species, *T. dubius* Scop. (*T. major* Jacq.), *T. porrifolius* L., and *T. pratensis* L. These are coarse herbs from thick biennial taproots which in *T. porrifolius* furnish the familiar salsify or vegetable oyster. The three species are widespread on this continent. In southeastern Washington and adjacent Idaho, all are found. *T. dubius* is the most common here, having successfully invaded waste places, roadsides, fields, and pastures, until its occurrence is practically continuous throughout the region. *T. porrifolius* and *T. pratensis* are more restricted in distribution, being almost wholly confined to towns, and the latter is absent from some towns. It grows abundantly, for instance, in Moscow, Idaho, but has not been found in Pullman, Washington, 10 mi. away. Like *T. dubius*, both *T. porrifolius* and *T. pratensis* can withstand considerable competition, and although their ecological amplitude is not so great, it is strange that they have not become more generally established. According to herbarium records, *T. porrifolius* was established in Pullman prior to 1916 (*Pickett 314*) and *T. dubius* in Pullman prior to 1928 (*Jones 2066*). Local botanists remember the sudden appearance of the latter in great abundance about 1930. *T. pratensis* was collected in Spokane County, Washington, as early as 1916 (*Suksdorf 8729, 8911*).

THE PARENTAL SPECIES.—Each of the three species as it occurs in our area is sharply defined by a combination of qualitative and quantitative characters (table 1) and, with the exceptions to be discussed below, there is never the slightest difficulty in recognizing a given individual as a member of one of the three discrete populations. The genetic hiati are broad, sharp, and absolute, and there simply is no biological intergradation between the entities. The species differ in habit; in the color, shape, crisping, curling, and indument of the leaves; in the color, number, and shape of involucral bracts; in the number of flowers per head; in the relative lengths of involucral bracts and ligules; in the color of ligules; in the shape and relative length of the beak and body of the fruit; in the color of fruit and pappus; and in other ways.

Tragopogon dubius.—This species (fig. 1) is easily recognized by its pale lemon-yellow ligules, all shorter than the involucral bracts. The habit is low

and bushy, the branches originating from near the base of the stem. The leaves taper uniformly from base to apex, and are neither crisped on the margins nor curled backward at the tip. They are usually conspicuously floccose when young, becoming glabrate and somewhat glaucous with age. The peduncles of well developed heads are strongly inflated and fistulose toward the apex. The flowers of the head are many, ranging in number from 104–180 in the heads counted, with an average of well over 100 flowers per head. The bracts of the involucre are usually thirteen per head—exactly this number in 75 per cent of the first heads of a random sample of forty plants of a pure colony growing under favorable conditions. Occasionally a particularly robust plant may have as many as seventeen bracts in the first head, and frequently the number may be as few as eight on depauperate plants or in late heads. The bracts are long and narrow—always longer than the longest ligules—and are not margined with purple. The expanded mature heads range from 8–12 (av. 10.5) cm. in diameter. The achenes are slender, ranging from 25–36 (av. 33) mm. long, including the beak. The body is gradually narrowed to and not strongly differentiated from the beak. The outer achenes are pale brown, the inner ones straw colored, and the pappus is whitish.

Meiosis was studied in pollen mother cells from two plants, using the aceto-carmine smear technique (employed throughout this study). In both plants, six bivalents regularly were formed at metaphase I (fig. 14). Mature pollen grains, stained with iodine throughout this study, appeared to be 99 per cent good. The mean diameter of mature pollen protoplasts was 29.3 μ. The calculated diameter of mature pollen protoplasts of mean volume was 29.2 μ. These values are identical with those for *T. pratensis*. The species is highly fertile (fig. 5), not more than 2 or 3 per cent of the flowers failing to produce fruits under normal conditions, and even the poorly developed achenes toward the center of the head uniformly contain apparently viable embryos.

Tragopogon porrifolius.—This species is distinguished at once by its pale to dark violet ligules. Lengths of the longest ligules in the wild species are grouped around two means. In the form with long outer ligules, these are nearly as long as the involucral bracts. In the other form, the outer ligules are short, like the inner ones, averaging about half the length of the involucral bracts or less. Both forms are frequent in our populations, the former being more abundant than the latter. The habit is stout but strict, with the branches fewer in number

[1] Received for publication November 12, 1949.
This investigation was supported in part by funds provided for biological and medical research by the State of Washington Initiative No. 171.

TABLE 1. *A morphological comparison of the three introduced diploid species of Tragopogon.*

Dubius	Porrifolius	Pratensis
Leaves tapering uniformly from base to apex, neither crisped on margins nor curled backward at tip, usually conspicuously floccose when young, glabrate and somewhat glaucous with age.	*Leaves* tapering uniformly from base to apex, neither crisped on margins nor curled backward at tip, glabrous and glaucous, somewhat broader than in *T. dubius.*	*Leaves* narrowed more abruptly below, the margins concave and crisped, the tips recurved, obscurely floccose when young, later glabrate, pale green, not glaucous.
Peduncles strongly inflated toward apex.	*Peduncles* strongly inflated toward apex.	*Peduncles* scarcely at all inflated, even in fruit.
Heads averaging well over 100 flowers (up to 180 counted), in fruit 8–12 cm. in diameter (av. 10.5).	*Heads* averaging about 90 flowers (up to 117 counted), in fruit 9–11 cm. in diameter (av. 10.0).	*Heads* averaging about 75 flowers (up to 96 counted), in fruit 5–6 cm. in diameter.
Bracts usually 13 (sometimes as many as 17 on the first head of vigorous plants or as few as 8 on the latest heads or on depauperate plants), long and narrow, not margined with purple, longer than the outer ligules.	*Bracts* usually 8 or commonly 9 on the first head, rarely as many as 12, broader than in *T. dubius,* not margined with purple, longer than the outer ligules.	*Bracts* usually 8 or commonly 9 on the first head, rarely as many as 13, broad and short, margined with purple, about equaling the outer ligules in length.
Ligules pale lemon yellow, all shorter than the bracts.	*Ligules* pale to deep violet-purple, all shorter than the bracts, the longest sometimes less than half as long.	*Ligules* chrome yellow, the outer ones about equaling the bracts in length.
Achenes slender, 25–36 mm. long (av. 33), gradually narrowed to the not strongly differentiated beak, outer pale brown, inner straw colored; pappus whitish.	*Achenes* thicker, 29–35 mm. long (av. 32), abruptly tapering to a slender beak longer than the body, outer usually dark brown, inner paler; pappus brownish.	*Achenes* thicker, 20–25 mm. long, abruptly tapering to a slender beak which is often shorter than the body, outer usually dark brown, passing to straw colored inwardly; pappus whitish.

and usually originating higher on the stem than in *T. dubius.* The leaves taper uniformly from base to apex, and are neither crisped on the margins nor curled backward at the tips. They are somewhat broader than in *T. dubius* and glabrous and glaucous from the beginning. The peduncles of well developed heads are strongly inflated and fistulose toward the apex. The number of flowers per head ranged from 84 to 117 (av. 93) in the heads counted. The bracts of the involucre are usually eight per head—exactly this number in 62 per cent of the first heads on a random sample of twenty-nine plants growing under favorable conditions, and in 90 per cent of the second heads of the same plants. Heads with nine bracts are common, this number accounting for 34 per cent more of the first heads and the remaining 10 per cent of the second heads in the population sampled. Heads with as many as twelve bracts, however, are occasionally found. No head with fewer than eight bracts has been noted. The bracts are relatively broader and shorter than in *T. dubius.* They are usually longer than the longest ligules—often twice as long—and are not margined with purple. The expanded mature heads range from 9–11 (av. 10) cm. in diameter. The achenes are stout, ranging from 28–36 (av. 31.7) mm. long including the beak. The thick body is abruptly narrowed to and clearly differentiated from the somewhat longer beak, which is stouter than in *T. dubius.* The outer achenes are dark brown or rarely paler, the inner ones paler, and the pappus is brownish.

Pollen mother cells of three plants were examined. In all three plants, six bivalents regularly were formed at metaphase I of meiosis (fig. 15). Mature pollen grains appeared 97.5 per cent good. The mean diameter of mature pollen protoplasts was 30.8 μ. The calculated diameter of mature pollen protoplasts of mean volume was 31.0 μ. In comparison with those for the other diploid species, and the tetraploid involving *T. porrifolius* and *T. dubius,* these values appear a little high. This species is highly fertile (fig. 6), not more than 2 or 3 per cent of the flowers failing to produce fruits under normal conditions, and even the poorly developed achenes toward the center of the head uniformly contain apparently viable embryos.

Tragopogon pratensis.—This species (fig. 2) is marked by chrome-yellow ligules, the longest about equaling the involucral bracts in length. The *forma minor* with all ligules much shorter than the involucral bracts has not been found in our area. The habit is slender and much branched. The leaves are abruptly narrowed below, resulting in concave margins which are conspicuously crisped. The long acuminate tips are curled backward. The herbage is obscurely floccose when young, later glabrate, and pale green, not glaucous. The slender peduncles are scarcely at all inflated, even in fruit. The number of flowers per head ranged from fifty-one to ninety-six (av. seventy-five) in the heads counted. The number of involucral bracts per head is usually eight (70 ± per cent) or nine (30 ± per cent),

94

rarely as many as thirteen (two observed). The bracts, which about equal the outer ligules in length, are short, broad, and margined with purple. The expanded mature heads average between 5 and 6 cm. in diameter. The achenes are stout, ranging from 20–25 mm. in length, including the beak. The thick body is abruptly narrowed to and clearly differentiated from the usually somewhat shorter beak. The outer achenes are usually dark brown, passing to straw colored inwardly, and the pappus is whitish.

Pollen mother cells of four plants were examined. In three plants, meiosis was regular with six bivalents at metaphase I (fig. 16). In one, a chromatin bridge (but no fragment) was observed in some anaphase I configurations. Since no fragment was found, this may have been a delayed separation of one of the longer chromosome pairs. It was not found in any of the other plants. Mature pollen grains, appeared 98.5 per cent good. The mean diameter of mature pollen protoplasts was 29.3 μ. The calculated diameter of mature pollen protoplasts of mean volume was 29.2 μ. These values are identical with those of *T. dubius*. The species is highly fertile (fig. 7), not more than 2 or 3 per cent of the flowers failing to produce fruits under normal conditions, and even the poorly developed achenes toward the center of the head uniformly contain apparently viable embryos.

INTERSPECIFIC DIPLOID HYBRIDS.—Wherever any two of the three introduced diploid species grow together, natural hybrids can be expected. These hybrids are not found except in patches including both of their parents. All three possible hybrids have been found. They combine certain dominant characteristics derived from the parents involved, and on this basis form three additional classes. In most features, they are not intermediate, but display a re-combination of the characteristics which mark their parents. In two cases, those involving *T. porrifolius* and the two yellow-flowered species, a striking "new" character, bicolored ligules, appears through the interaction of genes for anthocyanin coloration derived from *T. porrifolius* and for yellow plastids derived from the yellow-flowered parent. There is also a gene involved which restricts the anthocyanin to the distal portion of the ligule. As a result, the ligules in these two hybrids are reddish brown to violet brown distally and yellow proximally.

The frequency of the hybrids varies from place to place, presumably depending on the relative opportunities for cross pollination between the parents 2 or more years previously. They usually can be found wherever the two parental species are growing together, and sometimes form a very considerable percentage of the individuals in a patch. The only actual frequency count was made by Dr. Gerald B. Ownbey at Pullman in 1946. He found in one patch, extending for 750 ft. along a roadside, 782 individuals of *T. dubius*, 123 of *T. porrifolius*, and 20 *T. dubius* × *porrifolius* hybrids. His results

by 50-ft. intervals are presented as table 2. This appears to be a relatively typical situation. The total number of individuals of each hybrid combination flowering annually in the Pullman-Moscow area runs into the thousands.

The hybrid individuals as a whole are strikingly uniform for each parental combination. There is some variation in color intensity of the ligules in the hybrids involving *T. porrifolius*, but this is no greater than in this parental species. The factor governing ligule length, also, is passed from *T. porrifolius* to its hybrid offspring, producing long- and short-liguled individuals in about the same proportions as in the parental species.

All three hybrid combinations are extremely sterile. This sterility is usually obvious at a glance (fig. 8, 9, 10). The heads, after flowering, do not continue to develop normally, even though a few fruits with embryos are produced. The bracts of the involucre do not grow as in the species, and the peduncle does not enlarge appreciably. Almost all of the ovaries abort at the flowering stage or shortly thereafter. In all three hybrid combinations, however, individual plants have been observed which appear to develop normal heads of achenes, and at maturity these heads of achenes expand normally. Three such plants were observed in *T. dubius* × *porrifolius*, five in *T. dubius* × *pratensis*, and eleven in *T. porrifolius* × *pratensis*. Each set of these plants was found in only one very limited area, suggesting close genetic relationship between the mem-

TABLE 2. *Frequency of Tragopogon dubius, T. porrifolius, and F$_1$ hybrids by 50-ft. intervals along roadside, Pullman, Washington, 1946.*

Interval	dubius	porrifolius	hybrids
1	11	18	0
2	56	20	0
3	129	6	0
4	114	1	0
5	44	4	1
6	11	0	0
7	8	2	0
8	19	3	1
9	31	7	0
10	65	14	1
11	99	11	0
12	43	8	8
13	26	5	4
14	51	6	4
15	75	18	1
Totals	782	123	20
Per cent	84.5	13.3	2.2

bers of the set. Paradoxically, these quasi-fertile hybrid individuals were the most sterile of any examined. Although from 15–28 per cent of the flowers produced mature fruits that superficially appeared to be fully developed, direct observation showed that at most only 0.4 per cent produced fruits with embryos.

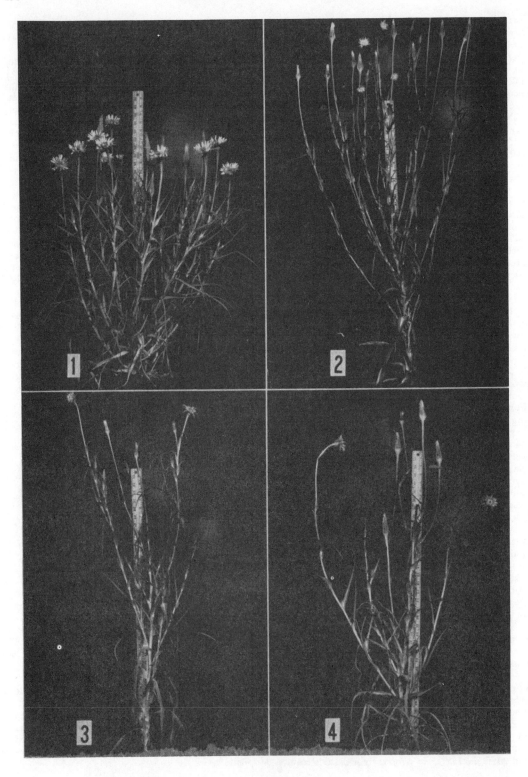

The sterile hybrids are often taller, more branched, and more floriferous than the diploid species, particularly with age. They do not, however, possess the marked "gigas" characteristics of the amphiploids to be discussed later.

From the uniformity of the hybrids within each of the three classes, and their sterility, it is inferred that most, of not all, of those observed are F_1 individuals. Evidence for back-cross or F_2 generations is presented in a later section. The characteristics of hybrid individuals of each of the parental combinations follow:

Tragopogon dubius $\times$ *porrifolius* F_1.—Hybrids of this parentage are marked by bicolored ligules, violet brown (or infrequently reddish brown) distally and yellow at the base. They are distinguished from *T. porrifolius* $\times$ *pratensis* hybrids by their uniformly tapering leaves with neither crisped margins nor recurved tips, the number and shape of the involucral bracts which are not margined with purple, and the generally more violet cast of the ligules. The leaves are obscurely floccose when young, but soon become glabrate and glaucous. The habit is generally more strict than in *T. dubius*. The few mature fully developed achenes show a close resemblance to those of *T. porrifolius* in size, shape and color.

The number of bracts per head was determined on the first heads of forty-six plants growing under favorable conditions. Of these, thirty-six (78 per cent) had thirteen bracts, six (13 per cent) had twelve bracts, and the remaining four (9 per cent) had eleven bracts. As in *T. dubius*, late heads and those of depauperate plants were noted commonly to have as few as eight involucral bracts. The bracts resemble those of *T. dubius* in shape as well as number.

These F_1 hybrids are highly sterile (fig. 8). One hundred eighty-seven heads collected at maturity yielded only 161 fruits with embryos. The number of flowers per head was found to average 130 in ten heads. This figure is likely high, since these heads were from particularly robust plants. These same ten heads also produced an average of 1.7 fruits with embryos per head *vs.* the general average of 0.87 for the entire lot. On the basis of 130 flowers per head, the fertility of the hybrid is 0.67 per cent.

Pollen mother cells of two plants were examined. Metaphase I of meiosis was irregular in both plants. In the first, two bivalents and eight univalents were found; in the second a ring of four, two bivalents, and four univalents were observed (fig. 17). Notwithstanding these irregularities, spore-tetrad formation was not conspicuously abnormal. Mature pollen grains, however, were found to be 92 per cent visibly abortive. No mature pollen grains exceeding the diploid size range were observed.

Three quasi-fertile plants, otherwise indistinguishable from the usual F_1 type, were found. Nineteen heads from these plants yielded only seven fruits with embryos. The average number of flowers for 13 heads was 105, and fertility, calculated on this basis, 0.35 per cent. Pollen mother cells were not examined.

Tragopogon dubius $\times$ *pratensis* F_1.—Hybrids of this parentage (fig. 3) are yellow-flowered, the shade being intermediate between those of the parents. They are easily recognized by their recombination of characteristics marking the parents, together with their usually obvious sterility (fig. 9). The relative lengths of involucral bracts and ligules are those of *T. dubius*, as are the number of bracts per head and their shape. The bracts, however, are margined with purple as in *T. pratensis*, and the leaves resemble this parent in their shape, color, crisped margins, and recurved tips. The influence of *T. pratensis* is also evident in the size and shape of the few fully developed achenes matured, although in color these approach those of *T. dubius*.

The number of bracts per head was determined in sixty heads collected for seed. Of these, twenty-three (38 per cent) had thirteen bracts, thirteen (21 per cent) had twelve bracts, eleven (18 per cent) had eleven bracts, five (8 per cent) had ten bracts, five (8 per cent) had nine bracts, and three (5 per cent) had eight bracts. These same 60 heads yielded 101 fully developed fruits, some of which, it was subsequently discovered, lacked embryos. On a basis of an assumed average of 100 flowers per head —no more reliable figure is available—this is a maximum fertility of less than 1.7 per cent. Taking the achenes without embryos into consideration, and the possibility that the assumed number of flowers per head is low, the actual fertility may be as low as 1 per cent.

Five quasi-fertile plants, otherwise indistinguishable from the usual F_1 type, were observed. One of these was studied in considerable detail. As usual in this hybrid (fig. 9), only one or two fruits per head developed beyond anthesis in the earlier heads, but later heads appeared fertile. Twelve later heads from this plant yielded only three fruits with embryos, although many fruits appeared fully developed until they were broken in two. Four heads averaged 114 flowers per head, and on this basis, fertility was 0.22 per cent. It is likely that this plant furnished some of the pollen mother cells studied.

Pollen mother cells of three plants were examined. All three plants were nearly regular with the usual six bivalents at metaphase I of meiosis (fig. 18), occasionally with a pair of univalents, or possibly a ring of four. Evidence of pairing between heteromorphic chromosomes was observed in four cells. A few microspore groups at the tetrad stage contained five and six cells, sometimes of uniform

Fig. 1–4. *Tragopogon.*—Fig. 1. *T. dubius.*—Fig. 2. *T. pratensis.*—Fig. 3. Diploid *T. dubius* $\times$ *pratensis.*—Fig. 4. Amphiploid *T. dubius* $\times$ *pratensis* (*T. miscellus*).

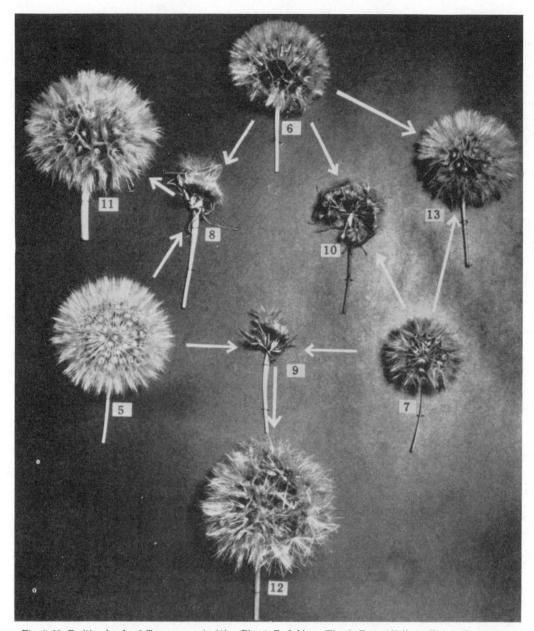

Fig. 5–13. Fruiting heads of *Tragopogon* (×⅓).—Fig. 5. *T. dubius.*—Fig. 6. *T. porrifolius.*—Fig. 7. *T. pratensis.*—Fig. 8. Diploid *T. dubius* × *porrifolius.*—Fig. 9. Diploid *T. dubius* × *pratensis.*—Fig. 10. Diploid *T. porrifolius* × *pratensis.*—Fig. 11. Amphiploid *T. dubius* × *porrifolius* (*T. mirus*).—Fig. 12. Amphiploid *T. dubius* × *pratensis* (*T. miscellus*).—Fig. 13. Quasi-fertile diploid *T. porrifolius* × *pratensis.*

size and sometimes of varying sizes. Mature pollen grains were found to be 93 per cent visibly abortive. No mature pollen grains exceeding the diploid size range were observed.

T. porrifolius × *pratensis* F_1.—Hybrids of this parentage exhibit bicolored ligules which are red-

dish brown (or infrequently violet brown) distally and yellow at the base. They are distinguished from *T. dubius* × *porrifolius* hybrids by their abruptly tapering leaves with crisped margins and recurved tips, the number and shape of the involucral bracts, which are margined with purple, and the generally more reddish cast to the ligules. The leaves are glabrous or nearly so from the beginning, and are not very glaucous. The habit is generally strict. Mature, fully developed achenes show a close similarity to those of *T. porrifolius* in size, shape, and color.

The number of involucral bracts per head was counted in sixty-two heads collected for seed. Of these, fifty-two (84 per cent) had eight bracts, eight (13 per cent) had nine bracts, one (1.6 per cent) had ten, and one (1.6 per cent) had seven. Occasionally, a head is found with as many as eleven bracts. These same 62 heads yielded 120 fully developed achenes, a few of which, it was subsequently discovered, lacked embryos. On the basis of an assumed average of ninety flowers per head—no more reliable figure is available—this is a maximum fertility of about 2 per cent.

Eleven quasi-fertile plants (fig. 13), otherwise indistinguishable from the usual F_1 type (fig. 10), were observed. Eight heads from eight of these plants, averaging ninety-five flowers per head, yielded only three fruits with embryos, although 28 per cent of the potential fruits in them appeared fully developed otherwise. This is a fertility of 0.4 per cent.

Pollen mother cells or four plants, including one of the quasi-fertile ones, were examined. Chromosome pairing at metaphase I of meiosis varied greatly. Although six bivalents were found in a number of cells (fig. 19), there were often from two to ten univalents—this last number in the quasi-fertile plant. Microspore groups of three (one much larger than the other two) and six, as well as the usual four were observed. Mature pollen grains from one of the normal hybrids, were found to be only 60 per cent visibly abortive; from one of the quasi-fertile plants, 96 per cent were abortive. No mature pollen grains exceeding the diploid size range were observed in either.

Two AMPHIPLOID SPECIES.—In the season of 1949 four small colonies were detected in which the members differed from the corresponding diploid hybrids in their very evident fertility and in the possession of conspicuous "gigas" features. These four colonies were immediately suspected to represent two newly originated allotetraploid species ($n = 12$), and this chromosomal constitution has since been confirmed. These allotetraploids differ in none of their characters from the corresponding diploid hybrids, but they are much larger in every way. The stems and leaves are thicker, coarser, more massive and succulent. The heads are larger both in flower and in fruit, and the fruits larger and thicker. The mean volume of the spherical pollen grains is almost precisely the sum of the mean volumes of those of the parental species. The fertility averages between 52 and 66 per cent, although there is wide variation in individual plants beyond these limits. In all four colonies, the amphiploids occurred with both the parental species, and the diploid F_1 hybrid. In one patch, three species, three F_1 hybrids and one amphiploid species grew together.

Tragopogon dubius $\times$ *porrifolius Amphiploid* (*T. mirus*).—Two colonies of this amphiploid (fig. 11) were studied, one in Pullman, Washington, and one in Palouse, Washington, 15 mi. away. Extensive search revealed no other individuals. Only *T. dubius* and *T. porrifolius*, with frequent diploid F_1 hybrids between them, are found in either of these two towns. The Pullman colony was studied more intensively. It consisted of fifty-six flowering individuals growing close together in fertile bottom land along a railroad track. The Palouse colony consisted of twenty-five or more individuals growing on a dry hillside. Both these and the associated parental species and diploid F_1 hybrids were considerably smaller than at the Pullman site, which is attributable to the less favorable habitat.

In the Pullman colony, the bract number was thirteen in ten of the twelve heads—one per plant —examined, eleven and nine, respectively, in the remaining two heads. Of five heads (from five plants) from the Palouse population, one had twelve bracts, one had ten, two had nine, and one had eight. The number of flowers was found to average 147 per head in the 12 heads from Pullman, 89 per head in 4 heads from Palouse. The twelve heads from Pullman averaged seventy-six fruits with embryos per head, a fertility of 52 per cent; those from Palouse, fifty-nine fruits with embryos per head, a fertility of 66 per cent.

Pollen mother cells of three plants from the Pullman colony were examined. In all three plants, twelve bivalents were observed at meiotic metaphase I of some cells (fig. 20). Multivalent formation was frequent, however, and many cells could not be analyzed completely. Among the multivalents studied, the maximum number of associated chromosomes discerned was six (fig. 21). It will be recalled that the F_1 hybrid showed a ring of four. Where multivalents were not formed, a strong secondary association between similar bivalents was frequently noted. Spore-tetrad formation appeared normal, with the four microspores notably larger than in the diploid species and hybrids. Mature pollen grains appeared to be 92 per cent good. The mean diameter of mature pollen protoplasts was 37.0 μ. The calculated diameter of protoplasts of mean volume was 36.8 μ. The mean volume of the pollen protoplasts was 8 per cent less than the sum of the mean volumes of those of the parental species. This difference is probably not significant.

Tragopogon dubius $\times$ *pratensis Amphiploid* (*T. miscellus*).—Two colonies of this amphiploid (fig. 4, 12) were studied, both in Moscow, Idaho. Extensive search in Moscow and other towns where the parental species and F_1 hybrids between them occur, revealed no other individuals. All three diploid species and the three diploid hybrids are frequent in Moscow, but only this amphiploid occurs here. Each colony included between thirty and thirty-five individuals, scattered over a few hundred square yards. They were separated by about a mile. At one

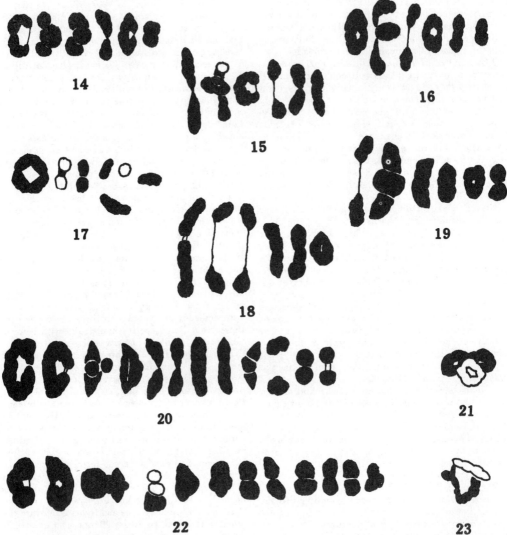

Fig. 14–23. Meiotic chromosomes of *Tragopogon* (×2700).—Fig. 14. *T. dubius.*—Fig. 15. *T. porrifolius.*—Fig. 16. *T. pratensis.*—Fig. 17. *T. dubius* × *porrifolius.*—Fig. 18. *T. dubius* × *pratensis.*—Fig. 19. *T. porrifolius* × *pratensis.*—Fig. 20. Amphiploid *T. dubius* × *porrifolius* (*T. mirus*).—Fig. 21. Multivalent from same interpreted as VI.—Fig. 22. Amphiploid *T. dubius* × *pratensis* (*T. miscellus*).—Fig. 23. Multivalent from same interpreted as heteromorphic IV.

site, in bottom land along a railroad track, conditions were favorable for full development. At the other, along a roadside away from the creek, conditions were not so favorable, and the plants were smaller and less vigorous, as were also those of the parental species and diploid hybrids which occurred there.

In the first colony, the bract number was thirteen in nine of twenty-six heads, twelve in seven heads, eleven in four heads, nine in one head, and eight in one head. Twenty-three heads averaged 115 flowers and 64 fruits with embryos per head, a fertility of 56 per cent.

In the second colony, the bract number was thirteen in two of eighteen heads, twelve in one, eleven in two, ten in seven, nine in five, and eight in one.

Plate 1. Flowering heads of *Tragopogon*. Arranged as in fig. 5–13. *T. dubius* (lower left), *T. porrifolius* (top center), and *T. pratensis* (lower right) form a triangle, along the sides of which are arranged the diploid hybrids, *T. dubius* × *porrifolius* (left center), *T. dubius* × *pratensis* (below center), and *T. porrifolius* × *pratensis* (right center). Adjacent to the diploid hybrids are the corresponding amphiploids, *T. mirus* (upper left), and *T. miscellus* (bottom center). At upper right is the head of the quasi-fertile diploid *T. porrifolius* × *pratensis* hybrid.

[*Editors' Note:* In the original, Plate 1 appears in color.]

Sixteen heads averaged ninety-one flowers and fifty-three fruits with embryos per head, a fertility of 58 per cent.

Pollen mother cells of two plants from the first population were examined. In both, regular or nearly regular plates of twelve bivalents were observed at metaphase I of meiosis (fig. 22). Quadrivalents and some univalents, however, were frequent, and some cells could not be analyzed completely. One configuration interpreted as a quadrivalent involving two heteromorphic chromosome pairs (fig. 23) is of probable significance. It will be recalled that there was evidence also of heteromorphic pairing in the corresponding diploid hybrid. Sporetetrad formation appeared normal, with the microspores notably larger than in the diploid species and hybrids. Mature pollen grains appeared 91.5 per cent good. The mean diameter of mature pollen protoplasts was 36.3 μ. The calculated diameter of protoplasts of mean volume was 36.4 μ. The mean volume of the pollen protoplasts was only 3 per cent less than the sum of the mean volumes of those of the parental species. This difference is not significant.

F_2 AND BACKCROSS GENERATIONS.—Although the high degree of sterility, both of the pollen and ovules, of the three diploid interspecific hybrids would impose a limitation of major importance on the occurrence of F_2 and backcross individuals, these should nevertheless be expected to appear in small numbers wherever interspecific hybridization is extensive. Furthermore, the well marked dominant characters evident in the F_1 generation should provide "markers" for the study of introgression into the parental species. The absence of such evidence of gene flow across the interspecific barriers, together with the sterility of the individuals showing recombination of characters, was the basis for the earlier statement that most, if not all, of these individuals represent the F_1 generation. Among the thousands of plants examined, however, three individuals appeared to represent a later hybrid generation. Two of these resembled *T. pratensis* very closely, except for the color of the ligules, which in one were pale orange, and in the other, deep red. That these individuals were of hybrid origin was confirmed by their sterility, which approximated 70 per cent. The other species involved must have been *T. porrifolius*. The third individual involved *T. dubius* and *T. porrifolius*. It was characterized by very pale ligules, and may have represented either a backcross to *T. dubius* or an F_2 segregate. It was about 99 per cent sterile. In the absence of positive evidence to the contrary, it seems improbable that introgression into any of the three species is taking place.

DISCUSSION.—The genus *Tragopogon* has furnished a classic example of interspecific hybridization since Linnaeus in the summer of 1759 obtained what is usually considered to be the first interspecific hybrid produced for a scientific purpose,

that between *T. pratensis* and *T. porrifolius* (Linnaeus, 1760; Focke, 1881, 1890; Lotsy, 1927; Winge, 1938).[2] This genus now likewise supplies the second and third well documented examples of the origin of a species through amphiploidy in natural populations in historic time. The other example is that of *Spartina Townsendii* (Huskins, 1931), but both parents of this species are themselves undoubtedly also polyploid.

Linnaeus, after rubbing the pollen from the flowers of *Tragopogon pratensis* early in the morning, sprinkled the stigmas with pollen *T. porrifolius* at about eight o'clock. The heads were marked, and the seeds harvested and planted in a separate place. The F_1 hybrids flowered in 1759, producing purple flowers, yellow at the base. Seeds of these F_1 hybrids, along with an essay describing this and other experiments and observations bearing on sex in plants, were submitted in a competition sponsored by the Imperial Academy of Sciences at St. Petersburg. The essay was awarded the prize on September 6, 1760, and the seeds planted in the botanical garden at St. Petersburg, where the F_2 flowered in 1761. These were observed by Kölreuter who recorded (1761) his conclusion that "the hybrid goat's-beard . . . is not a hybrid plant in the real sense, but at most only a half hybrid, *and indeed in different degrees*" (italics added). This record of segregation in the F_2 of an experimental hybrid, a century before Mendel, has escaped recent notice, even of Roberts (1929), who brought together the pertinent facts. I am indebted to Dr. Jens Clausen for calling it to my attention.

This cross was repeated by Focke (1890), Lotsy (1927), and Winge (1938). Focke's detailed, point by point comparison of the parental species and F_1 hybrids has been overlooked by later workers. Lotsy's contribution is in the form of a color plate illustrating the flowering heads of the parental species, F_1, and F_2 segregates. No attempt at analysis of the spectacular segregation in the F_2 is attempted in the paper. Winge's investigations covered a period of 15 years, and carried the hybrids through the F_7 generation. He was particularly concerned with the genetic bases of specific differences, and gives a detailed account of five independent segregating pairs of genes affecting flower color. It was possible by selection to recover both parental species in apparently pure form from the segregating hybrids. Full fertility was regained in the F_2 and subsequent generations. The chromosomes of the two parental species and hybrids of different generations were thoroughly investigated. Both species and hybrids were diploid, $2n = 12$, with the regular formation of six bivalents at metaphase I of meiosis. No meiotic irregularities were noted which

[2] Zirkle (1935) maintains that the first artificial hybrid was produced by Thomas Fairchild, a London horticulturist, prior to 1717, between the carnation (*Dianthus Caryophyllus*) and the sweet william (*D. barbatus*). The evidence that this hybrid resulted from a deliberate experimental cross pollination is conflicting and inconclusive.

would explain the low fertility of the F_1 hybrids. Root-tip mitoses revealed differences in morphology of the somatic chromosomes of the two species, and these differences were found in the reconstituted parental types of hybrid parentage. No tetraploids were found among the 113 plants of Winge's F_2 cultures, although root-tips of 82 were examined. The observation of limited sectors of tetraploid tissue in root-tips of two of these plants, however, is perhaps of significance.

Aside from the work of Winge, chromosome numbers of five species of *Tragopogon* have been reported casually by Poddubnaja-Arnoldi *et al.* (1935) as follows: *T. brevirostris*, $2n = 12$; *T. Cupani*, $2n = 24$; *T. major (dubius)*, $2n = 12$; *T. marginatus* $2n = 12$; and *T. porrifolius*, $2n = 12$. As the only previously known tetraploid species, *T. Cupani* is of considerable interest. Examination of a single specimen so named, preserved in the Herbarium of the Missouri Botanical Garden, suggests the possibility that this may be the amphiploid involving *T. porrifolius* and *T. pratensis* which has not been found in either Winge's cultures or our wild populations.

In later papers, Focke (1897, 1907) reports additional hybrids including *T. orientalis* $\times$ *porrifolius*, *T. dubius* $\times$ *porrifolius*, and the triple hybrid (*T. pratensis* $\times$ *porrifolius*) $\times$ *orientalis*. The first and last of these were sterile, but the second always matured about a quarter of the usual number of fruits. From it was obtained a fertile, constant line of plants with brownish-purple flowers, which was grown for about eight generations. Focke considered this line to represent a newly originated species, *T. phaeus*, but it does not seem to be our amphiploid of the same parentage. A second constant, fertile form, *T. hortensis*, of uncertain origin, also appeared in his cultures. The first plant of this line was unusually robust, but this characteristic was lost in later generations. It is neither of our amphiploids.

There are numerous brief references to natural hybridization in *Tragopogon*. Linnaeus (1760) records the spontaneous appearance in 1757 of *T. porrifolius* $\times$ *pratensis* in a part of his garden where he had planted its parental species. Schultz-Bipontii (1846) noted this hybrid and also *T. major* $\times$ *pratensis* in his garden. The occurrence of *T. porrifolius* $\times$ *pratensis* in wild populations has been reported in Denmark (Lange, 1864), Germany (Focke, 1887), France (Rouy, 1890), and Sweden (Rouy, 1890). In the United States, it has been found in Illinois (Sherff, 1911) and Michigan (Farwell, 1930). That the hybrid occurred with its parents is definitely stated in most instances. In central Germany, Haussknecht (1884, 1888) found all three possible hybrid combinations of the species occurring there, *T. major* $\times$ *orientalis*, *T. major* $\times$ *pratensis*, and *T. orientalis* $\times$ *pratensis*. The sterility of the first and last was noted. *T. major* $\times$ *orientalis* is also recorded for Austria by Dichtl (1883) and Waisbecker (1897). Chenevard (1899) reports *T. crocifolius* $\times$ *major?* growing with its presumed

parents in France [?] and Cockerell (1912) found sterile *T. dubius* $\times$ *porrifolius* growing with its parents at Boulder, Colorado. Further search of the literature would probably reveal many additional records, but these are sufficient to show that natural hybridization in *Tragopogon* is extensive and involves several species.

The chromosome studies reported in the present paper were directed primarily toward the determination of the ploidy levels of the entities involved and the detection of gross meiotic irregularities which might explain the high degree of sterility in the F_1 together with the success of the amphiploids. The material is suitable for much more detailed analysis, which modify considerably these preliminary observations.

The chromosome complement of all three diploid species consists of three longer and three shorter pairs. The three longer pairs are further generally distinguishable at meiosis by the number and position of chiasmata. At first metaphase, one long pair usually forms a ring with two terminal chiasmata, or through absence of one chiasma, a chain. The second pair is characterized by a submedian localized chiasma, and sometimes by one or two others. The chiasma number in the third long pair is more variable. There are often probably three, but these chromosomes may form a ring with only two, or a chain with only one. The three short pairs are less easily distinguished at metaphase, although one may be a little larger than the other two. Generally in these there is a single terminal chiasma.

The bivalents formed in the F_1 hybrids usually correspond closely to those of the diploid species, indicating a rather high degree of homology between the chromosomes of the different species, at least as far as pairing is concerned. Univalents, when formed, come mostly from the three short pairs. Conclusive evidence as to which two of the long pairs form the ring of four in the *T. dubius* $\times$ *porrifolius* F_1 has not been obtained.

The bivalents of the amphiploids also correspond closely to those of the diploid species, except that there are twice as many. Often this correspondence is obscured by some multivalent formation. Where multivalents are not formed, a strong secondary association between similar bivalents was sometimes noted. Although the chromosomes, at least in *T. dubius* $\times$ *pratensis* F_1 and *T. porrifolius* $\times$ *pratensis* F_1, are able to form normal-appearing allosynaptic bivalents, pairing in the amphiploids, on the whole, seems to be strongly autosynaptic. It should be observed that residual allosynapsis might be expected again to occur in the progeny of such amphiploids as *T. miscellus* when crossed with a third species, and that meotic pairing in this hypothetical hybrid might not indicate the third species to be an ancestor of the amphiploid.

The high degree of sterility in the F_1 could be caused by evolutionary differentiation of the chromosomes of each species brought about by translocation or interchange of segments between the non-homologous chromosomes of the genome. Ring formation in *T. dubius* × *porrifolius* and evidence of heteromorphic pairing in *T. dubius* × *pratensis* indicate that differences in homology do exist. Around these structural differences, with the resultant interference with random chromosome recombination (because of non-viability of deficient gametes), could be built the association of distinctive genes which mark each species. If all six chromosomes of each parental genome were non-homologous for deficiencies caused by translocation or interchange of essential segments, as compared with the corresponding member of the other parental genome, only those gametes containing a reconstituted parental genome with respect to these structural differences would be viable. Disregarding crossing-over, and given random distribution of the chromosomes of each of the six pairs, one genome of each of the parental species should be reconstituted in each $2^6 =$ sixty-four gametes. In other words, approximately 3.1 per cent of the gametes should contain a parental set of chromosomes. The maximum fertility, if only reconstituted gametes were viable, would be 3.1 per cent, and the F_2 would fall into three classes, reconstituted parental species, 25 per cent for each, and reconstituted F_1 hybrids, 50 per cent. Since any deficient chromosome segments of one parental species might be compensated for by the addition through crossing-over of non-deficient segments of the other parent, the net effect of crossing-over would be an increase in the variability of the F_2, and the genes on the crossover segments would behave in the manner which Winge has described. Some such mechanism might explain the restored fertility in the F_2 and subsequent generations of Winge's hybrids, and the infrequency of detectable later generations in our wild populations.

The mechanism of origin of the amphiploid species, whether by somatic or gametic doubling in the F_1 is obscure. Winge's observation of tetraploid sectors in root-tip tissues would favor the former explanation, as would the absence of pollen grains exceeding the diploid size range in all of the diploid hybrids examined in the present study. Supporting the latter explanation would be. the presence of spore triads in some of our hybrids, which suggests that diploid pollen grains might be produced, and the lack of extended vegetative growth. It should be noted that positive evidence favoring either of these mechanisms was observed only in the one of the three hybrid combinations for which no amphiploid is known, that between *T. porrifolius* and *T. pratensis*.

Whatever the mechanism of origin, it is apparent that the amphiploids do not originate with great frequency. The four known colonies probably represent four independent instances of chromosome doubling, and the subsequent establishment of the resultant tetraploid. Considering the frequency of all three F_1 hybrid combinations, however, chromosome doubling must be an exceedingly rare event.

For theoretical reasons, its frequent occurrence in species hybrids with essentially regular meiotic pairing is not to be expected, and amphiploids derived from such should be unsuccessful.

In spite of these theoretical handicaps, the amphiploids of *Tragopogon* have appeared, and have attained a degree of success. Although the populations are still small and precarious, fertility is good, and these species are competing successfully with their parents. Crossing-over has not led to deterioration, presumably because each chromosome usually pairs with its exact homologue, and the consequences of crossing-over, therefore, are not deleterious. Fertility ought to improve with succeeding generations, since any genetic factor which will increase fertility—and there is wide variation in this respect—will enjoy a real selective advantage.

The ecological characteristics of the new amphiploids are not yet apparent. In all instances, they occur within the ecological amplitude of the most restricted parental species. The ecological requirements of natural amphiploids are often such that they have achieved an ecological and geographical distribution somewhat different from the species from which they are presumed to have been derived (Clausen *et al.*, 1945). Since, in both instances, the present amphiploids combine genomes from species with significantly different ecological requirements, it will be interesting to follow their ecological development. At the present time, it is apparent that they have not spread far from their point of origin.

TAXONOMIC CONSIDERATIONS.—The two newly originated amphiploids are to be considered taxonomic species for the following reasons: (1) They are natural groups characterized by a combination of distinctive morphological features. (2) They are reproducing themselves under natural conditions. (3) Gene interchange between the amphiploids and the parental species is prevented by a genetic barrier (ploidy level), and presumably residual sterility factors—evident in the F_1 hybrids—would prevent free interbreeding between the two.

Search of the systematic literature has not revealed the existence of these amphiploid species in Europe, although it would be surprising if they do not occur there. The identification of many obscure species which have been proposed in *Tragopogon* must await a comprehensive taxonomic and cytogenetic study of the genus. Accordingly, these two amphiploids are here described as new species.

Tragopogon mirus Ownbey, sp. nov.—Herbae biennes primum obscure floccosae deinde glabrae glaucaeque. Folia lineari-lanceolata semi-amplexicauliausque ad 5 cm. lata paulatim attenuata, marginibus non crispis, apicibus non cirrosis. Capitula multiflora, pedunculis inflatis fistulosis usque ad 15 mm. crassis. Bracteae involucri lineari-lanceolatae ubique virides, in plantis robustioribus plerumque 13. Ligulae bicoloratae ad apicem lilacinae ad basem flavae bracteis paulum breviores. Achenia rostraque conjuncta 25–35 mm. longa, exteriora fusca, interiora straminea, rostro corpore subaequilongo, pappo cervino.

Type: Washington. Whitman County: in fertile bottom land, Pullman, June 9, 1949, *Ownbey 3195,* in Herbarium of the State College of Washington, Pullman.

*Tragopogon **miscellus*** Ownbey, sp. nov.—Herbae biennes primum obscure floccosae deinde glabrae viridesque. Folia lineari-lanceolata semi-amplexicaulia usque ad 3 cm. lata abrupte attenuata, marginibus crispis, apicibus cirrosis. Capitula pluriflora, pedunculis inflatis fistulosis usque ad 10 mm. crassis. Bracteae involucri lineari-lanceolatae in plantis robustioribus plerumque 13, marginibus purpureis. Ligulae flavae bractea dimidia subaequilongae. Achenia rostraque conjuncta 25-35 mm. longa, exteriora fusca, interiora straminea, rostro corpore subaequilongo vel longiore, pappo cinereo.

Type: Idaho. Latah County: in fertile bottom land, Moscow, June 10, 1949, *Ownbey 3196,* in Herbarium of the State College of Washington, Pullman.

SUMMARY

Three diploid ($n = 6$) species of the Old World genus *Tragopogon* (Compositae), *T. dubius, T. porrifolius,* and *T. pratensis,* have become widely naturalized in North America. In southeastern Washington and adjacent Idaho, where all three occur, extensive natural hybridization is taking place. Each species crosses readily with both of the others, and wherever two or more grow together, easily detected F_1 hybrids are frequent. These diploid hybrids for all three species combinations are highly sterile, not more than 1–2 per cent of the flowers producing fruits with embryos. They are intermediate only in the sense that they recombine certain dominant characteristics of the parental species involved. F_2 and back-cross individuals are absent or nearly so. Meiosis in the hybrids is fairly regular, although some multivalents and univalents are formed, particularly in *T. dubius* $\times$ *T. porrifolius.* Four small amphiploid populations were discovered in 1949. These represent apparently four recent and independent instances of the doubling of the chromosome sets, two cases each for the *T. dubius* $\times$ *porrifolius* and *T. dubius* $\times$ *pratensis* hybrids. These two tetraploid entities ($n = 12$) are fairly regular meiotically, usually forming bivalents at metaphase I in pollen mother cells. They are moderately fertile, on the average from 52–66 per cent of the flowers producing fruits with embryos. They are established and true-breeding entities, although population size is still precariously small. Morphologically, they are like the corresponding diploid hybrids except for conspicuous "gigas" features and their very evident fertility. Their cell volume, as revealed by measurement of the spherical pollen grains, is almost precisely the summation of the cell volumes of the two parental genomes. They are accorded species rank, described and named *T. mirus* (amphiploid *T. dubius* $\times$ *porrifolius*) and *T. miscellus* (amphiploid *T. dubius* $\times$ *pratensis*).

DEPARTMENT OF BOTANY,
    STATE COLLEGE OF WASHINGTON,
        PULLMAN, WASHINGTON

LITERATURE CITED

CHENEVARD, P. 1899. Notes floristiques. Bull. Trav. Soc. Bot. Genève 9: 130.

CLAUSEN, J., D. D. KECK, AND W. M. HIESEY. 1945. Experimental studies on the nature of species. II. Plant evolution through amphiploidy and autoploidy, with examples from the Madiinae. Carnegie Inst. of Washington. Publ. No. 564.

COCKERELL, T. D. A. 1912. *Tragopogon* in Colorado. Torreya 12: 244–247.

DICHTL, P. A. 1883. Ergänzungen zu den "Nachträgen zur Flora von Nieder-Österreich." (Fortsetzung). Deutsch. Bot. Monatschrift 1: 187–188.

FARWELL, O. A. 1930. Botanical gleanings in Michigan. VI. Amer. Midl. Nat. 12: 113–134.

FOCKE, W. O. 1881. Die Pflanzen-Mischlinge. Gebrüder Bornträger. Berlin.

———. 1887. *Tragopogon porrifolius* $\times$ *pratensis.* Abhandl. Naturwiss. Ver. Bremen 9: 287–288.

———. 1890. Versuche und Beobachtungen über Kreuzung und Fruchtansatz bei Blütenpflanzen. Abhandl. Naturwiss. Ver. Bremen 11: 413–421.

———. 1897. Neue Beobachtungen über Artenkreuzung und Selbststerilität. Abhandl. Naturwiss. Ver. Bremen 14: 297–304.

———. 1907. Betrachtungen und Erfahrungen über Variation und Artenbildung. Abhandl. Naturwiss. Ver. Bremen 19: 68–87.

HAUSSKNECHT, C. 1884. Botanischer Verein für Gesamtthüringen. I. Sitzungsberichte. Mitteil. Geogr. Ges. (für Thüringen) Jena 2: 211–217.

———. 1888. Kleinere botanische Mitteilungen. Mitteil. Geogr. Ges. (für Thüringen) Jena 6 (Bot. Ver. Gesamtthüringen): 21–32.

HUSKINS, C. L. 1931. The origin of *Spartina Townsendii.* Genetica 12: 531–538.

KÖLREUTER, J. G. 1761. Vorläufige Nachricht von einigen das Geschlecht der Pflanzen betreffenden Versuchen und Beobachtungen. Gleditsch. Leipzig. (Not seen). Reprinted by W. Pfeffer in Ostwald's Klassiker der exakten Wissenschaften 41: 3–37. 1893.

LANGE, J. 1864. Haandbog i den danske Flora. Ed. 3. C. A. Reitzel. Copenhagen.

LINNAEUS, C. 1760. Disquisitio de quaestione ab Academia imperiali scientiarum Petropolitana in annum MDCCLIX pro praemio proposita: "Sexum plantarum argumentis et experimentis novis. . . ." Academy of Sciences, St. Petersburg. (Not seen.) Reprinted as "Disquisitio de sexu plantarum. . . ." in Amoenitates Academicae 10: 100–131. 1790; in English translation as "A dissertation on the sexes of plants," by J. E. Smith. Nichol. London. 1786.

LOTSY, J. P. 1927. What do we know of the descent of man? Genetica 9: 289–328. Plate II.

PODDUBNAJA-ARNOLDI, W., N. STESCHINA, UND A. SOSNOVETZ. 1935. Der Charakter und die Ursachen der Sterilität bei *Scorzonera tausaghys* Lipsch. et Bosse. Beih. Bot. Centralblatt 53A: 309–339.

Roberts, H. F. 1929. Plant hybridization before Mendel. Princeton University Press. Princeton, N.J.

Rouy, M. G. 1890. Remarques sur la synonomie de quelques plantes occidentales. Bull. Soc. Bot. France 37: XIV–XX.

Sherff, E. E. 1911. *Tragopogon pratensis* × *porrifolius*. Torreya 11: 14–15.

Shultz-Bipontii, K. H. 1846. *Tragopogon*. In P. B. Webb and S. Berthelot's Historie Naturelle des Iles Canaries. 3²(2): 469.

Waisbecker, A. 1897. Beiträge zur Flora des Eisenburger Comitates. Österreich. Bot. Zeitschr. 47: 4–9.

Winge, Ö. 1938. Inheritance of species characters in *Tragopogon*. A cytogenetic investigation. Compt. Rend. Trav. Lab. Carlsberg Série Physiol. 22: 155–193. Plates I and II.

Zirkle, C. 1935. The beginnings of plant hybridization. University of Pennsylvania Press. Philadelphia, Pa.

Reprinted from *Heredity* **2**:119-129 (1948)

POLARISED SEGREGATION IN THE POLLEN MOTHER CELLS OF A STABLE TRIPLOID

S. SMITH-WHITE

Museum of Technology and Applied Science, Sydney

INTRODUCTION

THE Australasian family of the Ericales, the Epacridaceæ, are known to the Author in their chromosome numbers by only a single report, that given by Samuelson (1913) for *Epacris impressa*, cited by Gaiser (1930). A difference of opinion in papers by Hagerup (1928) and Wanscher (1934) concerning the probable basic chromosome number in the Ericales, suggested that a survey of the Epacridaceæ might be of interest, and the results of such a survey are being published elsewhere (Smith-White, 1947*a*). In this survey a triploid species of *Leucopogon* was discovered in which chromosome behaviour at meiosis is peculiar and of general cytological interest.

MATERIALS AND METHODS

Leucopogon juniperinus. R. Br. is a species occurring abundantly on shale-soil areas in the vicinity of Sydney. Bentham (1869) also records it from Moreton Bay, Queensland, and doubtfully from the Upper Macalister River, Victoria. These occurrences, however, would appear to be geographically isolated from one another, and the present study concerns only the species in the Port Jackson area. Plants have been examined from four localities within the area, viz. Lane Cove River, Gordon, Kuringai and Yagoona, all within 12 miles of Sydney.

Mitotic chromosomes were drawn from aceto-lacmoid crushes of young ovules, following maceration in HCl-lacmoid as in the schedule given by Darlington and La Cour (1942), and made permanent in Euparal. Meiosis in pollen mother cells was studied from aceto-carmine crushes, which have also been made permanent. For studies of fertility, pollen from mature anthers, ready to dehisce, was mounted in a special dextrin-sorbitol-acid fuchsin stain mountant (*cf.* Smith-White, 1947*b*). Chromosome drawings were made with a camera lucida at an initial magnification of ca. 3900, and have been reduced for reproduction to ca. 2600.

OBSERVATIONS

The mitotic chromosomes. The somatic chromosome number for the species is 12 (fig. 1 ; pl. I, fig. 1). At metaphase in ovule mitoses, the chromosomes are rod-shaped bodies, of uniform length (ca. 2·7-

$3 \cdot 0 \mu$), and all show median centric constrictions. So far it has not been possible to observe any characteristic morphological differences between the chromosomes.

Although the genus *Leucopogon* shows considerable variation in chromosome number ($n = 4$, 6, 10, 12, 24 (Smith-White, 1947a)), two closely related species, *L. setiger* and *L. esquamatus*, have a haploid complement of 4, so that *L. juniperinus* is apparently triploid in constitution.

Meiosis in the pollen mother cells. At the first metaphase the pollen mother cells regularly show 4 bivalents and 4 univalents. No multivalents, and never more or less than 4 bivalents, were found in 586 pollen mother cells from five plants (fig. 2 ; pl. I, fig. 2). Pairing of the bivalents is very uniform, with a rather high chiasma-frequency, a majority being united on both sides of the centromere. In one plant (No. 0) in which 122 1-M cells were studied (488 bivalents), 412 bivalents, or 84·4 per cent., were associated on both sides of the centromere. Uniform association and regular high chiasma-frequency is an indication of homozygosity and lack of difference between the pairing chromosomes.

The extraordinary regularity of the chromosome associations suggests that the species is a triploid on a basic haploid set of 4 chromosomes, and that its constitution may be represented by the formula—

$$AA \quad BB \quad CC \quad DD \quad E F G H$$

in which each letter represents a whole chromosome. This means that the species is probably of hybrid origin, and that the paired chromosomes A, B, C and D are derived from one parent, and that the unpaired chromosomes E, F, G and H are derived from another parental species, sufficiently remote phylogenetically from the first to exclude the occurrence of multivalent associations between the two sets. Such complete failure of chromosome pairing has been demonstrated in many artificial diploid species hybrids, and the condition is essentially similar to that found in triploid hybrids, in which the chromosome set from one parent is present in the diploid number. (*cf.* Darlington, 1937, pp. 210-212). Such a doubling of one of the two chromosome sets could be derived either from an unreduced diploid gamete, from backcrossing to one parent, or if one of the parents was autotetraploid in constitution.

The behaviour of the univalents. In general, the behaviour of univalents on the spindle during the first meiotic division is well understood (Darlington, 1937, p. 410 ; Kihara, 1931 ; Ribbands, 1937). Usually they are late in movement to the equatorial plate at M1, and if the anaphase division of the bivalents occurs before their arrival at the plate, they may be included in one or other anaphase group at the poles. In this case their distribution is random. On the other hand, if they reach the plate, they tend to remain there

FIGS. 1-9.—× ca. 2000.   1. Somatic chromosomes in ovule tissue.   2. M1 in polar view, showing the bivalent plate.   3. M1, in side view, showing a 2-2 distribution of the univalents. Note how the univalents are forced to the extremities of the poles of the spindle.   4. M1, in side view, showing a 1-3 distribution of the univalents. One univalent at the bottom of the picture is out of focus.   5. M1 in side view, showing all 4 univalents at one end of the spindle.   6. M2, showing unequal plates of 4 and 8 chromosomes.   7. T2, showing a second division bridge.   8. T2, showing laggard chromosomes, possibly the result of misdivision.   9. A pollen mother cell showing the rearrangement of the four microspore nuclei.

FIGS. 10 and 11.—× ca. 900.   10. The pollen mother cell shortly after cytokinesis.   11. Young pollen grain in the binucleate stage (section).

FIG. 12.—× ca. 210.   Mature pollen.

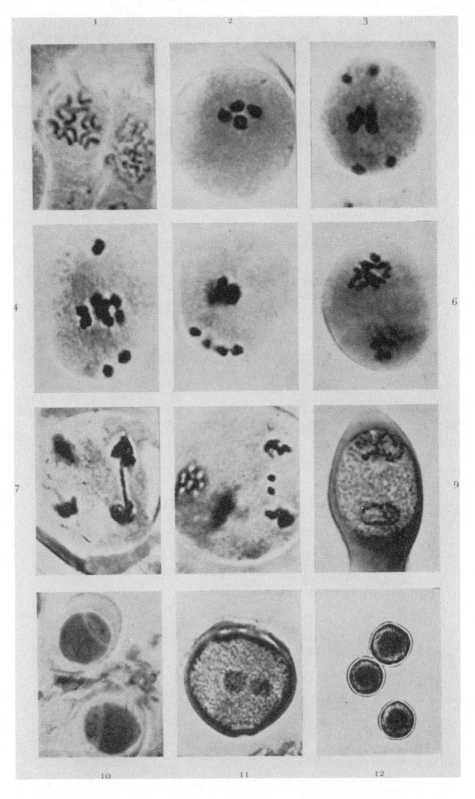

after the anaphase separation of the bivalents, and to divide into daughter chromatids, which follow the daughter chromosomes of the bivalents to the poles.   At the second division such divided uni-

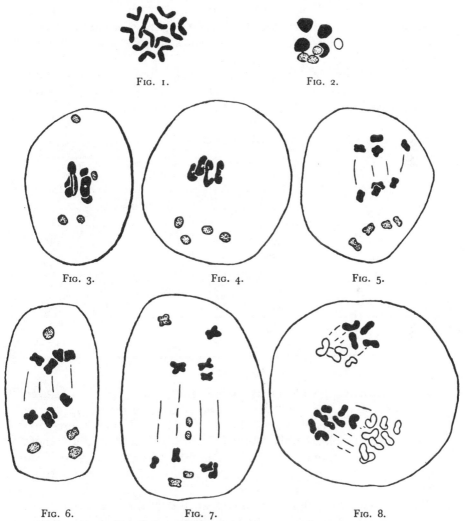

Fig. 1.

Fig. 2.

Fig. 3.

Fig. 4.

Fig. 5.

Fig. 6.

Fig. 7.

Fig. 8.

Figs. 1-8.—*Leucopogon juniperinus* × 2600.   1. Somatic chromosomes in ovule crush $2n = 12$.
2. 1-M in pollen mother cell, polar view, three univalents are above, and one is below the bivalent plate. 3. 1-M, showing a 1-1-2 distribution of univalents. 4. 1-M, with the more frequent 0-0-4 distribution of univalents. 5. 1-A, with the four univalents moving to the one pole ahead of the paired chromosomes. 6. 1-A, with a 1-3 distribution of the univalents. 7. 1-A, showing the (comparatively rare) division of a univalent. 8. 2-A, showing small and large spindles of 4 and 8 chromosomes respectively.

valents are usually randomly distributed to the poles of the second anaphase spindles.   Where univalents are present, the regularity of the meiotic division is upset, and a majority of the microspore nuclei are of unbalanced constitution.

In *L. juniperinus* the four univalent chromosomes show an unusual behaviour. At M1 they usually lie off the equatorial plate, towards the poles of the spindle (figs. 3 and 4 ; pl. I, figs. 3, 4 and 5). In a minority of cells one or more univalents may have reached the equator at this stage, when such univalents lie on the edge of the bivalent plate. Of 586 cells at M1, for which data are given in table 1, only 152, or 25·9 per cent., showed univalents in such a position.

The most remarkable feature in the behaviour of the univalents is their non-random distribution (table 1). In a majority of cells,

TABLE 1

Distribution of the univalents at M1 in L. juniperinus

Plant no.	Type of distribution				Total
	0-4	1-3	2-2	Incomplete	
0	52	23	7	44	126
1	54	32	12	39	137
3	21	12	7	9	49
4	44	33	19	32	128
5	71	36	11	28	146
Total	242	136	56	152	586
Expected nos. (random basis)	54·25	217·00	162·75	...	...

$$\chi^2 = 750 \cdot 02. \qquad n = 2. \qquad p \ll 0 \cdot 001.$$

all four univalents pass to the one pole (fig. 4 ; pl. I, fig. 5), and even at mid-metaphase they are then placed at the extreme end of the spindle, near the edge of the cell, but usually lie well apart. Ignoring the 152 cells in which univalents lie near the equator, and whose final distribution cannot be determined, the distribution is seen to differ widely from that expected on a random basis. ($\chi^2 = 750 \cdot 02$, $p \ll 0 \cdot 001$.)

The cause of this polarity in univalent distribution is not apparent, but its result is clear. A majority of the pollen mother cells at M2 show unequal plates, most frequently one with 4 and the other with 8 chromosomes, the larger plate containing all the unpaired chromosomes (table 2 ; pl. I, fig. 6).

Division of univalents on the equator of the A1 spindle, after the separation of the bivalent chromosomes (fig. 7) is rare, as indicated by the frequency of chromosome numbers at M2, given in table 2. This would be expected as an incidental result of the polarity in univalent behaviour.

At anaphase, after the separation of the bivalent chromosomes, the univalents are first to reach the poles (figs. 5 and 6). At this stage

112

the repulsion between the chromatid-arms is very evident in all chromosomes.

TABLE 2

Distribution of chromosomes at M2 in L. juniperinus

	Type of distribution					Total
	4-8	5-7	5½-6½ *	6-6	5-6-(1) †	
Plant no. 1 observed frequency . .	15	8	2	3	1	29
Expected frequency (random basis) .	3·25	13·00	...	9·75	...	...

* Due to division of univalent at M1.
† Chromosome lost in cytoplasm.

$$\chi^2 = 49·07.   n = 2.   p \ll 0·001.$$

Apart from the uneven size of the two plates, the second meiotic division is usually normal (fig. 8). Occasionally, however, bridges (pl. I, fig. 7) or laggards (pl. I, fig. 8) are seen. The former indicate that structurally the bivalent chromosomes are not completely homozygous, but that they differ at least in respect of segmental inversions. The laggards shown in pl. I, fig. 8, may represent a case of misdivision of a centromere, since they are smaller than the other chromosomes. Darlington (1939a, 1940) has shown that the centromeres of univalents which have divided at first anaphase are unable to divide normally at second anaphase, and are consequently liable to misdivision. Several cases of such possible misdivision have been observed in *L. juniperinus.*

Pollen development. Pollen development is of a peculiar type, also found in other species of *Leucopogon*, in *Styphelia*, and other genera of the Stypheleæ. This type of development has been described elsewhere (Smith-White, 1947a), as the " Styphelia " or " S " type, as distinct from the " tetrad " type of pollen development characteristic of the Ericales generally. The four telophasic microspore nuclei do not show uniform behaviour. At first, having a tetrahedral or quartite arrangement, they assume, prior to cytokinesis, a grouped arrangement, in which three nuclei are clustered at one end of the cell, and a single nucleus at the other (pl. I, fig. 9). The pollen mother cell is then divided into three small and one large microspore (pl. I, fig. 10). The large microspore is the only one to continue development. The other three degenerate, or become crushed out of existence (pl. I, fig. 11), so that the mature pollen consists of " single " grains, of the type described by Brough (1924) for *Styphelia* (pl. I, fig. 12).

It has not yet been determined whether the functional microspore of each " tetrad " is of any particular chromosome constitution, but from size comparisons of the nuclei, it is suggested that it may be one of the 8 chromosome nuclei. The rapid thickening of the wall of the

developing pollen grain offers some difficulty in the examination of
the pollen grain mitosis. Very occasionally " double " pollen grains,
resulting from the development of two microspore nuclei, are found.
Their frequency, however, is almost negligible, only three cases being
observed in a slide containing over 2000 pollen grains.

Pollen fertility. Pollen fertility is variable, with a considerable
proportion of aborted grains, but it must be regarded as unusually
high for a triploid. The data given in table 3 indicate that significant
variations in pollen fertility may occur between plants, or even between
flowers on the same plant, and it is probable that pollen development
and fertility are very susceptible to environmental, and particularly,
temperature conditions.

L. juniperinus is abundantly seed-fertile. In fact it sets a much
higher proportion of fruits than its diploid relatives *L. setiger* and
L. esquamatus, and it is morphologically a uniform and distinctive
species.

TABLE 3.

Pollen fertility in Leucopogon juniperinus

Per cent. good pollen

Plant	Slide (a)					Slide (b)					M
No.	1	2	3	4	*m*	1	2	3	4	*m*	
1	31	31	45	33	35·00	46	42	47	41	44·00	39·50
2	36	43	40	46	41·25	58	60	76	47	60·25	50·75
3	8	18	30	21	19·25	10	5	12	9	9·00	14·13
4	65	76	67	67	68·75	64	71	65	63	65·75	67·25
5	70	78	67	60	68·75	68	70	74	61	68·25	68·50
General mean	.	.	.	.	...	.	.	.	.	...	48·25

Analysis of variance

	SS	DF	MS	F	SE	Sign difference at 5 per cent. point
Total . .	18,542·75	39	...	...	...	...
Between plants.	15,858·875	4	3,964·719	17·8 †	±5·31	19·3
Slides *	1,112·625	5	222·525	...	...	...
Error . .	1,571·250	20	78·5625	...	...	...

* Used as error control.
† Exceeds the 1 per cent. point of F.

DISCUSSION

The high fertility of the species may simply mean that it is fully
apomictic, as are so many triploids and odd-numbered polyploids in
nature (Darlington, 1937, p. 468). Apomictic polyploids, however,

usually show extreme irregularity of meiosis, particularly in the pollen mother cells, resulting in polysporous tetrads and highly infertile pollen. With such a method of reproduction, it is difficult to imagine the evolution of the unusual stabilised meiosis which has been described. Such stabilisation could offer selective evolutionary advantage only if the pollen is involved in seed production, and it could be developed only by adaptation in the course of sexual reproduction (Darlington, 1937, p. 414). The only known comparable case is the condition of " semi-apomixis " found in the *Caninæ* section of *Rosa* (Täckholm, 1922 ; Hurst, 1931), in which the egg carries a number of univalents, which have been distributed in a polarised manner. The present case may be similar, except that it concerns pollen formation, and the extra univalents may be carried by the pollen. Genetically the system has a similar result, in permitting a hybrid species to breed true, but the species does not seem to be divided into the numerous interrelated but true-breeding forms that occur in the *Caninæ*.

Several cases of polarity in chromosome behaviour have been reported in the past. Morgan, Bridges and Sturtevant (1925) showed that in triploid *Drosophila* the distribution of univalents may not be according to chance, but that certain combinations of sex chromosomes and autosomes are formed more often than others. Sturtevant (1936) found that in triplo-IV females of *Drosophila melanogaster* the segregation of the three chromosomes IV was not random, and that the segregation of the third chromosome was determined by a property of the other pair. Darlington (1940) has shown that the segregation of supernumerary sex chromosomes in *Cimex* may be preferential with respect to the Y chromosome.

In these cases, the differential segregation appears to be determined by the chiasma relationships or other properties of the chromosomes themselves, or by a property of the spindle mechanism, rather than by any polarisation of the cell as a whole. Rhoades (1942) found that in the embryo-sac mother cell in maize heterozygous for an abnormal heterochromatic segment of chromosome 10, the abnormal chromosome passes to the lower (chalazal) end of the spindle most frequently, and Catcheside (1944) has shown a clear case of non-random segregation in the ascomycete *Bombardia lunata*, which he ascribes to the existence of a gradient in the cell. Renner (1940) has also discussed the relation of heterogamy and polarised segregation in *Oenothera*.

There are several possible explanations of the polarity in behaviour of the univalents in *Leucopogon juniperinus*. First, the distribution might be a mechanical effect consequent on the persistence through meiotic prophase of the chromosome arrangement derived from the preceding mitotic telophase. (*a*) Chromosome movement during meiotic prophase, which must be considerable, would tend to eliminate any such polarisation. (*b*) A similar behaviour would be expected in other plants with univalents, and this is not so. (*c*) With such a derivation, the polarity of the pollen mother cells would be in opposed pairs which are not found.

Secondly, the polarity in movement might be the consequence of a gradient in the cell. Catcheside (*loc. cit.*) has pointed out that the orientation and differential segregation of chromosomes involves two kinds of inequality : (i) a gradient in the cell as a whole, and (ii) a difference possessed by the chromosomes showing the bias, and capable of directing them with reference to the cell gradient. It is significant that a morphological polarity in the pollen mother cells is apparently normal for most species of *Leucopogon*, for *Styphelia* and for related genera. This polarity controls the unusual type of pollen development described. It may be in some way analogous to the gradient which determines the development of the embryo-sac from a particular one, usually that at the chalazal end, of the row of four megaspores. This gradient provides the first of Catcheside's requirements. On the other hand, it appears to be intracellular in nature, since the pollen mother cells show a random arrangement in the anther as a whole, as seen in longitudinal section. The functional pollen grain of each " tetrad " may face in any direction with reference to the axis of the anther. Since the pollen mother cells have no obvious morphological " base," it is difficult to fit the explanation offered by Catcheside for *Bombardia*. He was dealing with a situation in which there was a definite linear arrangement of the nuclei, but here no such arrangement exists.

The condition in *L. juniperinus* also shows interesting similarities and differences as compared with the case in maize, in which the non-random segregation of a heterozygous bivalent is controlled by an abnormal heterochromatic segment in one of the chromosomes. It is possible that in both plants a similar action of heterochromatin, in directing the movement of chromosomes along the gradient present in the cell, is being manifested. In *L. juniperinus* the univalents are similar in staining behaviour to the bivalents at all stages. There is, however, in all chromosomes, a distinct region close to the centromere which frequently fails to stain, especially at second metaphase, and a relation between these heterochromatic segments and the segregation of the univalents cannot be excluded. In maize, the effect of the heterochromatin is shown in bivalents heterozygous for the heterochromatic segment. Differential segregation could not be shown by a bivalent homozygous either for the presence or absence of the segment, since it is dependent upon unbalance between the component chromosomes. With the univalents of *Leucopogon*, it is possible that heterochromatic or other segments, not being balanced against similar segments as in bivalents, may act in the same way, directing the univalents in a definite direction with respect to the cell gradient. It is impossible to say whether any particular segment is responsible for this direction, but a comparison of the behaviour of univalents with different distributions of heterochromatin in this or other related species might provide evidence of whether this directional property is characteristic of heterochromatin. In order to test this hypothesis it is proposed to undertake the synthesis of new allotriploids in *Leucopogon* and *Styphelia*.

It is not known whether the four univalents tend to be associated with any particular daughter bivalent or group of daughter bivalents. Any polarity in the behaviour of the bivalents could be determined cytologically only in the presence of visible differences between the chromosomes, or genetically by the use of plants heterozygous for marker genes. Genetically the species is totally unknown.

L. juniperinus demands further cytological study. It is necessary to determine the exact constitution of the functional pollen grains, and the meiotic behaviour in the embryo-sac mother cell. Brough (1924) has shown that the normal development of the embryo-sac from the chalazal megaspore, typical of most Angiosperms, is replaced in *Styphelia longifolia* by development of the megaspore at the micropylar end of the linear tetrad ; the same condition is to be expected in *Leucopogon*. An attempt at the resynthesis of the species would be desirable, and its probable parents may be found amongst those species of *Leucopogon*, or even *Styphelia*, which have a diploid number of 8. It would be of interest to know the cytological constitution of the species in its occurrence near Brisbane, Queensland, especially if its origin there should prove independent of its origin in the Sydney district.

SUMMARY

1. An unusual case of a stabilised triploid species, *L. juniperinus*, with a somatic number of 12, has been described. The reduction divisions in the pollen mother cells show great regularity, with the formation of 4 bivalents and 4 univalents.

2. Polarity in the distribution of the univalents is pronounced, all four usually going to the same end of the pollen mother cell. This polarity, correlated with an unusual type of pollen development, and with pollen of moderate fertility, suggests that the species is semiapomictic. It presents a condition analogous with that known for *Rosa canina* except that the extra univalent chromosomes may be carried by the pollen.

3. The polarity in univalent distribution is considered to be a direct consequence of allotriploidy superimposed upon the intracellular polarity which controls the method of pollen development.

4. The species is considered to be of hybrid origin, and to have achieved a true-breeding condition and reasonable fertility by this form of semiapomixis.

Acknowledgment.—The author wishes to express his gratitude to Dr H. N. Barber for helpful discussion of the problem, and for criticism of the text. Acknowledgment is also due to the Trustees and the Director of the Museum of Technology and Applied Science for the provision of facilities for the work.

REFERENCES

BENTHAM, G.   1869.
Flora Australiensis 4.
London : Lovell, Reeve.

BROUGH, P.   1924.
Studies in the Epacridaceæ.   (i) The life history of *Styphelia longifolia*.   R. Br.
P. Linn. Soc. N.S.W. 49, 162-178.

CATCHESIDE, D. G.   1944.
Polarised segregation in an Ascomycete.
Ann. Bot. N.S. *8*, 119-130.

DARLINGTON, C. D.   1937.
Recent advances in cytology.
London : Churchill.

DARLINGTON, C. D.   1939*a*.
Misdivision and the genetics of the centromere.
J. Genet. 37, 341-364.

DARLINGTON, C. D.   1939*b*.
The genetical and mechanical properties of the sex chromosomes.   V. *Cimex* and
    the Heteroptera.
J. Genet. 39, 101-137.

DARLINGTON, C. D.   1940.
The origin of Isochromosomes.
J. Genet. 39, 351-361.

DARLINGTON, C. D., AND LA COUR, L. F.   1942.
The handling of chromosomes.
London : Allen and Unwin.

GAISER, L. O.   1930.
Chromosome numbers in the angiosperms.   II.
Bibliog. Genet. 6, 171-412.

HAGERUP, O.   1928.
Morphological and cytological studies in the Bicornes.
Dansk. Bot. Arkiv. 6, 1-26.

HURST, C. C.   1931.
Embryo-sac formation in diploid and polyploid species of Roseæ.
P.R.S. (B) *109*, 126-148.

KIHARA, H.   1931.
Genomanalyse bei *Triticum* und *Aegilops* II *Aegilotricum* and *Aegilops cylindrea*.
Cytologia 2, 106-156.

MORGAN, T. H., BRIDGES, C. H., AND STURTEVANT, A. H.   1925.
The genetics of *Drosophila*.
Bibliog. Genet. 2, 1-262.

RENNER, O.   1940.
Kurze Mitteilungen über *Oenothera*, IV.   Über die Beziehung zwischen Heterogamie
    und Embryo-sac entwicklung u.s.w.
Flora, N.F. *34*, 145-158.

RHOADES, M. M. 1942.
Preferential segregation in maize.
Genetics 27, 395-407.

RIBBANDS, C. R. 1937.
The consequences of structural hybridity at meiosis in *Lilium x testaceum*.
J. Genet. 35, 1-24.

SAMUELSON. 1913.
Studien über die Entwicklungsgeschichte der Bluten einiger Bicornes typen.
Svensk. Bot. Tidsk. 7, 97-188.

SMITH-WHITE, S. 1947a.
A survey of chromosome numbers in the Epacridaceæ.
P. Linn. Soc. N.S.W. (in press).

SMITH-WHITE, S. 1947b.
Cytological studies in the Myrtaceæ. II. Chromosome numbers in the Leptospermoideæ and Myrtoideæ.
P. Linn. Soc. N.S.W. (in press).

STURTEVANT, A. H. 1936.
Preferential segregation in Triplo-IV females of *Drosophila melanogaster*.
Genetics 21, 444-466.

TÄCKHOLM, G. 1922.
Zytologische Studien über die Gattung *Rosa*.
Acta Hort. Berg. 7, 97-381.

WANSCHER, J. H. 1934.
Secondary associations in the Umbelliferæ and Bicornes.
New Phyt. 33, 58-65.

8

FURTHER STUDIES ON POLYPLOID AMPHIBIANS
(CERATOPHRYDIDAE)
I. MITOTIC AND MEIOTIC ASPECTS

M. L. Beçak, W. Beçak, and M. N. Rabello

Abstract. Odontophrynus cultripes REINHARDT and LUTKEN, 1862 has 22 chromosomes in its diploid complement. Spermatocyte I contained 11 ring bivalents and metaphase II exhibited 11 chromosomes. *Odontophrynus americanus* (DUMÉRIL and BIBRON) 1882 has 44 chromosomes in somatic as well as germ cells, these can be sorted into 11 groups of homologues. Metaphase I showed varying numbers of quadrivalents and metaphase II exhibited 22 dyads. *Ceratophrys dorsata* WIED., 1824 has 104 chromosomes in somatic and germ cells; these 104 chromosomes comprise 8 each of 13 kinds of homologues. The spermatocyte I contained ring octovalents and other multivalents, and metaphase II 52 chromosomes. The above findings indicate that evolution by polyploidization occurred in South American frogs belonging to the family *Ceratophrydidae*.

Introduction

A previous report (BEÇAK, BEÇAK and RABELLO, 1966) revealed the South American frog (*Odontophrynus americanus*, $2n = 44$) to be a tetraploid species. This indicated the possibility that a series of polyploidization might have occurred during speciation of frogs belonging to the family *Ceratophrydidae*. The present report describes our findings on *Odontophrynus cultripes* ($2n = 22$), which is a diploid species, and on *Ceratophrys dorsata* ($2n = 104$), an octoploid species.

Material and Methods

The cytological study has been done in 3 males and 2 females of *O. cultripes* collected in two different localities of the State of Minas Gerais, 9 males and 1 female of *O. americanus* collected at three different localities of the State of São Paulo and 1 male and 1 female of *C. dorsata* collected at different localities of São Paulo.

The chromosomes were obtained from spleen, bone-marrow and gonads by the squash technique. Prior to sacrifice the animals were treated with a 1% solution of colchicine in the dosage of 0.1 ml per 10 g of weight. The slides were stained with Giemsa after hydrolysis in 1N HCl at 60° C for 10 min.

Observations
Odontophrynus cultripes

The diploid number of 22 chromosomes has been ascertained from countings of 128 cells (Table 1). The chromosomes can be grouped into

Table 1. *Number of chromosomes in somatic metaphases and of the first and second division of male meiosis of diploid species O. cultripes from Minas Gerais*

Number and sex of specimen	Mitoses			Metaphases I			Metaphases II			Total
	≤ 21	22	≥ 23	≤ 21	22	≥ 23	≤ 10	11	≥ 12	
994, ♀	—	3	—	—	—	—	—	—	—	3
1001, ♂	—	10	—	1	13	1	—	11	—	36
1074, ♂	—	7	—	—	10	—	—	11	—	28
1093, ♂	3	7	3	3	21	1	5	8	—	51
1133, ♀	—	10	—	—	—	—	—	—	—	10
Total	3	37	3	4	44	2	5	30	—	128

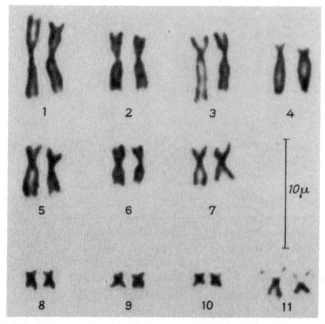

Fig. 1. Mitotic karyotype of *O. cultripes* ♀ showing 11 pairs of homologues

11 pairs of homologues. Pairs 1, 5, 6, 7, 8, 9, 10 and 11 are metacentrics; pairs 2 and 3 are submetacentrics and pair 4 is acrocentric. The karyotype revealed no sex difference (Fig. 1).

In male meiosis the chromosomes showed a "bouquet" configuration during zygotene with pairing at the distal ends. Eleven ring bivalents seen at metaphase I exhibited two terminal chiasmata. In metaphase II 11 chromosomes were present. No differential condensation was observed in any bivalent. Apparently, *O. cultripes* is a true diploid species.

Odontophrynus americanus

This tetraploid species has previously been described (Beçak, Beçak and Rabello, 1966). New specimens were examined in order to confirm the previous finding and to study the different phases of meiosis in material not treated with colchicine. This study permitted us to elucidate the behavior of homologues in the tetraploids during pairing as well as during segregation of the multivalents in anaphase I.

Table 2. *Number of chromosomes in somatic metaphases and of the first and second divisions of male meiosis of the tetraploid species O. americanus from São Paulo*

Number and sex of specimen	Mitoses			Metaphases I			Metaphases II			Total
	≤ 43	44	≥ 45	≤ 43	44	≥ 45	≤ 21	22	≥ 23	
857, ♂	—	12	—	—	17	—	—	5	—	34
858, ♀	—	15	1	—	—	—	—	—	—	16
859, ♂	—	16	—	—	11	—	—	4	—	31
860, ♂	—	13	—	—	12	—	—	6	—	31
975, ♂	1	26	1	—	—	—	1	5	—	34
1006, ♂	—	10	—	—	1	—	1	3	—	15
1007, ♂*	—	2	—	—	5	—	3	6	—	16
1018, ♂*	2	5	1	—	3	—	—	9	—	20
1019, ♂	—	5	—	—	—	—	—	—	—	5
1025, ♂*	—	2	—	—	6	—	2	15	—	25
Total	3	106	3	—	55	—	7	53	—	227

* Animals not treated with colchicine.

Table 2 shows the number of chromosomes counted in 227 cells. The 44 chromosomes can be classified into 11 types, each one being represented by 4 homologues. At metaphase I in the male 11 ring quadrivalents were most often seen, but some spermatocytes contained a number of trivalents and bivalents. At metaphase II, 22 chromosomes were invariably present.

For the study of earlier stages of male meiosis, colchicine treatment was avoided. During zygotene, the chromosomes occupied the entire nuclear sphere, with the extremities polarized in a peripheral region resembling a "bouquet" configuration. The 8 terminal ends of 4 homologues were paired two-by-two, resulting in a configuration with four loops. At the terminal ends, each pairing involved only two homologues (Figs. 2 and 3). Consequently, the number of synaptic terminals in zygotene is 44, representing 11 sets of 4 paired homologues.

The stage corresponding to pachytene was not observed. The longitudinal pairing along the entire length of the homologues apparently

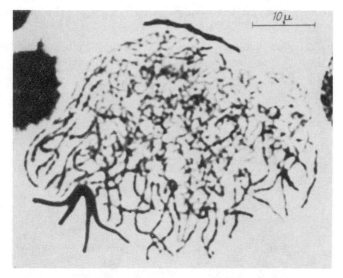

Fig. 2. Zygotene stage of *O. americanus* spermatocyte with "bouquet" aspect

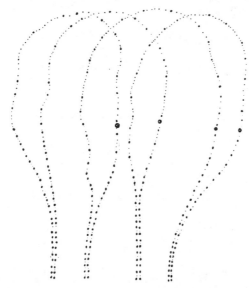

Fig. 3. Schematic drawing of the *O. americanus* zygotene, showing two-by-two
pairing of the ends of four homologues

did not occur. This lack of complete longitudinal pairing appeared to
be the very reason why interstitial chiasmata were not observed
even during diplotene. Figure-of-eight configurations often observed

123

resulted from a twist at the middle rather than from the presence of an interstitial chiasma (Figs. 4 and 5).

The observation of several cells in zygotene, diplotene and metaphase I indicated that the occasional formation of bivalents rather than

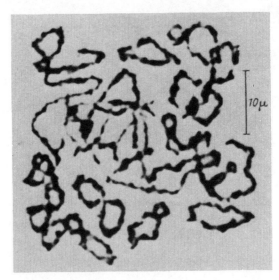

Fig. 4. Diplotene stage of *O. americanus* spermatocyte showing ring quadrivalents

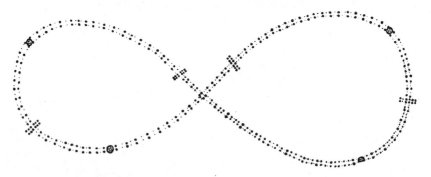

Fig. 5. Schematic drawing of a quadrivalent of *O. americanus* in diplotene stage

quadrivalents was not due to the breakdown of quadrivalents at the advanced stage of meiosis I, for bivalents were already seen in some of the diplotene figures. The occasional formation of them is expected, because synapsis at each chromosome end always involved only two homologues (Fig. 3).

At anaphase I, the quadrivalents oriented themselves in such a way that the movement of two bivalents to the opposite division pole was insured.

Thus, in this tetraploid species, there apparently was no formation of aneuploid gametes. Univalents and trivalents occasionally observed may be regarded as artifacts produced as a result of severe squashing employed during the preparation.

Table 3 shows the frequency of tri-, bi-, and univalents observed in 55 first meiotic metaphase figures.

Table 3. *Patterns of multivalent configurations in 55 metaphases I of O. americanus*

Numbers of M I	Configurations	Numbers of M I	Configurations
9	11 IV	1	8 IV, 2 II, 4 I
9	10 IV, 2 II	8	7 IV, 8 II
1	10 IV, 4 I	1	7 IV, 6 II, 4 I
1	10 IV, 1 III, 1 I	1	7 IV, 6 II, 1 III, 1 I
10	9 IV, 4 II	1	6 IV, 10 II
1	9 IV, 2 II, 4 I	1	6 IV, 8 II, 1 III, 1 I
10	8 IV, 6 II	1	4 IV, 14 II

Ceratophrys dorsata

In this species, each cell contained 104 chromosomes (Table 4). These chromosomes could be arranged into 13 groups of 8 homologues

Table 4. *Number of chromosomes in somatic metaphases and of the first and second spermatocyte divisions of the octoploid species C. dorsata from São Paulo*

Number and sex of specimen	Mitoses			Metaphases I			Metaphases II			Total
	≦103	104	≧105	≦103	104	≧105	≦51	52	≧53	
1088, ♀	2	10	1	—	—	—	—	—	—	13
1211, ♂	3	14	1	1	3	—	5	9	1	37
Total	5	24	2	1	3	—	5	9	1	50

each. Again, no sexual difference was observed. Groups 1, 4, 6, 10 and 11 were composed of metacentrics, groups 2, 3, 7, 8, 9, 12 and 13 were submetacentrics, and group 5 was made of acrocentrics. The chromosomes of group 6 were characterized by having an achromatic gap adjacent to the centromere. Metacentrics of group 11 presented a satellite at the distal end of one arm, but frequently not all homologues show this aspect simultaneoulsy (Fig. 6).

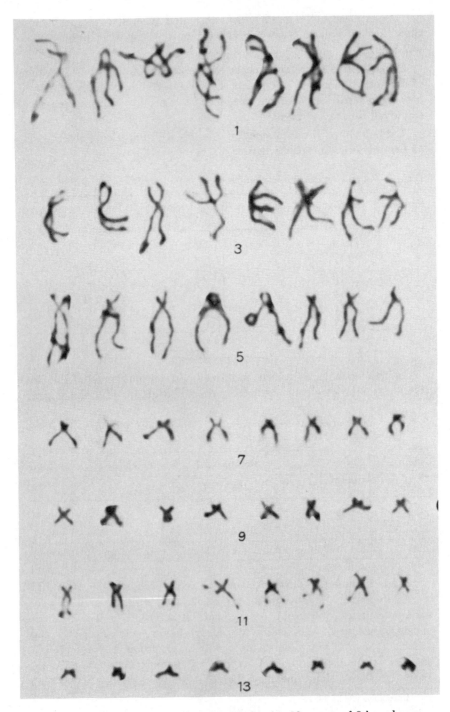

Fig. 6. Mitotic karyotype of *C. dorsata* ♀ showing 13 groups of 8 homologues

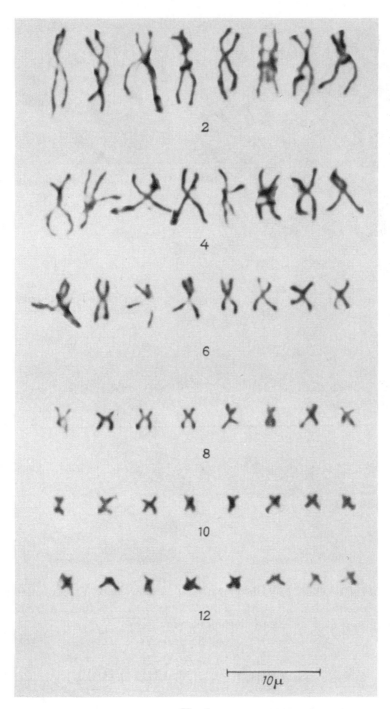

Fig. 6

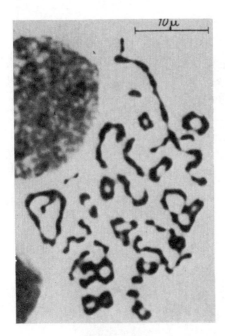

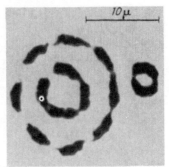

Fig. 8. Partial aspect of metaphase I of *C. dorsata* ♂ with rings of 8, 4 and 2 homologues

Fig. 7. Metaphase I of *C. dorsata* ♂ showing ring multivalents

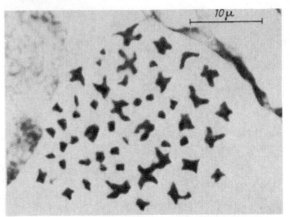

Fig. 9. Metaphase II of *C. dorsata* ♂ showing 52 chromosomes

Spermatocytes I contained an admixture of octovalents, hexavalents, quadrivalents and bivalents (Figs. 7 and 8). Three fully analyzable metaphases I showed the following combinations:

No. 1   3 VIII, 3 VI, 9 IV, 13 II
No. 2   1 VIII, 2 VI, 9 IV, 23 II, 2 I
No. 3   4 VIII, 2 VI, 7 IV, 1 III, 14 II, 1 I.

Metaphases II contained 52 chromosomes (Fig. 9). The analysis of the mitotic and meiotic chromosomes indicates that *C. dorsata* is an octoploid species.

There was no indication of heteromorphic sex chromosomes in mitoses of both sexes and in male meiosis. The scarcity of specimens prevented a complete study of meiotic events, especially of the segregation in anaphase I.

The material of all 3 species studied were normal males and females, possessing either functional testes or ovaries.

Discussion

SAEZ and BRUM (1959) were the first to observe the presence of multivalents in spermatocytes I of *Ceratophrys ornata*. The authors stated that the diploid chromosome number of this species is higher than $2n = 80$.

Our observations of mitotic and meiotic studies in the closely related species, *C. dorsata* suggested that this species is octoploid and evolved by polyploidization from an ancestral diploid of $2n = 26$. Indeed, J. P. BOGART (personal communication) found the diploid chromosome number of 26 in 2 members of the family *Ceratophrydidae*, namely *Chacophrys pierotti* and *Lepidobatrachus llanensis*. The comparison of the karyotype of *O. cultripes* (22 chromosomes) with that of *O. americanus* (44 chromosomes), on the other hand, suggested that a diploid ancestor of *O. americanus* had a karyotype nearly identical with that of *O. cultripes*.

The DNA content of these three species, measured by cytophotometry, confirms our cytologic findings (BEÇAK et al., in press). According to these measurements, *C. dorsata* cells have approximately twice the DNA content found in *O. americanus* cells and the values of the latter are approximately twice as much as those found in *O. cultripes* cells.

Acknowledgments. We are thankful to Dr. SUSUMU OHNO from City of Hope Medical Center, Duarte, California, for his suggestions and kind revision of the manuscript.

References

BEÇAK, M. L., W. BEÇAK, and M. N. RABELLO: Cytological evidence of constant tetraploidy in the bisexual South American frog *Odontophrynus americanus*. Chromosoma (Berl.) **19**, 188—193 (1966).

BEÇAK, W., M. L. BEÇAK, D. LAVALLE, and G. SCHREIBER: Further studies on polyploid amphibians (*Ceratophrydidae*) II. DNA content and nuclear volume (in press).

SAEZ, F. A., y N. BRUM: Citogenética de anfíbios anuros de America de Sud. Los cromosomas de *Odontophrynus americanus* y *Ceratophrys ornata*. An. Fac. Med. Montevideo **44**, 414—423 (1959).

Part III

CLASSIFICATION OF POLYPLOIDS

Editors' Comments
on Papers 9 and 10

9   **CLAUSEN, KECK, and HIESEY**
     Excerpt from *Experimental Studies on the Nature of Species.*
     II. Plant Evolution Through Amphiploidy and Autoploidy with
     Examples from the Madiinae

10   **STEBBINS**
     Types of Polyploids: Their Classification and Significance

Paper 9 is an excerpt from the exposition on plant evolution through amphiploidy and autoploidy by Clausen and his coworkers. Almost half of this treatise deals with research carried out by the authors in the 1930s and early 1940s. The discussion is thus founded on a large amount of their own experimental data plus literature findings, giving rise to their classification of polyploids based on biosystematic units. In retrospect, it is unfortunate that they used taxonomic rank as a criterion in a classification of polyploidy.

Stebbins has essentially set forth the classification of polyploidy still in use today (Paper 10). His article represents a synthesis of the data available at the time and was based on genetic principles then available. Unfortunately, many systematists and others have misinterpreted his system and have continued to be misled by the idea that taxonomic rank is of major importance in describing the kind of polyploid being studied. In fact, the data influencing such decisions are necessarily of a genetic and cytogenetic nature, as emphasized by Stebbins (1980) and more recently by Jackson (1982). The paper is important both for the system of polyploid classification proposed and for the incisiveness with which it examines the central genetic issues.

REFERENCES

Jackson, R. C., 1982, Polyploidy and Diploidy: New Perspectives on Chromosome Pairing and Its Evolutionary Implications, *Am. J. Bot.* **69:**1512–1523.
Stebbins, G. L., 1980, Polyploidy in Plants: Unsolved Problems and Prospects, in *Polyploidy: Biological Relevance*, Plenum Press, New York, pp. 495–520.

9

Reprinted from *Carnegie Inst. Washington Publ. No. 564*, pp. 62–73 (1945)

EXPERIMENTAL STUDIES ON THE NATURE OF SPECIES.
II. PLANT EVOLUTION THROUGH AMPHIPLOIDY AND AUTOPLOIDY WITH EXAMPLES FROM THE MADIINAE

J. Clausen, D. D. Keck, and W. M. Hiesey

[Editors' Note: In the original, material precedes this excerpt.]

V

BIOSYSTEMATIC RELATIONSHIPS AND AMPHIPLOIDY

The findings reported here represent a small part of a series of investigations designed to clarify current concepts regarding the organization of living things and their mode of evolution. This objective has been approached from the point of view of classical taxonomy, including morphology and distribution, as well as from the points of view of ecology, genetics, and cytology. It is therefore appropriate to examine the findings on amphiploidy in the perspective of the entire series of investigations.

PRINCIPLES GOVERNING BIOSYSTEMATIC INVESTIGATIONS. Experimental studies reveal an organization of living things based primarily upon genetic and ecologic principles. The genetic principles include those of the chromosomes, and the ecologic ones include that of physiologic fitness to the environment in which the organism is native. Genetic and ecologic factors control the morphological characters utilized in separating species and other natural units. Therefore, once this relation has been determined by experiment, these morphological characters can be successfully employed to trace the distribution of the species in the wild, and they become indicators of more basic genetic-physiologic differences.

Two fundamental principles govern the appearance and distribution of wild plants and probably of all living things, as follows:

1. Natural species consist of individuals whose genes are in internal balance so that a harmonious physiologic and morphologic development is assured generation after generation.

2. The individuals of wild species not only are balanced internally, but fit their natural environment; they are in rhythm with the seasons and adapted to the over-all conditions of temperature, moisture, wind, soil, light, and biotic elements. Individuals out of balance with their environment are unable to compete with others better fitted.

Considering the number of genes involved in the differentiation of kinds of plants and the intricacy of their balances, it is not remarkable that it is difficult to restore the balance once it has been broken through a rearrangement of genes following interspecific hybridization. The very intricacy of the interreactions of many genes—largely precluding as they do most of the possibilities of change—may account for the relative stability found in species, many of which have preserved their identity essentially unchanged through geologic periods of time. When species are closely enough related to permit the production of interspecific hybrids, the offspring are characterized by various combinations of gametic and zygotic sterility, poor germination, sublethal seedlings, and an assortment of subnormal plants. These effects are evidently closely related and indicate very severe disturbances in the gene-determined physiological balances.

It is therefore natural that species, which have to meet the exacting requirements of both the internal and external balances, change slowly, and that they are relatively few considering the large number of genes available. It is more remarkable that widely different subspecies and ecotypes of one species from geographically and climatically widely separated areas are able to interchange their genes freely without consequential disturbances in the intricate internal balance.

THE BIOSYSTEMATIC UNITS. From an abundance of experimental data it is now clear that natural units of various ranks exist. The higher units are separated by more distinct internal barriers than the lower units within them, hence they are more effectively prevented from exchanging genes. This is taken to indicate that the genetic balances among the units of higher rank are the more unlike. Many small and gradual steps are involved in the differentiation of the natural units from the local population to the genus. One therefore finds natural units on different levels, the better marked of which have received special recognition. These represent significant nodes, or departures, in evolutionary differentiation. Four major levels are recognized in the present experimental investigations. These are the foundation of a classification which is basically biological rather than morphological. Three of the biological units recognized are adopted from Turesson (1922a, 1922b), and the fourth and most inclusive from Danser (1929). This classification is a further development of one previously proposed by the writers (1939). These biological—or, better, biosystematic—units are as follows:

1. *Ecotype*. Species that occupy a series of contrasting environments develop genetically and physiologically distinct ecologic races, the ecotypes, which are suited to these environments. Ecotypes of one species have the same internal balance, for there is no genetic obstacle to a free interchange of their genes when they meet and hybridize. Each of such ecotypes, however, strikes a different balance with the environment and is prevented from free migration to other environments by natural selection.

Some species are monotypic, that is, they contain only one ecotype.

Then they occupy only one environmental zone, although they may have a wide geographical distribution within a climatically fairly uniform area. Also, some so-called species, not separated by genetic barriers but recognized only by morphology and ecology, are but ecotypes of one species. The ecotype is a biological unit defined in terms of ecology and genetics combined, but taxonomically it frequently approximates the geographical subspecies.

2. *Ecospecies.* Different species have evolved separate genetic systems that are balanced both internally and externally. These balances are so intricate that the genes of related systems cannot be freely interchanged without seriously impairing the ensuing development of the offspring. Such constitutional barriers between species are based in the genic structure, but are expressed through physiological development or morphology. These constitutional barriers are carried along wherever a species migrates. This situation differs from that in ecotypes, which depend exclusively for their separation upon their fitness to given environments. The species is thereby insured more permanence than the ecotype.

Species capable of a limited interchange of genes with one another are ecospecies of one cenospecies. Like ecotypes of one species, they usually occupy a series of different environments. The ecospecies approximates the species of moderately conservative taxonomists working along conventional lines.

3. *Cenospecies.* Species entirely unable to exchange genes with one another belong to different cenospecies. Their genetic balances have become so unlike that no interchange is possible, although sterile hybrids may at times be formed between them. It is still possible, however, to add all the chromosomes of members of distinct but related cenospecies by the process of amphiploidy. Some cenospecies are monotypic, that is, they consist of only one ecospecies. The evolutionary possibilities open to such cenospecies are few. In most genera the level of the cenospecies approximates that of the taxonomic section or of the species complex, but in small genera it may equal the entire genus. The monotypic cenospecies is usually a very distinct taxonomic species.

4. *Comparium.* Distinct cenospecies which are still able to produce first-generation hybrids with one another belong to one comparium. In accordance with the definition of the cenospecies, such hybrids are always sterile except when all the chromosomes of both parents are added together through amphiploidy. Cenospecies which are unable to intercross and produce even a sterile hybrid belong to different com-

paria unless they are linked through intermediaries. At this evolutionary level the genetic balances have differentiated so far that even the addition of all the chromosomes of members of two comparia is impossible.

In many complex groups of plants the comparium corresponds to the genus, but there are also examples of several adjacent genera composing one comparium. In less complex groups which approach a static condition, it may include only a single taxonomic species. The genus *Ginkgo*, for instance, of the monotypic order Ginkgoales, has evidently become depleted in biotypes to such a degree that today it constitutes a single comparium, with one cenospecies, one ecospecies, and possibly only one ecotype.

These four kinds of biosystematic units represent important evolutionary nodes, but many intermediate steps connect them. Briefly, therefore, ecotype differentiation may be considered to represent the evolutionary level at which fitness to more than one major environment evolves. The ecospecies is at that level at which separate units arise through constitutional barriers to successful interbreeding. Beyond the level of the cenospecies gene exchange is impossible, although the addition of all the genes of members of two cenospecies in the same comparium is still possible through amphiploidy. The comparium marks the limit for even this avenue of evolution, because neither exchange nor addition of genes of related comparia is possible.

In general terms one may say that evolution is reticulate from the level of the ecotype to that of the comparium, but beyond that level it is exclusively of the forked type. The comparium therefore represents a very important node in the evolutionary process. In biosystematic studies on plants, the ecotype is the basic ecological unit, the ecospecies is the basic systematic unit, and the cenospecies and comparium are units of major evolutionary significance.

THE EVOLUTIONARY PROCESSES. If the data from a study of hundreds of different hybrids between taxonomic units of various ranks are correctly interpreted, then it appears likely that one kind of biosystematic unit may evolve from another in successive order. The processes of mutation, recombination, and selection provide the machinery for this development. Distinct ecotypes of one species may successively become distinct ecospecies, cenospecies, and comparia. The process of mutation furnishes the raw materials or genes, and the recombining processes repattern them. As these new forms are subjected to the environment, selective processes eliminate the unfit. The balance between the processes of mutation and recombination on the one hand, and of selection on the other, determines whether a group of organisms is on its way up or on its way out.

Processes of evolution not necessarily activated by hybridization include gene mutation, chromosome multiplication (autoploidy), and chromosome repatterning (translocation and inversion). Presumably it should be possible for these processes to occur at all stages in the evolution of a group, even when, like *Ginkgo*, it has become a monotypic comparium.

Evolutionary processes initiated only by hybridization include chromosome exchange, gene interchange, and addition of sets of chromosomes through amphiploidy. These have their best opportunities in comparia consisting of many cenospecies, with many ecospecies and ecotypes. Such a comparium is possibly in its most active and expansive stage of development. Groups of this nature, with their confusing patterns of interrelations, are those which the taxonomists call critical. Under these conditions the gene interchanges gradually repattern the ecotypes and, to a certain extent, the ecospecies also. The addition of complete sets of chromosomes is most effective in hybrids between cenospecies of the same comparium; for hybrids between species with nonhomologous chromosomes produce the most stable combinations. Amphiploidy is a remarkable way of circumventing gene interchange in interspecific hybrids. Thereby the gene balance of the first-generation hybrid is perpetuated, and the new form emerges suddenly.

In comparia that are rich in ecospecies and ecotypes, nonhybrid evolutionary processes also continue. These occur alongside of gene interchanges and additions of sets of chromosomes, and it is even possible that hybridism accentuates their rate of appearance. Accordingly, systematic groups in this stage of development have their greatest evolutionary possibilities. Conversely, as through the ages a comparium becomes depleted and approaches the monotypic condition, its chances for change through hybridization decrease. This situation is well illustrated by the genus *Sequoia*, which at one time encircled the northern hemisphere and must have consisted of several ecospecies and many ecotypes, but today is represented by only two monotypic species in California. It follows from this discussion that the biosystematic approach to classification gives an evaluation of the stage of development reached by the group studied and of its evolutionary possibilities.

APPLICATIONS OF BIOSYSTEMATIC PRINCIPLES. The biosystematic units are explored by experimental methods. They are usually found to correspond fairly closely to taxonomic units based on careful morphologic and distributional studies. A few interesting discrepancies in critical groups of plants, however, indicate the shortcomings of a purely morphological approach to classification.

In chapter II attention was called to *Layia nutans* (Greene) Jepson, which proved to be a *Madia* when tested by experiments. This is an example of the correctives offered by this approach. Similarly, experiments now indicate that another *Madia*, known for years as *Hemizonia Wheeleri* A. Gray, is no more than an ecotype of *Madia elegans* D. Don. These are not very unlike morphologically, each has 8 pairs of chromosomes, their hybrids are fertile, the second generation is vigorous, and their chromosomes are homologous.

Likewise, the complete genetic compatibility of *Zea Mays* L. with *Euchlaena mexicana* Schrad. (Mangelsdorf and Reeves, 1939) indicates that these two "genera" belong to one ecospecies. Recently Reeves and Mangelsdorf (1942), recognizing the close relationship of these plants, have combined them in *Zea* but retained them as separate species. These and the genus *Tripsacum* belong to one comparium.

Another striking discrepancy is found in the agriculturally equally important genera *Triticum, Aegilops, Agropyron* (including *Haynaldia*), *Secale, Elymus, Hordeum*, and *Sitanion*. The combined results of the investigations of many authors indicate that these genera belong to one comparium. The chromosome homologies of their species transcend present generic lines. Although the first three genera are widely accepted today, they have, at one time or another, been included under *Triticum*, and from the genetic point of view this has not been without reason. The synonymies and many "intergeneric" hybrids point to their close relationship.

A still more glaring discrepancy is presented by *Lolium perenne* L. and *Festuca pratensis* Huds. (= *F. elatior* L.). These two grasses are placed in different tribes. They hybridize spontaneously, they are slightly interfertile on backcrossing (Jenkin, 1933), and their chromosomes are homologous, for the hybrid has 7 pairs like each parent (Peto, 1933). Accordingly, they belong to one cenospecies. Hybrids link them with the other species of *Lolium* and *Festuca*, which places both genera in the same comparium. The positive evidence of homology between their chromosomes is a forceful reminder of their close evolutionary relationship not lightly to be ignored, and is in line with similarities in their floral morphology reported by Stebbins (in litt.).

These examples of marked inconsistencies between purely morphological and biological classifications are taken from the Gramineae and the

Compositae, two families admittedly characterized by very complex relationships. The intricacies of these relationships sometimes make it necessary for the classifier to use only one or two characters as key indicators of tribal differences. Such a situation in itself suggests that such tribes may be entirely artificial.

The nature of the complexity within these and other families suggests that they are in a very active stage of evolution. Such situations call for a re-examination of the classification on the basis of adequate experimental data. A recent brief summary of the relationships in the genus *Layia* shows the application of some of the biosystematic principles in a fairly critical genus (Clausen, Keck, and Hiesey, 1941).

REQUIREMENTS FOR SUCCESS IN AMPHIPLOIDS. Amphiploidy is now known to play a very significant role in the evolution of higher plants. Our modern cultivated wheats and other crop plants probably originated through it, and doubtless the process will receive greater attention in future plant breeding. It is therefore of economic as well as of scientific importance to consider the principles that govern the development of successful amphiploids.

When amphiploids were first produced it was tacitly assumed that the simple doubling of the chromosome number would, in some miraculous way, render any sterile hybrid fertile and vigorous. Sufficient information has now been gathered on experimental amphiploids to show that many of them are partly sterile or otherwise unsuccessful. This is exactly what one should expect from the biological concept of species as presented above.

The first condition for a successful amphiploid is that the genomes of its parents fit together properly to insure a harmonious development of the F_1. Weak first generations are more likely to result from crossings between the remotest species it is possible to hybridize than from those between species a little more closely related.

If the F_1 hybrid is vigorous and healthy, a successful amphiploid may be produced, but then it must be remembered that the genes of distinct species are not freely interchangeable without detriment to the offspring. If, therefore, some gene exchange takes place before or after the duplication, as happened in *Layia pentaglossa*, the resulting progeny should be expected to show the same weaknesses that would be observed in F_2 populations of such hybrids with undoubled chromosome numbers. Since the balances that determine success or failure are intricate and delicate, they may be upset by slight genetic interchanges. The principles that govern the development of amphiploids may therefore be stated as follows:

1. Successful amphiploids arise from successful interspecific hybrids. It is essential that the parental genomes which enter the first-generation hybrid interact in such a way as to produce a harmonious and vigorous development.

2. If the amphiploid is to remain successful during succeeding generations, the original balance must remain unchanged. The best way to insure such perpetuation is to use species as parents with chromosomes so different that those of one are nonhomologous with those of the other. This prevents interspecific segregation in their amphiploid.

In terms of biosystematic units, an amphiploid is most likely to succeed if its parent species are members of different cenospecies of one comparium. They should be closely enough related to produce a vigorous F_1 hybrid, but remotely enough so that the balance between their combined genomes can be perpetuated.

VI

CLASSIFICATION OF EXPERIMENTAL POLYPLOIDS
ON BIOSYSTEMATIC PRINCIPLES

Polyploids in the plant kingdom have been classified as autoploids and amphiploids. Various criteria have been used in the attempt to designate the differences which such a classification implies, but none of the attempts have been fully satisfactory, because they failed to reckon adequately with the complex organization of plant relationships. It therefore appears desirable to attempt a classification on biosystematic principles.

Autoploids may either spring spontaneously from nonhybrid individuals of a given species, or arise from intra-ecospecific hybrids, which, of course, are fully fertile. Amphiploids, on the other hand, are derived only from inter-ecospecific and inter-cenospecific hybrids, and consequently the F_1 hybrids giving rise to them are either partially or wholly sterile.

Amphiploids in the strictest sense are derived from inter-cenospecific hybrids. Between this extreme and that of true autoploids there are possibilities for chromosome doubling in hybrids between plants of intermediate degrees of relationship. Such combinations are generally unstable, because the chromosomes are not homologous enough to be completely interchangeable, but are too closely related to preserve the original genomes intact. One would not expect many products of evolutionary importance to arise from these intermediate cases. The occurrence of such intergradations makes it clear that the line of demarcation between amphiploids and autoploids is not very sharp, even though the cytogenetic requirements for success are sharply contrasted in the extreme cases.

INTERPLAY BETWEEN CYTOGENETIC PROCESSES IN POLYPLOIDS. All the chromosomes in the autoploid have originated within one species, and each chromosome is therefore represented more than twice. No specifically new hereditary materials have been introduced through this chromosome addition, and, the chromosome homology being complete between sets in this newly created genome, the genic balance has not been disturbed and the autoploid can be essentially as fertile as its parent. The presence of more than one pair of each kind of chromosomes tends to obscure the effect of gene mutations and so is a stabilizing factor. On the other hand, the orderly distribution of more than two chromosomes of a kind during meiosis is frequently difficult.

In amphiploids the situation is much more complex, and the perpetuation of a successful balance between specifically distinct genomes is achieved in various ways. In general, the balance is attained through the interplay between two groups of antagonistic processes. One group tends to destroy the identity of duplicated genomes by (1) interchange of genes or chromosomes between parental genomes, and (2) chromosomal rearrangements. Either or both may take place either before or after doubling of the chromosomes. The other group tends to preserve the identity of the genomes by (1) preferential pairing (due to closer homology) between duplicate chromosomes of one parent, as compared with pairing between them and their nearest homologues from the other parent, and (2) elimination of unsuccessful combinations resulting from occasional interchange between unlike genomes. Interspecific sterility, both gametic and zygotic, and weakness of offspring are the most effective means of elimination. If the relationship of the parents permits, stability is obtained after these groups of processes have produced a new genome of successful character from which discordant rearrangements have been practically eliminated.

A classification of auto- and amphiploids is attempted in table 12, in which various influences that determine their success or failure have been coordinated and plotted one against another. Across the top are listed progressively the degrees of difference between the forms that give rise to auto- and amphiploids. These are given in terms of the biosystematic units. As a corollary, distinctions in genetic relationship are expressed through the degree of sterility in the undoubled F_1 generation. In the cases given in the central vertical column, in which the F_1 hybrid is partially sterile, the degree of sterility is often manifest through constitutional weakness of the undoubled F_2 generation. Also, it is to be noted that autoploids of nonhybrid origin are purposely omitted from this table.

In the two left-hand columns influences are considered that tend to preserve, versus those that tend to recombine, the parental genomes. Since the success of amphiploids largely depends upon the preservation of balanced genomes intact through succeeding generations, the mechanism that governs this process is given special consideration. As has been pointed out by Darlington (1929), the preservation of the initial balance is endangered if the chromosomes of the parents are so homologous that intergenomal pairing takes place. If such pairing is complete, one should expect in all cases to find the identity of the original genomes

TABLE 12

CIRCUMSTANCES OF ORIGIN OF HYBRID AUTOPLOIDS AND AMPHIPLOIDS

		AUTO-PLOIDS	AMPHIPLOIDS	
		INTRA-ECO-SPECIFIC: Undoubled F_1 fully fertile	INTER-ECO-SPECIFIC: Undoubled F_1 partially sterile	INTER-CENO-SPECIFIC: Undoubled F_1 wholly sterile
PARENTAL GENOMES: Identity lost in recombination	**INTER-GENOMAL PAIRING:** Complete	1. Fragaria bracteata × vesca rosea (also many non-hybrid cases)	2. Primula kewensis 3. Aquilegia Janczewskii	Probable
	Partial	. . . .	4. Crepis rubra × foetida	5. Crepis capillaris × tectorum
	None	. . . .	Improbable	6. Layia pentaglossa
Identity preserved	Partial	. . . .	7. Galeopsis Tetrahit 8. Erophila duplex × quadruplex	9. Digitalis mertonensis
	None	. . . .	Probable	10. Phleum pratense 11. Triticale 12. Raphanobrassica 13. Brassica Napus 14. Gossypium Thurberi × arboreum 15. Nicotiana Tabacum 16. Nicotiana digluta 17. Madia nutrammii 18. Madia citrigracilis

144

lost in recombination, so the alternative combination, complete inter-genomal pairing with the identity of parental genomes preserved, is omitted from the table. But even if the parental chromosomes do not pair, the identity of the genomes may nevertheless be lost through the random distribution of unpaired chromosomes, or through the reciprocal translocation of segments of chromosomes. Also, the identity of the genomes may be preserved even after partial intergenomal pairing; for a strong gametic selection may eliminate unfit combinations, or the at-traction between the chromosomes from one parent, after doubling, may outweigh the attraction of their nearest homologue from the other.

In table 12 genetic relationships of the parents are closest in the upper left-hand corner, and become progressively more remote down to the lower right-hand corner. In the various boxes well known examples are placed, to give our conception of this scheme of classification in opera-tion. Each example is reviewed in later paragraphs. Boxes for which no examples are as yet known to us are marked as to the probability of find-ing cases for them. The combinations in the autoploid column below the first instance appear obviously incongruous. In general, it will be noted that the autoploids fall in the box in the upper left-hand corner, and the most stable amphiploids in the lower right-hand corner, following the order from closest to most remote relationship of the parents.

[*Editors' Note:* Material has been omitted at this point. Only those references cited in the preceding excerpt have been included here.]

REFERENCES

CLAUSEN, J., DAVID D. KECK, and W. M. HIESEY. 1939. The concept of species based on experiment. Amer. Jour. Bot. 26:103–106.

CLAUSEN, J., DAVID D. KECK, and W. M. HIESEY. 1941. Experimental taxonomy. Carnegie Inst. Wash. Year Book No. 40:160–170.

DANSER, B. H. 1929. Ueber die Begriffe Komparium, Kommiskuum und Konvivium und ueber die Entstehungsweise der Konvivien. Genetica 11:399–450.

DARLINGTON, C. D. 1929. Polyploids and polyploidy. Nature 124:62–64, 98–100.

JENKIN, T. J. 1933. Interspecific and intergeneric hybrids in herbage grasses. Initial crosses. Jour. Genetics 28:205–264.

MANGELSDORF, P. C., and R. G. REEVES. 1939. The origin of Indian corn and its relatives. Texas Agric. Exper. Sta. Bull. No. 574. 315 pp.

PETO, F. H. 1933. The cytology of certain intergeneric hybrids between *Festuca* and *Lolium*. Jour. Genetics 28:113–156.

REEVES, R. G., and P. C. MANGELSDORF. 1942. A proposed taxonomic change in the tribe Maydeae (family Gramineae). Amer. Jour. Bot. 29:815–817.

TURESSON, G., 1922a. The species and the variety as ecological units. Hereditas 3:100–113.

TURESSON, G., 1922b. The genotypical response of the plant species to the habitat. Hereditas 3:211–350.

10

Reprinted from *Adv. Genet.* **1**:403–429 (1947)

Types of Polyploids: Their Classification and Significance

G. LEDYARD STEBBINS, Jr.

University of California, Berkeley

CONTENTS

I. INTRODUCTION

Although polyploidy has long been recognized as a very important factor in plant evolution, many differences of opinion have existed and still exist concerning the nature of its role, and the relative importance of different types of polyploidy. The development of the use of colchicine to produce polyploids freely has not only made possible a thorough reëxamination of these points of view using experimentally controlled material but, in addition, has made such a reëxamination of great practical as well as theoretical importance. There is no doubt that the more we know of how polyploidy has operated in the past to produce new species and races of plants, the better we shall be able to use experimental polyploidy as a means of improving our useful cultivated plants. The whole subject of polyploidy is much too large to be discussed in any one article, so that this contribution will be confined to those aspects in which the relation between natural and artificial polyploidy is of major theoretical and practical importance.

II. BASIS FOR THE CLASSIFICATION OF POLYPLOID TYPES

In most literature on cytogenetics, polyploids are divided into two types, autopolyploids (recently contracted to autoploids by Clausen, Keck and Hiesey, 1945), and allopolyploids or amphidiploids (amphiploids in Clausen, Keck and Hiesey, 1945). These two categories have been variously defined and redefined, but no system yet devised has proved satisfactory. There are three main reasons for this. First, the two categories usually recognized are not sharply distinct from each other, no matter on what basis the distinction between them is drawn. The usual criterion is whether the component genomes are similar or different from each other, and the common symbols used are AAAA for auto- and AABB for allopolyploids, the letters referring to a set of chromosomes or genomes, containing the basic haploid number for the group. Obviously, however, all degrees and types of differences can exist between two genomes, so that the decision as to whether a polyploid is auto- or allopolyploid must depend on the amount and type of differences that are considered significant in separating the two categories. Differences in both genic content and structural arrangement of the chromosomes are so widespread both within and between species, that the component genomes of a polyploid are not certain to be identical with each other in either of these respects unless the polyploid originated by somatic doubling from a homozygous diploid.

The second difficulty arises from the fact that genic and chromosomal differentiation are independent of each other, so that two genomes widely different in genic content may be very similar in structural arrangement, and *vice versa*. A polyploid may, therefore, contain, when first formed, two chromosomal sets that are highly dissimilar in genic content but are, nevertheless, structurally identical and form a fertile hybrid on the diploid level. Such a polyploid would be termed an autopolyploid under any definition of the term, but it could differ morphologically from any stabilized diploid form, and could produce many new, and still more different, types of segregation in its progeny. Furthermore, even when genic content is disregarded and similarities or differences between genomes are judged solely on the basis of chromosome structure or segmental arrangement, reliable estimates of the amount of difference between two genomes are hard to obtain. The type of structural changes best known, and those which are easiest to detect because of meiotic abnormalities in structural hybrids involving them, are interchanges and inversions of relatively large segments. There is, however, a good deal of evidence now available that a large proportion, if not a majority of the structural differences between genomes, involve segments so small that they do not give the typical meiotic configurations such as multivalents and bridge-fragments in structural hybrids. The writer has pointed out elsewhere (1945) that many of the numerous examples now known of diploid hybrids that are partly or wholly sterile in spite of nearly regular meiosis are best explained on the basis of this phenomenon of cryptic structural hybridity. Furthermore, this condition is the most common one in diploid hybrids between closely related species. It follows, therefore, that tetraploids derived from such hybrids will possess four genomes that have the majority of their chromosomal segments in common and are capable of pairing more or less completely with each other. The type of polyploid which has the strong genomic differentiation usually associated with allopolyploidy or amphidiploidy is in most cases derived from a hybrid between species belonging to different sections, subgenera or genera.

The third difficulty encountered in classifying polyploid types involves those polyploids higher than tetraploids, many of which are of very complex origin. A hexaploid, for instance, may contain only two different chromosome sets, one of them duplicated, or three distinct sets (AABBCC), while an octoploid may contain one, two, three or four different sets, variously differentiated from each other. At these higher levels, therefore, one and the same plant may possess autopolyploid characteristics, due to duplication of chromosome sets, in addition to the characteristics of allopolyploidy.

The obvious conclusion from these facts is that the classification of a plant simply as an auto- or an allopolyploid, based on such criteria as the resemblance in external morphology to certain diploids, the gross morphology of the chromosomes or the behavior of the chromosomes at meiosis, does not provide any basis for conclusions about the cytogenetic behavior or the phylogenetic origin of the species concerned and, therefore, is of little practical value.

Amplified classifications of polyploid types have been made by Simonet (1935), Lilienfeld (1936), Fagerlind (1937, 1941), and Clausen, Keck and Hiesey (1945). The first three are based primarily on the nature of chromosome pairing in the polyploid or its diploid ancestor, and the amount of differentiation between its component chromosomal sets, while the last is based on biosystematic principles. These involve principally the development of barriers to gene interchange between the component genomes of a polyploid or, in other words, whether the polyploid is derived from a fertile species or a more or less sterile diploid hybrid. This criterion was also used by Lilienfeld (1936) and the writer (1940), but its implications were not fully understood at that time. Since barriers to gene interchange often fail to affect visibly chromosome pairing at meiosis (Stebbins 1945), it is obvious that classifications based primarily on chromosome behavior will give very different results when applied to a particular group from those based on biosystematic principles which emphasize the origin of the polyploid.

The nearest approach to a complete classification of these various polyploid types is presented by Clausen, Keck and Hiesey (1945, Table 12). This, however, includes only the less complex types and tells little about their cytogenetic behavior. The present discussion, therefore, although based largely on their classification, aims to supplement and broaden it and, in doing so, to review some of the more important recent literature on the cytogenetics of polyploids.

III. AUTOPOLYPLOIDY

The number of autopolyploids artificially produced is now very large, so that their morphological and cytogenetic characteristics are well understood. Morphologically, the effects of chromosome doubling vary greatly with the nature of the original material (Kostoff 1939a, Barthelmess 1941, Randolph 1941, Pirschle 1942). The characteristics most consistently present are thicker leaves, larger flowers, and larger fruits. The plant as a whole, on the other hand, is as often smaller and less vigorous as it is larger and more vigorous than its diploid progenitor. It generally, but not always, flowers and fruits later than the diploid. Because of their lateness and partial sterility, artificial autopolyploids of crop plants are as a rule distinctly undesirable. Autotetraploids of garden flowers are likely to have desirable qualities of greater size, sturdiness, durability and lateness (Emsweller and Ruttle 1941), while valuable qualities associated with alterations in their chemical composition have been found in the autotetraploids of tomatoes, maize, sugar beets, and other crop plants (Randolph 1941).

Cytological studies of several artificially produced autopolyploids have changed the opinions once held on the cause of their sterility. Darlington (1937) considered that the sterility of autopolyploids is due to the formation of multivalents, which divide irregularly at meiotic anaphase and so produce gametes with abnormal chromosome combinations. Kostoff (1939b), following this opinion, postulated that, in plants with small chromosomes and low chiasma frequency, polyploids would be less sterile than in species with large ones, and later (1940) cited a few examples which seemed to bear out this opinion. That this is not always true, however, is evident from the sterility of some autopolyploids with small chromosomes, such as those of *Antirrhinum majus* (Emsweller and Ruttle 1941), *Gossypium herbaceum* and *G. hirsutum* (Beasley 1940), *G. arboreum* (Stephens 1942), and *Stipa lepida* (Stebbins 1941 and unpubl.). Myers (1943), and Myers and Hill (1942) concluded that the reduced fertility in the tetraploid *Dactylis glomerata*, which behaves cytologically as an autopolyploid, is due in large part to meiotic irregularities but that these irregularities do not depend on irregular segregation of multivalents. Variance between clones in frequency of univalents at first metaphase was positively correlated with that in frequency of tetrads with micronuclei but not with frequency of quadrivalents. Similar results were obtained by Sparrow, Ruttle and Nebel (1942) in *Antirrhinum majus*, and by Myers (1945) in autotetraploid *Lolium perenne*. Randolph (1941) concluded that the sterility in autotetraploid maize is, to a large extent, genically controlled and is largely physiological in nature. Nevertheless, irregular chromosomal distribution does occur in this tetraploid, giving rise to unbalanced, partly sterile, aneuploid types. There are apparently at least three different causes of sterility in autotetraploids: (1) irregular chromosomal distribution caused by unequal separation of multivalents; (2) irregular distribution caused by meiotic abnormalities of a physiological nature, presumably controlled genetically; and (3) genetic-physiological sterility of an unexplained nature, but not associated with meiotic irregularity. The relative importance of these three causes varies with the tetraploid in question, but in most examples the first is less important than the last two.

None of the artificial tetraploids has been carried through enough generations so that its future evolution can be predicted, except for the tetraploid *Lycopersicum esculentum* (Lindstrom 1941). This has not given rise to anything new, and no fully fertile lines have been selected. On theoretical grounds the evolutionary future of an autopolyploid from an essentially homozygous diploid is likely to be limited (Stebbins 1940, Huskins 1941). On the other hand, if an autotetraploid should arise by somatic doubling from a diploid intervarietal hybrid exhibiting hybrid vigor, the tetraploid would retain and might augment this vigor. Further-more, as was pointed out by Randolph (1941), the smaller amount of segregation in intervarietal autotetraploids should lead to a longer persistence of hybrid vigor in their progeny as compared with the offspring of a diploid intervarietal hybrid, according to any of the current hypotheses concerning the nature of hybrid vigor. We should expect, therefore, that the most successful autopolyploids, both for the plant breeder and in natural evolution, would be derived from hybrids between varieties or subspecies. This seems to be the case in *Antirrhinum majus* (Sparrow, Ruttle and Nebel 1942), in *Fragaria bracteata* × *F. vesca* var. *rosea* (*cf.* review in Clausen, Keck and Hiesey 1945), and in *Allium paniculatum-oleraceum* (Levan 1937).

This frequent occurrence of intervarietal autotetraploids adds to the difficulties in the way of interpreting the evolutionary significance of autopolyploidy *per se* on the basis of naturally occurring polyploids. Before the differences in external morphology and geographic distribution between a natural autopolyploid and its nearest known diploid relative can be ascribed to the polyploid condition, one must be certain that no diploid variety or subspecies exists which possesses the morphological and ecological characteristics in question. And in many discussions of the evolutionary significance of autopolyploidy, this point has not been considered. Clausen, Keck and Hiesey (1945) have already indicated (p. 130) that most of the supposed autopolyploids listed by Müntzing (1936) are actually allopolyploids, and state that "relatively few natural polyploids reported in the literature can be regarded as clear examples of autoploidy." They cite ten examples of possible or probable natural autopolyploids, but of one of these, *Cuthbertia graminea* (*cf.* Giles 1942), they say that

151

(p. 144) "this is a case in which the earliest known facts pointed to auto-ploidy, but where an assembling of additional data . . . is likely to revise the first impression." Of the remaining nine, of which the morphological differences and ecological preferences of the tetraploid are listed in Table 14 (p. 151), eight are, in the opinion of the writer, either doubtfully autopoly-ploid or else represent intervarietal autopolyploids which have acquired their supposedly divergent characteristics as a result of hybridization with a diploid variety or subspecies different from that to which their origin has commonly been ascribed. These eight cases are as follows:

1. *Zea perennis* (Hitchcock) Reeves et Mangelsd. has been found only once in nature, and its supposed diploid parent, *Z. mexicana* (Schrad.) Reeves et Mangelsd., is itself believed to be of hybrid origin (Mangelsdorf and Reeves 1939). There can be no certainty, therefore, that the presence of rhizomes in *Z. perennis* is due to the polyploidy, and not to admixture with a rhizomatous species of Tripsacum, or from some other source.

2. The strictly autopolyploid origin of *Dactylis glomerata* L. is doubted by Clausen, Keck and Hiesey. The writer has seen herbarium specimens, identified as *D. hispanica* Roth. or *D. juncinella* Bory, from Spain, Morocco, Sardinia and Asia Minor, which have to an extreme degree the various morphological characteristics by which, according to Müntzing (1937), tetraploid *D. glomerata* is distinguished from diploid *D. Aschersoniana*. These specimens from the Mediterranean region, moreover, have pollen grains as small as those of *D. Aschersoniana*, and may therefore be diploids. If so, nearly all of the morphological and ecological divergence of *D. glomerata* from *D. Aschersoniana* could be explained on the basis of hybrid-ization between *D. Aschersoniana* and these Mediterranean forms.

3. The example of Eragrostis, first given by Hagerup (1932) and widely cited in general references as an example of the effects of autopoly-ploidy, is particularly doubtful. Eragrostis is a very large and complex genus, in which good taxonomic characters for separating species are particularly hard to find. Furthermore, it is very poorly known cytologi-cally. Hence, until more is known about the cytology of other African desert species of this genus, the status of the three cited by Hagerup must be considered ambiguous.

4. *Biscutella laevigata* L. The fine analysis by Clausen, Keck and Hiesey of Manton's classic work on this example overlooked one point, namely, that two of the western European diploid "species" of the group, *B. arvernensis* and *B. Lamottii*, share with the tetraploid the ability to produce stolons, or "root buds" (Manton 1937, p. 449). It seems likely, therefore, that the stoloniferous character of the tetraploid was acquired through crossing with one of these forms, so that *B. laevigata* must be considered an intersubspecific autotetraploid, and its differences from the

diploids in both external morphology and distribution are due to gene recombination as well as to doubling of the chromosome number.

5. *Tradescantia canaliculata, T. occidentalis, T. humilis, etc.* The genus Tradescantia has served as a classic example of one in which auto-tetraploids exist side by side with diploids of the same species, and have enabled the species to spread into new territory (Anderson and Sax 1936). This situation must, however, be reëxamined, particularly in the light of the experiments of Skirm (1942) with doubling artificially the chromosome number of the natural hybrid ("Oakhill") between *T. canaliculata* and *T. humilis.* This hybrid possesses cytological abnormalities usually associated with hybridization (Anderson and Sax 1936), and that are probably caused by structural hybridity for small chromosomal segments. When doubled somatically it produces a tetraploid with very few multivalents, resembling cytologically an allotetraploid. This is due, as Skirm pointed out, to differential affinity, and preferential pairing of exactly similar chromosomes (see discussion below). On the other hand, if doubling is accomplished through the medium of gametes with the unreduced chromosome number, which were produced after meiosis and crossing over had taken place, the resulting tetraploid behaves cytologically like an auto-tetraploid. Such a tetraploid, although allotetraploid in that it was produced from an undoubted interspecific hybrid, could nevertheless segregate in the direction of one or other of its parents, so that some of its descendants could come to appear both morphologically and cytologically like autopolyploids. The probability of such an event is increased by the evidence of Giles (1941), who found that the triploid hybrid between tetraploid *T. canaliculata* and diploid *T. paludosa* was morphologically indistinguishable from *T. canaliculata* and had as many trivalents as an autotriploid. It showed more evidence of structural hybridity than its parents, but this is also greater in tetraploid *T. canaliculata* than in any diploid. There is no doubt that autotetraploids occur in Tradescantia. But diploid interspecific hybrids also occur, which could give rise to allotetraploids, and hybridization between tetraploids, followed by back-crossing, has been described by Anderson and Hubricht (1938) as introgressive hybridization. These introgressive types are, as Clausen, Keck and Hiesey pointed out, partial allopolyploids. Since the evidence of Skirm and Giles has shown that neither morphological nor cytological evidence is reliable in this genus for distinguishing between partial allo- and true autopolyploids, it is not certain whether the widely distributed "weedy" tetraploids are strict autotetraploids, or whether they are partial allopolyploids, the new characteristics of which are due to the presence of genes from other species as well as to the doubling of the chromosome number.

6. *Galium mollugo, G. verum, et aff.* Fagerlind (1937, pp. 342–345) recognizes that in this genus certain known allopolyploids, such as *G. lucidum-mollugo*, are nearly indistinguishable from one of their diploid ancestors and, furthermore, that many of the "intraspecific" tetraploids, particularly those of *G. mollugo* and *G. verum*, show definite signs of admixture with genes from another species. Furthermore, quadrivalent frequency is low in all of the tetraploids, both putative auto- and known allotetraploids. Nevertheless, because of the fact that the diploid species are unable to hybridize with each other, Fagerlind correctly concludes that true intraspecific polyploidy, *i.e.*, autopolyploidy, must be present. The important question from the evolutionary point of view is, however, what proportion of the tetraploids are true autopolyploids, and what proportion of them, particularly those which diverge from the diploids in their ecological requirements and geographical distributions, owe their new characteristics wholly or in part to an admixture of genes from another species. Since the diploid species, and consequently the newly arisen tetraploids, often occur together, and since the latter intercross freely, there is plenty of chance for such admixture on the tetraploid level. Furthermore, the intermediate allopolyploid between *G. mollugo* and *G. verum* is less vigorous than the apparent autopolyploids and would, in nature, give way to backcross types, which produce segregates indistinguishable from the supposed autopolyploids. Hence, although in Galium, as in Tradescantia, true autopolyploids certainly exist, the supposed autopolyploid types which have greatly extended the range of the species may actually be partial allopolyploids which owe their new characteristics in large part to the presence of combinations of genes derived from different ancestral diploid species.

7. *Vaccinium uliginosum* L. The subgenus to which this species belongs has its center of variability in western North America, where large-leaved forms similar to the European tetraploid also occur. Hagerup's (1933) evidence must therefore be considered definitely incomplete, and the status of the European tetraploid is doubtful until its American relatives have been studied.

8. *Empetrum nigrum* L. and *E. hermaphroditum* (Lange) Hagerup. This famous case, which has been widely cited in discussions of the relationship between polyploidy and sex, must also be considered doubtful. The genus Empetrum is predominantly a North American one, and the forms on this continent are mostly bisexual and unknown cytologically. Until a good series of chromosome counts has determined whether these North American forms are diploid, tetraploid, or both, the origin of *E. hermaphroditum* must be considered doubtful. Significantly, Blackburn

(1938) has found diploid hermaphroditic plants of *E. nigrum* to occur occasionally in Great Britain.

The removal of these eight from Clausen's list of undoubted natural autopolyploids would leave only *Galax aphylla*, representing a monotypic genus (Baldwin 1941). In this genus, the morphological differences between the diploids and autotetraploids are about like those seen in artificial autotetraploids, and the tetraploid race differs only slightly from the diploid in geographic distribution. Beside those cited by Clausen *et al.*, still others originally interpreted as autopolyploids have turned out to be entirely or partly allopolyploid when more fully investigated. Some of the most notable of these are *Nasturtium officinale* (Manton 1935, Howard and Manton 1940), *Lilium tigrinum* (Stewart and Bamford 1943), *Rubus caesius* and its relatives (Thomas 1940a, Gustafsson 1943), *Oxycoccus quadripetalus* (Hagerup 1940, Camp 1944), and *Solanum tuberosum* (Lamm 1945). *Allium oleraceum* and *A. carinatum* remain as true autopolyploids which have acquired a geographic distribution distinct from that of their diploid ancestors (Levan 1937), but these are probably derived from intervarietal hybrids, and they maintain their heterozygosity and hybrid vigor by means of asexual reproduction, which is unknown in the diploids of this group. The evolutionary future of such asexually reproducing autopolyploids is of course decidedly limited. The evidence is mounting, therefore, that autopolyploidy by itself rarely produces morphologically distinct species. Furthermore, the divergence of an autopolyploid from its diploid ancestor by means of mutation and other genetic changes without hybridization has taken place seldom if at all. The role of autopolyploidy in evolution has been primarily as a means of preserving vigorous intraspecific hybrid combinations and, secondarily, as a means of enabling species to hybridize which are incompatible on the diploid level.

IV. Typical Allopolyploids

This term will be used here in the same sense as the term amphiploid of Clausen, Keck and Hiesey (1945), though not as a synonym of amphidiploid in the sense of Navashin (1927), as will be explained further below. The difficulty of using the widely accepted criterion of structural similarity or dissimilarity in the chromosomes as the primary distinction between auto- *vs.* allopolyploidy does not lie only in the fact that the categories based upon this criterion cannot be used for evolutionary studies. In addition, there is no way of measuring quantitatively the amount of structural difference between two chromosome sets, so that the only possible dividing line between the two categories on this basis would have to be, as Müntzing (1936) has suggested, whether or not the component genomes are structurally identical. Since, however, structural differences, both inversions and translocations, are commonly found in wild plants of pure species (cf. Dobzhansky 1941, p. 126), and can in most cases be detected only if they are so large or numerous that they affect chiasma formation and metaphase pairing, this structural identity could never be determined with certainty. Fagerlind's criterion (1941a), namely, whether the diploid ancestor of the polyploid has or has not perfect pairing, presumably at metaphase, is even more unreliable. A single large translocation or inversion will undoubtedly affect metaphase pairing more than several small ones, so that many polyploids derived from interracial hybrids showing quadrivalents or bridge-fragment configurations would, on the basis of Fagerlind's criteria, be considered more nearly allopolyploid than those derived from interspecific hybrids having cryptic structural hybridity (Stebbins 1945). For instance, Bergner (1944) found, as expected, many different types of configurations in meiosis of an artificial tetraploid produced from a hybrid between prime types 1 and 2 of *Datura stramonium*, but the designation of such a polyploid as an allopolyploid or even part allopolyploid would be very misleading.

Among allopolyploids at the tetraploid level two general types may be recognized, although these are, of course, connected by a whole series of intermediates. The best known type is defined on the chart of Clausen, Keck and Hiesey (1945, p. 72) as inter-cenospecific with no intergenomal pairing and the identity of the parental genomes preserved. To this type belong the classic examples Triticale, Raphanobrassica, *Gossypium hirsutum*, *barbadense et aff.*, and *Nicotiana Tabacum*. These are the only type which Fagerlind (1941a) recognizes as allopolyploid, and are the typical allopolyploids or amphidiploids of textbook accounts. An extensive review of the literature on these types is that of Goodspeed and Bradley (1942). They emphasize the fact that most representatives of this type are highly constant because chromosome pairing is between similar chromosomes derived from the same species. This type of pairing is termed allosyndesis by Darlington (1937, p. 199), following the original definition of Ljungdahl (1924). He refers to the fact that when it occurs in an established allopolyploid the chromosomes which pair have been derived from different parental gametes. Sharp (1943) uses for it the similar term allosynapsis. On the other hand, Lawrence (1930), Sansome and Philp (1932, p. 178), Dobzhansky (1941, p. 232) and Goodspeed and Bradley (1942, p. 287), call this type of pairing autosyndesis because it is pairing between chromosomes derived from the same species. Waddington (1939, p. 73) has suggested that Ljungdahl's definitions be followed very strictly, and that the terms auto- and allosyndesis be used only for pairing between chromosomes derived from the same or different immediate parents of the plant involved, whether it be an autopolyploid, a newly formed allopolyploid or an allopolyploid species of long standing. Used in this sense, these terms have no reference to either the structural similarity or the phylogenetic relationship of the chromosomes concerned. In an allopolyploid newly formed from a diploid hybrid by somatic doubling, pairing between chromosomes derived from the same parental gamete, or autosyndesis, is the pairing of similar chromosomes, while in the later progeny of this allopolyploid autosyndesis in this strict sense is the pairing of dissimilar chromosomes. Waddington has introduced the terms homogenetic and heterogenetic association to replace auto- and allosyndesis as used in the phylogenetic sense. The writer believes that the use of these terms will eliminate the confusion that has centered around the use of the older ones.

If we follow the system of Waddington, therefore, we have two different series of terms with different uses. Autosyndesis and allosyndesis are purely genetical terms without phylogenetic connotations. In a diploid species, pairing is always allosyndesis and free genetic segregation is, therefore, possible. In an established allopolyploid autosyndesis is predominant and segregation is restricted. In an autopolyploid auto- and allosyndesis occur with equal frequency, while in intermediate polyploid types and in hybrids between polyploids these two types of pairing occur with various relative frequencies. Homogenetic and heterogenetic association are terms with definite phylogenetic connotations and cannot be used unless something is known, or can be inferred, about the origin of the polyploid in question. In diploid and autopolyploid species only homogenetic association can occur. In F_1 hybrids between distinct diploid or allopolyploid species only heterogenetic association can occur, unless the parental species have identical chromosome arrangements and are separated from each other by isolation barriers other than chromosomal sterility. A hybrid between two distinct partial allopolyploids can have two different types of heterogenetic association, namely, allosyndesis, or pairing between chromosomes derived from different parents, or autosyndesis between the different genomes derived from the same parental gamete. Within an established partial allopolyploid species, on the other hand, heterogenetic association will be mainly allosyndesis and homogenetic association will be the commonly occurring autosyndesis.

Heterogenetic association has long been known to occur as an occasional anomaly in otherwise true breeding allopolyploids (Darlington 1937, p. 200) and to be responsible for genotypic aberrations in these species. In new allopolyploids, even between widely different parents, a small percentage of heterogenetic association may occur regularly, as in Howard's (1938) strain of Raphanobrassica and in *Gossypium Thurberi — arboreum* (Beasley 1942). Since even a small amount of this type of pairing usually leads to some sterility as well as to inconstancy, its absence or rarity in old, established allopolyploids is probably due to selection in the past of mutations and other genetic changes in this direction. From the cytogenetic point of view, therefore, the raw allopolyploid becomes progressively "diploidized" until its behavior resembles that of a· diploid species. The nature of this diploidization has been accurately determined by R. E. Clausen (1941) for certain chromosomes of *Nicotiana Tabacum.* This species was derived from a hybrid between the diploids *N. sylvestris* and *N. tomentosiformis* or a close relative, as has been demonstrated by Greenleaf (1941), through comparison of the experimentally produced allopolyploid between these two species with *N. Tabacum.* Nevertheless, *N. sylvestris—tomentosiformis* has in duplicate certain factors, such as MM (dominant allele for mammoth growth), the normal allele to an asynaptic factor, and another to a recessive white-seedling character, which are all present only singly in *N. Tabacum.* Since the F_1 between the raw amphidiploid and *N. Tabacum* has perfectly normal meiosis, the elimination of these duplicate alleles appears to have been either through mutation or the loss of very small chromosomal segments. In self-pollinated plants, this diploidization apparently proceeds differently in different inbred lines, as evidenced by the fact that hybrids between different pure lines of such polyploids as hexaploid wheat and oats often have multivalents and other cytological irregularities (Huskins 1941).

Although the presence of heterogenetic pairing in allopolyploids usually causes some sterility, its absence by no means assures fertility. Sterility due to the interaction of genic factors in upsetting one of the developmental processes necessary for seed production, or genic sterility (*cf.* Dobzhansky 1941, p. 293) may be superimposed on chromosomal sterility in hybrids between distantly related species, and only the latter is removed by doubling the chromosome number. This condition was first noted by Greenleaf (1941, 1942) in allopolyploids between *N. sylvestris* and various members of the *N. tomentosa* complex. In these, the sterility involves the abortion of the female gametophyte, or embryo sac, at the 2 to 4 celled stage. In *Aegilops umbellulata — Haynaldia villosa* (Sears 1941) there is genic sterility affecting meiosis and producing partial asynapsis, in spite of the presence of an exact homologue for each chromosome.

All of the allopolyploids which show genic sterility have been produced from a diploid hybrid by doubling in the somatic tissue. On the other hand, a fertile allopolyploid of *Nicotiana sylvestris* × *tomentosiformis* was produced by Kostoff through gametic doubling, using as an intermediate stage a triploid derived from backcrossing the F_1 diploid to *N. sylvestris*, and crossing this triploid to *N. tomentosiformis*. Greenleaf (1942), based on his analysis of an F_1 hybrid between this allopolyploid and the one which he produced by somatic doubling, concluded that during the process of gametic doubling in the Kostoff allopolyploid one of the two complementary factors for genic sterility was transferred by heterogenetic association and crossing over onto the chromosome that carried the other factor and, consequently, the homologue of this chromosome was neutral and viable in the female gametophyte. It seems likely, therefore, that if heterogenetic association occurs to any degree, this can act as a sieve to eliminate genic sterility from allopolyploids produced by gametic doubling. On the other hand, the presence in a diploid hybrid of several different bivalents formed by heterogenetic association between chromosomes that have only certain segments in common will produce unreduced as well as reduced gametes that differ from each other widely in the arrangement of chromosomal segments. The union of two such unreduced gametes, therefore, will produce an allopolyploid with considerable structural hybridity and consequent chromosomal sterility, as has been found in Triticum-Agropyron (Love and Suneson 1945). It can be said, therefore, that, if any pairing at all occurs in the diploid hybrid, the allopolyploid produced from it by somatic doubling may have genic sterility but not chromosomal sterility, while that resulting from gametic doubling will rarely if ever have genic but is very likely to have chromosomal sterility. The sterility reported by Clausen, Keck and Hiesey (1945) in *Layia pentaglossa*, which resulted from gametic doubling, is undoubtedly chromosomal. Allopolyploids produced by somatic doubling may have chromosomal sterility in later generations, but by means of differential affinity and preferential pairing (Darlington 1937, p. 185) this may be reduced enough to permit the survival of the line until the diploidization process has eliminated heterogenetic association.

Heterogenetic association in the ancestral diploid hybrids may be responsible for part of the peculiar phenomena found by Müntzing (1939) in hybrids between different strains of *Triticum aestivum — Secale cereale* (Triticale). This allopolyploid has been produced several times by different workers, and each of the initial allopolyploids has given rise after several generations of selfing to a distinct strain of Triticale. All of these strains have a somewhat irregular meiosis and are partially sterile as to both pollen and seeds. Interstrain hybrids, moreover, are harder to obtain than cross pollinations within a strain. The resulting F_1 plants are less fertile than their parents, indicating that these different strains of Triticale have developed new barriers of partial isolation. Müntzing explained these results as due to physiological sterility and incompatibility resulting from inbreeding of the rye set of chromosomes. Rye is a self-incompatible, normally cross-pollinated species, in which inbreeding is known to produce partial sterility and a reduction in chromosome pairing. Wheat, on the other hand, is normally self-fertilized, and carries this characteristic into the Triticale allopolyploids, thus enforcing inbreeding of the rye as well as the wheat genome. There is no doubt that some of the sterility, which is apparently physiological in nature, is due to this cause, but the high number of univalents (up to 18 in some strains and interstrain hybrids) must be due in part to reduced homology between some of the chromosomes, which must also account for some of the haplontic sterility.

This evidence shows that many typical allopolyploids have a very different cytogenetic behavior from that of diploid species. Further evidence of this fact is the ability of all which have been so tested to tolerate much larger chromosomal deficiencies than can diploid species. Clausen (1941; Clausen and Cameron 1944) has been able to obtain monosomic plants deficient for one of each of the 24 different chromosomes found in the haploid set of *Nicotiana Tabacum*. Furthermore, although nullisomic (23-paired) plants are never viable in Nicotiana, in the case of one chromosome (the F) the vital portion is only a small region near the centromere. In *Triticum aestivum*, on the other hand, plants nullisomic for any one of the 21 pairs are viable (Sears 1944), and in the case of some chromosomes these nullisomics are fairly fertile. Sears has shown that the use of nullisomics provides a new rapid method for analysis of the gene content of individual chromosomes of this species. He also was able to show with striking clarity the presence of duplications in chromosomes belonging to different genomes. Plants which were tetra-II, nulli-XX were nearly normal and highly fertile, indicating that these chromosomes, one homologous to a chromosome of *T. durum*, and the other a chromosome of the Aegilops (*C*) genome in *T. aestivum*, have many genetic factors in common, in spite of the fact that they do not pair, even when both are monosomic. Even allopolyploids which behave cytologically as diploids under normal conditions are therefore actually quite different from them and can be expected to show more complex genetic ratios as well as reacting less strongly to cytogenetic disturbances of various sorts.

For allopolyploids of this usual type the term amphidiploid is often used. This term, however, is not synonymous with allopolyploid. It was first used by Navashin (1927) for the hypothetical doubled hybrid of *Crepis capillaris* ($n=3$) $\times C.$ *setosa* ($n=4$). Since the haploid number ($n=7$) of such a doubled hybrid is not a multiple of any basic number, it could not be called a polyploid in the strictest sense of the word. On the other hand, the type of allopolyploid to be discussed below, which undergoes segregation because of regular heterogenetic association, could not be termed an amphidiploid, since it does not behave like a diploid in any respect. Therefore, according to the original definitions and connotations of the two terms they are overlapping but not synonymous in meaning. The introduction by Clausen, Keck and Hiesey (1945) of the new, abbreviated term amphiploid, which they have defined exactly according to modern biosystematic concepts, which covers the meaning of both of the old terms, and has no connotations, is a simplification of terminology which has much in its favor.

V. SEGMENTAL ALLOPOLYPLOIDS

The second type of allopolyploid or amphiploid is that defined in the chart of Clausen, Keck and Hiesey as inter-ecospecific or inter-cenospecific with intergenomal pairing partial or complete, and the identity of the parental genomes lost in recombination. The best known example of this type is *Primula kewensis*, and Clausen, Keck and Hiesey list *Aquilegia Janczewskii*, *Crepis rubra — foetida*, *Crepis capillaris — tectorum* and *Layia pentaglossa* as additional artificially produced examples. To this list may be added *Nicotiana glauca — Langsdorffii* (Kostoff 1938), several combinations in Aegilops and Aegilotriticum (Sears 1941), *Allium cepa — fistulosum* (Jones and Clarke 1942), *Tradescantia canaliculata — humilis* (Skirm 1942), *Solanum Douglasii — nodiflorum* (Paddock 1942), *Lycopersicum esculentum — peruvianum* (Lesley and Lesley 1943), *Nicotiana paniculata — solanifolia* (Bradley and Goodspeed 1943), *Melica imperfecta — Torreyana* and *M. californica — imperfecta* (Joranson 1944), *Triticum durum — Timopheevi* and *T. vulgare — Timopheevi* (Zhebrak 1944 a,b), *Elymus glaucus — Sitanion jubatum* (Stebbins, unpubl.), and several combinations in Bromus, sect. Ceratochloa (Stebbins, unpubl.). Natural tetraploids of this type are *Zauschneria californica* (Clausen, Keck and Hiesey 1940), *Galium mollugo — verum* (= *G. ochroleucum*, Fagerlind 1937), *Aesculus carnea* (Upcott 1936) and *Lilium tigrinum* (Stewart and Bamford 1943). In external morphology, these allopolyploids usually differ from typical ones in resembling more closely one or both parents. This is both because their parents are more closely related and therefore differ less from each other in appearance, and because segregation of interspecific differences occurs, so that an initial intermediate allopolyploid of this type may in later generations produce segregates resembling more or less closely one or the other of its original parents. Cytologically, they are characterized by the presence of multivalents in varying numbers, so that in meiosis they often resemble autopolyploids more than true allopolyploids.

Clausen, Keck and Hiesey (1945, pp. 68–73) have advanced two reasons why these allopolyploids between closely related species might be expected to be unsuccessful and therefore infrequent in nature. In the first place, segregation of interspecific differences, particularly the incompatibility and sterility barriers which existed between the parental diploids, would result in the appearance of many weak or sterile types in their progeny. Secondly, this segregation would prevent the polyploid from breeding true and, therefore, of maintaining the proper physiological balance with its environment. They recognize, however, that such types might be successful if they became stabilized in later generations through the elimination of weak, sterile and unfit combinations. As a test of this hypothesis, the cytogenetic behavior of the above mentioned artificially produced examples of this type will be summarized. Of the 14 different examples or groups of examples, 5 — those in Crepis, Layia, Allium, Lycopersicum and Solanum — were either themselves weak and completely sterile or produced offspring entirely of this type, so that they fulfilled in every respect the prediction of Clausen, Keck and Hiesey. Six — those in Aquilegia, Nicotiana, Aegilops, Melica, Elymus-Sitanion and Bromus — segregated or varied in respect to both morphological characteristics and fertility, while the remaining 3 — *Primula kewensis, Tradescantia canaliculata — humilis* and *Triticum durum — Timopheevi* — are highly fertile, but later generations, when they have been produced, show considerable segregation for morphological characteristics. Only two of these segregating types — *Nicotiana glauca — Langsdorffii* and *Triticum durum — Timopheevi* — have been carried on for a sufficient number of generations so that their ultimate fate can be ascertained. From both of these, relatively constant, highly fertile types have been secured after four to six generations of inbreeding and selection. Furthermore, the great amount of segregation for morphological characteristics in the early generations permitted the production of a whole series of different lines, the number of which is limited only by the number of plants which the breeder is able to grow. There is good reason to believe, therefore, that these segregating allopolyploids are a valuable source of new variants for the plant breeder and well worth his attention, although the production of useful types from them obviously requires considerable time. Another valuable feature of these allopolyploids is that, in contrast to non-segregating allopolyploids derived from distantly related parents, they can be crossed to autopolyploids derived from their parental species and the resulting hybrids will in many cases be vigorous and reasonably fertile (Zhebrak 1946). This makes it possible to transfer genes or groups of genes from one species to another on the tetraploid level when this transfer is impossible on the diploid level because of the sterility of the F_1 hybrid. These same qualities give the segregating allopolyploids certain unique evolutionary possibilities. In the first place, the variability of these allopolyploids in early generations would give them an opportunity of exploring and occupying "adaptive peaks" in the sense of Wright (*cf.* Dobzhansky 1941) that might lie between or apart from those occupied by the parents (*cf.* Müntzing 1932). Secondly, if autotetraploids of the parent species should exist in nature, these could, by hybridization with the segregating allopolyploid, increase their variability and adaptability to potential new habitats and, in addition, lose some of the well-known drawbacks of most new autopolyploids, such as slow growth, irregular meiosis and consequent sterility.

For these reasons, allopolyploids of this type are sufficiently important so that they should be designated by a distinctive name. The name applied by Fagerlind (1937, 1941) interspecific autopolyploid, is inappropriate, as has been indicated above. A more satisfactory term is segmental allopolyploids. A segmental allopolyploid may, therefore, be defined as an allopolyploid of which the component genomes bear the majority of their chromosomal segments in common, so that the diploid hybrid from which it is derived has good pairing at meiosis, but in which these genomes differ from each other by a large enough number of chromosomal segments or gene combinations so that free interchange between them is barred by partial or complete sterility on the diploid level. The examples of segmental alloploids, both artificial and natural, have been cited above.

From the standpoint of both evolution and plant breeding, it is important to know what factors contribute to the success of segmental allopolyploids and which ones are responsible for their failure. Our knowledge of these factors is as yet very imperfect, but certain considerations are undoubtedly of paramount importance. These are, first, the nature of pairing, as determined by the amount of differential affinity between the chromosomes as well as their size and genically determined factors of chiasma frequency and distribution; second, the amount and nature of the sterility in the diploid hybrid from which the polyploid arose; and third, certain physio-ecological features of the plant group concerned which determine its ability to pass through the "bottleneck" of partial sterility which must intervene between the formation of the "raw" allopolyploid and the stabilization of constant, fertile lines from it.

The phenomenon of differential affinity (Darlington 1937, pp. 198–200) is the most characteristic feature of segmental allopolyploids and the degree to which it is developed is one of the most important factors determining their success or failure. As Darlington has pointed out, it is caused by the fact that chromosomes pair segment by segment, so that those which are completely homologous have a greater affinity for each other than those which differ in respect to large or small non-homologous segments. In diploid hybrids having nearly complete pairing and regular distribution of the chromosomes at meiosis, as in *Primula verticillata—floribunda*, and various hybrids in Galeopsis (Müntzing 1938), the sterility seems to be produced chiefly by the random segregation of small non-homologous segments, so that the gametes come to possess non-viable duplications or deficiencies. This is the chromosomal sterility of Dobzhansky (1941, p. 293), and it is partly eliminated in polyploids from such hybrids by means of homogenetic pairing, which results from differential affinity. It does not follow from this, however, that the fertility of segmental allopolyploids is directly correlated with the lack of heterogenetic association. Sears (1941) has clearly shown in a series of 21 allotetraploids of the Triticinae that this is not the case. Although in this group there is a general correlation between the amount of heterogenetic association, as measured by the frequency of multivalents and the degree of pollen and seed sterility, there are striking exceptions. The example of *Aegilops umbellulata — Haynaldia villosa*, a true allopolyploid in which genic sterility is found, has been mentioned above. Three other allopolyploids obtained by Sears which have a high number of univalents in spite of perfect homology of the chromosomes are *Ae. speltoides ligustica* II—*Ae. uniaristata*, *Ae. caudata-Ae. speltoides lig.* II, and *Ae. speltoides lig.* II—*Ae. umbellulata*. The first two of these had much lower seed fertility than any of the other allopolyploids except for the Aegilops—Haynaldia example mentioned above and, therefore, genic sterility connected with partial asynapsis or desynapsis may exist in them also. A more significant exception is the difference between two different allopolyploids involving the same two species, *Ae. caudata* and *Ae. umbel-lulata*. One of these (produced in 1938), obtained from a diploid hybrid characterized by a relatively low amount of pairing, had the high average of 5.81 chromosomes per cell in multivalents, while the other (produced in 1939), of which the diploid hybrid had closer pairing, had only 3.86 chromosomes per cell in multivalents. Nevertheless, the pollen and seed fertility in the two allopolyploids was nearly identical. A similar example is the pair of allopolyploids involving two different strains of *Ae. speltoides ligustica* (I and II) and *Ae. umbellulata*. Finally there is the example of *Ae. comosa—uniaristata*, which had the relatively high seed fertility of 78% in spite of the fact that the number of chromosomes in multivalents, 6.64%, was the next to the highest recorded. It is perhaps significant that *Ae. comosa* and *Ae. uniaristata* are placed by all monographers in the same taxonomic section (*cf.* Kihara 1940), while all but one of the other allopolyploids are intersectional. This evidence from Aegilops shows that, even within groups that are relatively homogeneous as to chromosome size and chiasma frequency and distribution in the diploid species, one cannot predict accurately the chromosome behavior or the fertility of an allopolyploid on the basis of the pairing in the diploid hybrid from which it is to be derived.

The factors of chromosome size and chiasma distribution, as they affect multivalent formation and fertility in polyploids, have already been discussed above in connection with autopolyploidy. These same factors obviously hold, with similar qualifications, for segmental allopolyploids. For instance, the low frequency of multivalents in *Primula kewensis*

(Upcott 1939) and *Galeopsis tetrahit* (Müntzing 1932), in spite of the high degree of pairing in their ancestral diploid hybrids, is probably due to the small size of the chromosomes and the low chiasma frequency in these genera. Nevertheless *Lycopersicum esculentum—peruvianum* var. *dentatum* has as high a multivalent frequency as autotetraploid *L. esculentum*, in spite of the fact that chromosome size and chiasma distribution are approximately the same in Lycopersicum as in Primula. Undoubtedly, therefore, unknown factors, in addition to the recognized ones, affect the frequency of multivalent formation and heterogenetic association in segmental allopolyploids.

The final factor determining the success of all new polyploids, and particularly segmental allopolyploids, is the character of the plants themselves. The writer (1938) pointed out that the chance of chromosome doubling in a sterile hybrid is much greater if the plant is a long lived perennial than if it is an annual. This chance would be increased still more if the hybrid had an efficient means of vegetative propagation, such as rhizomes, tubers or bulbs. Furthermore, the partly sterile descendants of an unstabilized autopolyploid or segmental allopolyploid would also have a much greater chance of survival if they were perennials with vegetative means of reproduction. Botanists are well aware that many wild perennial species which produce rhizomes, tubers or bulbs, such as those of Acorus, Agropyron, Elymus, Ammophila, Fritillaria and Tulipa, often set little or no seed and propagate themselves in the main vegetatively. In such species the partial sterility which accompanies the segmental allopolyploid condition is only a slight, or even a negligible, selective disadvantage which could easily be counterbalanced by the vigor and evolutionary possibilities of polyploids of this type. We should, therefore, not expect either autopolyploidy or segmental allopolyploidy to occur commonly in annual species, because of their difficulty in passing through the bottleneck of partial sterility which always accompanies these conditions in their initial stages. In perennials, on the other hand, these conditions should be more common, and perhaps as frequent as true allopolyploidy. This agrees with the evidence previously obtained by the writer (1938) that perennial groups have a higher percentage of polyploid types than annual ones, and that wherever they have been sufficiently investigated, these polyploid perennials can be seen to be descended from perennial diploid ancestors.

At levels of polyploidy higher than tetraploidy, types can occur which combine completely the characteristics of auto- and allopolyploidy. If, for example, an autotetraploid is crossed with a different diploid species, and the resulting triploid hybrid is doubled, a hexaploid will be produced which will be autopolyploid with respect to one genome, but allopolyploid in that it contains a different genome. This type has been called by Kostoff (1939c) an autoallopolyploid, and the example given by him is *Helianthus tuberosus*, which has a diploid chromosome number of 102. When it is crossed with *H. annuus* ($2n=34$), the resulting tetraploid hybrid has usually 34 bivalents. Kostoff interprets this result as due to the fact that the haploid complement of *H. tuberosus* has one genome, *Bt*, homologous with that of *H. annuus* (*Ba*), and two, *AtAt*, that are entirely different. *Phleum pratense* apparently has a similar constitution. Clausen, Keck and Hiesey (1945), after a review of most of the literature on this much disputed case, agree with the original opinion of Gregor and Sansome that it is an allopolyploid, and consider that the parental forms belong to different cenospecies. Nordenskiold (1941), however, concludes that *P. pratense* is an autopolyploid of *P. nodosum*, while Myers (1944), after finding in *P. pratense* both multivalents and tetrasomic genetic ratios considers it to be at least partly an autopolyploid. Critical evidence, in the writer's opinion, is provided by two different haploids of this species described by Nordenskiöld (1941) and Levan (1941). Both have typically fourteen bivalents and seven univalents, indicating that their genomic formula is *AAB*, and that normal *P. pratense* is *AAAABB*. This interpretation would reconcile the apparently conflicting evidence which has suggested on the one hand an autopolyploid and on the other an allopolyploid origin for *P. pratense*.

Another type of autoallopolyploid is the autopolyploid produced by somatic doubling from the allopolyploid species *Nicotiana tabacum* (Clausen 1941) and the similar one produced from *Gossypium hirsutum* (Beasley 1940). A natural octoploid of this type is *Rubus ursinus* (=“*R. vitifolius*”, Thomas 1940a, b). There are, moreover, many examples of high polyploid species which have some autopolyploid characteristics, such as the presence of multivalents and a close morphological resemblance to certain diploid species, but which are known more or less definitely to contain genomes derived from more than one species. Typical of these are *Iris versicolor* (Anderson 1936), *Agropyron elongatum* (Wakar 1935), *Pentstemon nectericus* (Clausen 1933), *Rubus lemurum* (Brown 1943) and *Bromus arizonicus* (Stebbins, Tobgy and Harlan 1944). These do not fit the definitions of either auto- or allopolyploids, and had best be designated either as partial allopolyploids, or undefined secondary polyploids. The observations of Love and Suneson (1945) on hybrids between *Triticum* and *Agropyron* have shown that interspecific hybrids involving these higher polyploids may produce fertile derivatives with euploid numbers between the undoubled and the doubled one. Thus the 41-chromosome F_1 *T. macha*$\times$*A. trichophorum* gave rise to a fertile plant with 70 chromosomes, presumably through the functioning of a partially reduced gamete with 28 chromosomes. It is possible, therefore, that some of these higher polyploid F_1 hybrids may give rise in later generations to several distinct species all descended from the same hybrid combination. The limitless possibilities and complexities of such a situation can only be imagined. In many genera containing these high polyploids we must, therefore, be content with assuming that most of these species contain various combinations of autopolyploidy, segmental allopolyploidy and true allopolyploidy, and that their phylogenetic relationships will be difficult or impossible to unravel.

VI. SUMMARY AND CONCLUSION

The various polyploid types may now be summarized as follows. On the tetraploid level there are autopolyploids, segmental allopolyploids and true allopolyploids. Autopolyploids usually are characterized by the presence of multivalents at meiosis, of tetrasomic ratios and, in the examples artificially produced, of slower development and reduced fertility. They may be descended from relatively homozygous diploids, or from hybrids between varieties or subspecies of a diploid species. The latter are more likely to be successful in nature, so that differences between wild autopolyploids and their nearest diploid relatives may be genetic in nature as well as the result of chromosome doubling *per se*.

Segmental allopolyploids in which, by definition, all of the component genomes have a majority of chromosomal segments in common, will resemble autopolyploids to a greater or lesser degree in possessing multivalents and tetrasomic ratios, but these will be less common. They may also come to resemble morphologically one or other of their diploid ancestral species, as a result of heterogenetic association and the consequent segregation for interspecific differences, as well as of the proven adaptive value of the gene combination possessed by these ancestral diploids. There is, therefore, no certain way of distinguishing between autopolyploids and segmental allopolyploids except by finding out through systematic studies and experimental verification the actual origin of the polyploid in question.

True allopolyploids may rarely have multivalent associations and tetrasomic ratios, but they usually do not, and they, therefore, resemble diploids to a large extent in their cytogenetic behavior. All of them, however, differ from diploids in that they can tolerate much more easily deficiencies of chromosomal material, in particular monosomic and nullisomic types. This doubtless is an important cause of the fact that fertility is usually higher in interspecific hybrids of polyploid species than it is between diploid species of about the same degree of relationship.

On levels of polyploidy higher than tetraploidy, complete autopolyploidy may exist, but since, on these higher levels, experimental autopolyploids are nearly always weak and aberrant, the success of such types in nature is highly problematical and no unquestionable examples are known to the writer. True allopolyploids, which contain three or more strongly differentiated genomes, derived from as many sharply distinct species, definitely exist in nature (*e.g.*, *Madia citrigracilis*, Clausen, Keck and Hiesey 1945), and have been synthesized artificially (*e.g.*, *Nicotiana digluta*, Triticale). It is likely, however, that a large proportion, if not a majority, of hexaploids, octoploids and higher polyploids represent some variant of the autoallopolyploid condition, in other words, that they have resulted from autopolyploids, segmental allopolyploids and true allopolyploids combined in different ways. The complete cytogenetic and phylogenetic analysis of such higher polyploids will probably be made in only a few clear and important examples.

These considerations lead to a revaluation of the importance of polyploidy in evolution and plant breeding. From the scientific point of view, it is important to know to what extent evolution in polyploid groups has been affected by chromosome doubling *per se*, with its attendant morphological and physiological alterations of the genotype and its creation of an isolation barrier which would permit divergent evolution, and to what extent divergence of polyploids from their diploid ancestors has been due to hybridization, with polyploidy acting as a stabilizer of hybrid combinations and as a means of obtaining fertility. The practical breeder needs to know how much he can expect from autopolyploidy and subsequent selection within a single variety as compared with the use of polyploidy as a tool in hybridization, either for transferring gene complexes from one species to another (Clausen 1941), or as a means of either fixing or rendering fertile hybrid combinations.

The best answer that can at present be given to the evolutionary question is that earlier estimates of the importance of autopolyploidy must definitely be revised. When we recognize the fact, first that true allopolyploids may often resemble diploid relatives so closely that taxonomists place the two in the same species (Clausen, Keck and Hiesey 1945, p. 130), and second, that segmental allopolyploids may resemble autopolyploids in cytogenetic as well as morphological characteristics, we must be more critical of most previously assumed examples of autopolyploidy. Much more work will have to be done before the extent of this necessary revision will be clear. Two facts, however, supply indirect evidence unfavorable to the assumption that polyploidy *per se* has played a large part in the differentiation of plant species. In the first place, the only known natural sexually reproducing polyploid which does not belong to a group of more or less closely related subspecies or species, including various diploids or putative diploids from which it could be descended, is the autotetraploid

Galax aphylla (Baldwin 1941). This strain resembles closely the diploid of the species in morphological and physiological characteristics as well as in distribution and, from the evolutionary point of view, is of distinctly minor importance. Secondly, all polyploid complexes fully analyzed follow the pattern suggested by the writer (1940, 1942), namely, that the morphological, physiological and ecological characteristics of their polyploid members are entirely, or almost entirely, recombinations of characteristics present in the ancestral diploids, with little or no evidence of divergence along a new evolutionary path (*cf.* Gustafsson 1943). The importance of polyploidy in fixing and spreading hybrid combinations is undoubted, and this is very likely its major role in evolution.

This conclusion leads directly to a suggestion to plant breeders that they will find much more profit in the use of polyploidy as a tool, combined in various ways with hybridization and selection, than in the creation of polyploids which might be expected to have immediate value. Viewed in this light, the importance of polyploidy in both plant evolution and plant breeding is by no means diminished, but it must be considered as one factor which is integrated with many others in producing various types of change.

The conclusion that polyploidy in nature is nearly always associated with hybridization, either intervarietal or interspecific, is of far-reaching evolutionary significance. About half of the species of Angiosperms have chromosome numbers that clearly indicate polyploid origin, and in some families, like the Gramineae, three-fourths of the species are polyploids. These polyploids are distributed through about two-thirds of the genera of Angiosperms. Hence in this proportion of genera, species formation has been associated in part with hybridization and the phylogenetic "tree" of the genus is partly reticulate in nature. And in genera like Rumex, Dianthus, Thalictrum, Mentha and Salix, in which the majority of the species are polyploids, the interrelationships of the species and the phylogenetic pattern of their evolution must be largely reticulate.

Of even broader significance is the possibility, which has arisen from recently counted chromosome numbers, that a large proportion of genera and even higher groups of Angiosperms may be of polyploid and therefore principally of hybrid origin. The writer pointed out (1938) that the basic chromosome number of woody genera is on the average higher than that of herbaceous ones, and that, of the nearly 150 genera that were well enough known to be listed, 35 had basic numbers of $x = 16$ or higher, numbers which strongly suggest a polyploid origin. Furthermore, 78 more of these woody genera had basic numbers of $x = 10$ to $x = 15$, which is higher than the modal numbers found in herbaceous genera. The prevailing evidence was and is against the hypothesis that these woody genera originated through polyploidy from herbaceous types. However, there is some reason for revising the estimate made by the writer of the relative probability of the two remaining hypotheses, namely, first, that the woody genera with basic numbers of $x = 11, 12, 13$ and 14 were derived from more ancient woody types with numbers of $x = 5, 6$ and 7, and second, that the numbers $12, 13$ and 14 are themselves primitive. In 1938, the writer favored the latter hypothesis. The following facts, however, now favor the former. In the family Anonaceae, which is entirely woody, predominantly tropical and phylogenetically primitive, basic numbers of $x = 7, 8$ and 9 have been found (Bowden 1945, Asana and Adatia 1945). In the primitive, tropical subfamily Caesalpinoideae of the large and relatively primitive family Leguminosae one ancient, woody genus, Cercis, has the basic numbers $x = 7$ and 6, while the related, also ancient genus Bauhinia has $x = 14$ (Senn 1938, Pantulu 1942). Finally the opinion of Avdulov (1931), that the original basic number of the Gramineae is $x = 12$, is now questionable. This opinion was based on the fact that this number is prevalent in the tribe Bambuseae and the series Phragmitiformes, in which the numbers 7 and 6 were unknown. Recently, however, these latter numbers have been determined as basic for Danthonia (Calder 1937, Stebbins and Love 1941), one of the clearly primitive and ancient genera of the Phragmitiformes and one which connects this series with the Festuciformes, containing the familiar northern genera with $x = 7$. It seems likely, therefore, that the number 12 in the Phragmitiformes is of very ancient polyploid origin. The well-established phylogenetic sequence by which Avdulov derived the series Sacchariferae from the Phragmitiformes through a gradual reduction in the basic number can still be accepted, so that the important tribes Oryzeae, Eragrosteae, Chloridae, Paniceae, Andropogoneae and Maydeae may well be derived secondarily from ancient polyploids. It therefore seems likely that reticulation is characteristic not only of the phylogenetic pattern within genera of this important family, but also of the pattern of relationships between genera and tribes.

If future evidence continues to favor the hypothesis that basic numbers of 10 and higher in the woody Angiosperms are originally or secondarily of polyploid derivation, we must conclude that within this group the pattern of relationships between genera of a family, and even between the families themselves, is to a large extent reticulate. If this is true, an explanation is at hand for the fact that plant systematists have never been able to construct a satisfactory system of relationships for the flowering plants. Although the genera and families of insects, vertebrates and other groups of animals have been arranged according to orders and classes that have become relatively stabilized and have met with general approval, the arrangement of the Angiosperms into groups higher than families and even the grouping of genera into families has been interpreted in widely different ways by such competent authorities as Bentham, Engler, Wettstein, Bessey and Hutchinson, and none of these systems has been considered satisfactory by more than a small fraction of modern systematists. The cytogeneticist will never be able to resolve this confusion, but he may be able to point to its cause. This is that any genus or family of flowering plants may share some genes with two, three or more other genera or families which otherwise have little in common. Hence the system will depend on the characters emphasized by its maker and, from the cytogenetic and phylogenetic point of view, any one of several systems is as nearly correct as another.

These speculations do not, however, alter the concept expressed by the writer (1940) and affirmed by a number of other cytogeneticists, that polyploidy is a conservative rather than a progressive force. Even if reticulation through allopolyploidy has played a major role in the origin of genera and families, this role has probably involved chiefly the production of new combinations of characters, rather than the origin of the characters themselves. Such fundamental changes as from polypetaly to sympetaly, from hypogyny to epigyny, and from actinomorphy to zygomorphy have probably been produced by successive genetic changes on the diploid level or in secondarily diploidized polyploids, while hybridization and polyploidy have acted mainly to put together the various resulting conditions in innumerably different ways, and the favorable combinations have been preserved and stabilized by selection. We can, therefore, conclude that, in the flowering plants as in other organisms, mutation, recombination, selection and isolation have been the chief agents of evolution, but that the prevalence of polyploidy has given recombination a more predominant and basic role than it plays in most other organisms.

REFERENCES

Anderson, E., *Ann. Mo. bot. Gdn.* **23**, 457–509 (1936).
Andérson, E., and Hubricht, L., *Amer. J. Bot.* **25**, 396–402 (1938).
Anderson, E., and Sax, K., *Bot. Gaz.* **97**, 433–476 (1936).
Asana, J. J., and Adatia, R. D., *Curr. Sci.* **14**, 74–75 (1945).
Avdulov, N. P., *Bull. appl. Bot. Leningrad Supp.* **44**, 428 pp. (1931).
Baldwin, J. T., Jr., *J. Hered.* **32**, 249–254 (1941).
Barthelmess, A., *Z. indukt. Abstamm.- u. VererbLehre* **79**, 153–170 (1941).
Beasley, J. O., *J. Hered.* **31**, 39–48 (1940).
Beasley, J. O., *Genetics* **27**, 25–54 (1942).
Bergner, A. D., *Proc. Natl. Acad. Sci., Wash.* **30**, 302–308 (1944).
Blackburn, K. B., *J. Bot.* **76**, 306–307 (1938).
Bowden, W. M., *Amer. J. Bot.* **32**, 81–92 (1945).
Bradley, M. V., and Goodspeed, T. H., *Proc. Natl. Acad. Sci., Wash.* **29**, 295–301 (1943).
Brown, S. W., *Amer. J. Bot.* **30**, 686–697 (1943).

Calder, J. W., *J. Bot. linn. Soc.* **51**, 1-9 (1937).

Camp, W. H., *Bull. Torrey bot. Club* **71**, 426-437 (1944).

Clausen, J., *Hereditas, Lund* **10**, 65-76 (1933).

Clausen, J., Keck, D. D., and Hiesey, W. M., *Publ. Carnegie Instn. No.* **520.** 432 pp. (1940).

Clausen, J., Keck, D. D., and Hiesey, W. M., *Publ. Carnegie Instn. No.* **564.** 174 pp. (1945).

Clausen, R. E., *Amer. Nat.* **75**, 291-306 (1941).

Clausen, R. E., and Cameron, D. R., *Genetics* **29**, 447-477 (1944).

Darlington, C. D., Recent Advances in Cytology, 2d ed. (1937). 671 pp.

Dobzhansky, T., Genetics and the Origin of Species. Rev. ed. (1941). 446 pp.

Emsweller, S. L., and Ruttle, M. L., *Amer. Nat.* **75**, 310-326 (1941).

Fagerlind, F., *Acta Hort. berg.* **11**, 195-470 (1937).

Fagerlind, F., *Chron. Bot.* **6**, 251-252 (1941a).

Fagerlind, F., *Chron. Bot.* **6**, 320-321 (1941b).

Giles, N., *Bull. Torrey bot. Club* **68**, 207-221 (1941).

Giles, N., *Amer. J. Bot.* **29**, 637-645 (1942).

Goodspeed, T. H., *Proc. Calif. Acad. Sci.* **25**(12), 291-306 (1944).

Goodspeed, T. H., and Bradley, M. V., *Bot. Rev.* **8**, 271-316 (1942).

Greenleaf, W. H., *Genetics* **26**, 301-324 (1941).

Greenleaf, W. H., *J. Genet.* **43**, 69-96 (1942).

Gustafsson, Å., *Acta Univ. Lund, K. fysiogr. Sälsk.*, N.F. **54**, No. 6, 200 pp. (1943).

Hagerup, O. *Hereditas, Lund* **16**, 19-40 (1932).

Hagerup, O., *Hereditas, Lund* **18**, 122-128 (1933).

Hagerup, O., *Hereditas, Lund* **26**, 399-410 (1940).

Howard, H. W. *J., Genet.* **36**, 239-273 (1938).

Howard, H. W., and Manton, I., *Nature, Lond.* **146**, 303-304 (1940).

Huskins, C. L., *Amer. Nat.* **75**, 329-344 (1941).

Jones, H. A., and Clarke, A. E., *J. Hered.* **33**, 25-32 (1942).

Joranson, P. N., Thesis (Ph.D.), Univ. of Calif. (1944).

Kihara, H., *Züchter* **12**, 49-62 (1940).

Kostoff, D., *J. Genet.* **37**, 129-209 (1938).

Kostoff, D., *Nature, Lond.* **144**, 868-869 (1939).

Kostoff, D., *Biodynamica* (Normandy, Mo.) *No.* **51**, 14 pp. (1939).

Kostoff, D., *Genetica* **21**, 285-300 (1939).

Kostoff, D., *J. Hered.* **31**, 33-34 (1940).

Lamm, R., *Hereditas, Lund* **31**, 1-128 (1945).

Lawrence, W. J. C., *Genetics* **12**, 269-296 (1930).

Lesley, M. M., and Lesley, J. W., *J. Hered.* **34**, 199-205 (1943).

Levan, A., *Hereditas, Lund* **23**, 317-370 (1937).

Levan, A., *Hereditas, Lund* **27**, 243-252 (1941).

Lilienfeld, F. A., *Jap. J. Bot.* **8**, 119-149 (1936).

Lindstrom, E. W., *Genetics* **26**, 387-397 (1941).

Ljungdahl, H., *Svensk bot. Tidskr.* **18**, 279-291 (1924).

Love, R. M., and Suneson, C. A., *Amer. J. Bot.* **32**, 451-456 (1945).

Mangelsdorf, P. C., and Reeves, R. G., *Bull. Texas agric. Exp. Sta. No.* **574**, 315 pp. (1939).

Manton, I., *Z. indukt. Abstamm.- u. VererbLehre* **69**, 132-157 (1935).

Manton, I., *Ann. Bot., Lond. N. S.* **1**, 439-462 (1937).

Müntzing, A., *Hereditas, Lund* **16**, 105-159 (1932).

Müntzing, A., *Hereditas, Lund* 21, 263–378 (1936).
Müntzing, A., *Hereditas, Lund* 23, 113–235 (1937).
Müntzing, A., *Hereditas, Lund* 24, 117–188 (1938).
Müntzing, A., *Hereditas, Lund* 25, 387–430 (1939).
Myers, W. M., *Bot. Gaz.* 104, 541–552 (1943).
Myers, W. M., *J. agric. Res.* 68(1), 21–33 (1944).
Myers, W. M., *Bot. Gaz.* 106, 304–316 (1945).
Myers, W. M., and Hill, H. D., *Bot. Gaz.* 104, 171–177 (1942).
Navashin (Nawaschin), M., *Z. Zellforsch.* 6, 195–233 (1927).
Nordenskiöld, H., *Bot. Notiser* 1941, 12–32.
Paddock, E. F., *Ohio J. Sci.* 42, 147–148 (1942).
Pantulu, J. V., *Curr. Sci.* 11, 152–153 (1942).
Pirschle, K., *Z. indukt. Abstamm-. u. VererbLehre* 80, 247–270 (1942).
Pirschle, K., *Z. indukt. Abstamm-. u. VererbLehre* 80(1), 126–156 (1942).
Randolph, L. F., *Amer. Nat.* 75, 347–363 (1941).
Sansome, F. W., and Philp, J., *Recent Adv. Plant Genet.* 1932, 414 pp.
Sears, E. R., *Res. Bull. Mo. agric. Exp. Sta. No.* 337. 20 pp. (1941).
Sears, E. R., *Genetics* 29, 232–246 (1944).
Senn, A., *Bibliogr. genet.* 12, 175–336 (1938).
Sharp, L. W., Fundamentals of Cytology. 270 pp. (1943).
Simonet, M., *Bull. Biol.* 69, 178–212 (1935).
Skirm, G. W., *Genetics* 27, 635–640 (1942).
Sparrow, A. H., Ruttle, M. L., and Nebel, B. R., *Amer. J. Bot.* 29, 711–715 (1942).
Stebbins, G. L., Jr., *Amer. J. Bot.* 25, 189–198 (1938).
Stebbins, G. L., Jr., *Amer. Nat.* 74, 54–66 (1940).
Stebbins, G. L., Jr., *Amer. J. Bot.* 28, Suppl. 6s (1941).
Stebbins, G. L., Jr., *Amer. Nat.* 76, 36–45 (1942).
Stebbins, G. L., Jr., *Bot. Rev.* 11, 463–486 (1945).
Stebbins, G. L., Jr., and Love, R. M., *Amer. J. Bot.* 28, 371–382 (1941).
Stebbins, G. L., Jr., Tobgy, H. A., and Harlan, J. R., *Proc. Calif. Acad. Sci.* 25, 307–322 (1944).
Stephens, S. G., *J. Genet.* 44, 272–295 (1942).
Stewart, R. N., and Bamford, R., *Amer. J. Bot.* 30, 1–7 (1943).
Thomas, P. T., *J. Genet.* 40, 119–128 (1940).
Thomas, P. T., *J. Genet.* 40, 141–156 (1940).
Upcott, M., *J. Genet.* 33, 135–149 (1936).
Upcott, M., *J. Genet.* 39, 79–100 (1939).
Waddington, C. H., An Introduction to Modern Genetics. 441 pp. (1939).
Wakar (Vakar), B. A., *Züchter* 7, 199–207 (1935).
Zhebrak, A., *Nature* 153, 549–551 (1944).
Zhebrak, A., *Acts Timiriazev agric. Acad. (Genet. div.)* 6, 5–54 (1944).
Zhebrak, A. R., *Amer. Nat.* 80, 271–279 (1946).

Part IV

VICINISM

Editors' Comments
on Papers 11 Through 14

The papers in this section were chosen because they analyze the mechanisms associated with the isolation of diploids and polyploids and the methods of gene exchange between the ploidy levels and among polyploids. It is no accident that two of the four papers involve the wheat group. This group has been and continues to be the subject of more intense genetical analyses than any other natural assemblage of plant species.

Paper 11 by Hagberg and Ellerström gives the most comprehensive treatment of vicinism of diploids and tetraploids available. It is important also for its applied aspects in agriculture, and it furnishes a framework for further studies of autopolyploid and diploid relationships in both artificial and natural populations.

One-way gene flow from diploids to tetraploid through triploid F_1 hybrids is demonstrated by Zohary and Nur in Paper 12. On theoretical grounds, this gene flow would seem highly unlikely because the F_1 is male sterile, but unreduced eggs provide partial female fertility. This study is important because it helps explain some of the variation found in polyploid complexes that appears to blend with diploids when they come into contact. Stebbins had previously suggested that such polyploid complexes were formed by hybridization of different varieties or ecologically differentiated populations followed by chromosome doubling to give highly variable polyploid groups.

In contrast to the previous paper, Vardi (Paper 13) shows experimentally that it is possible for highly male sterile triploid hybrids to act as bridges for the transfer of genes from a tetraploid to a diploid species. It is important theoretically and for applied breeding programs that complete diploidy is obtained in the third backcross generation despite an initial F_1 pollen fertility of only about 1 percent.

Zohary examines in Paper 14 the possible causes of widespread morphological and ecological diversity of polyploids in the wheat group. Diploid progenitor species are isolated by strong sterility barriers, but the polyploids are able to exchange genes even though they differ by one genome. A very interesting part of this paper is the demonstration of the importance of a pivotal genome shared by successfully hybridizing polyploids. The pivotal genome acts as a buffer in hybrids because it assures bivalent formation with its homologue in interspecific hybrids. The other genomes may behave variously and intergenomal crossing over may occur, but backcrossing and return of pollen fertility allows normal selfing and possible fixation of recombinant traits.

Additional papers bearing on problems discussed in the selected articles have been written by Cavanah and Alexander (1963), Carroll and Borrill (1965), and Estes (1969).

REFERENCES

Carroll, C. P., and M. Borrill, 1965, Tetraploid Hybrids From Crosses Between Diploid and Tetraploid *Dactylis* and Their Significance, *Genetica* **36:**65–82.

Cavanah, J. A., and D. E. Alexander, 1963, Survival of Tetraploid Maize in Mixed 2N-4N Plantings, *Crop Sci.* **3:**329–331.

Estes, J. R., 1969, Evidence For Autoploid Evolution in the *Artemesia Ludoviciana* Complex of the Pacific Northwest, *Brittonia* **21:**29–43.

11

Reprinted from pages 369–370, 382–391, and 406–415 of Hereditas **45**:369–416 (1959)

THE COMPETITION BETWEEN DIPLOID, TETRAPLOID AND ANEUPLOID RYE

THEORETICAL AND PRACTICAL ASPECTS

By *ARNE HAGBERG* and *SVEN ELLERSTRÖM*

SWEDISH SEED ASSOCIATION, SVALÖF, SWEDEN

(Received October 13th, 1958)

I. INTRODUCTION

ONE of the first artificial autotetraploids to be induced among agricultural plants was the tetraploid rye, described by DORSEY in 1936. Since then, breeding work with tetraploid rye has been carried out by several different institutions. MÜNTZING (1951) has reported in detail on his work, during the late 1930'ies and the 1940'ies, with the production of populations of tetraploid rye and with testing their agricultural value and their quality. The work with these populations has resulted in the marketing of one new variety, "Dubbelstål" (Double Steel). Similar work in Germany has resulted in the variety »Tetra-Petkus». v. SENGBUSCH (1940 and 1941) described the preliminary work, of producing the tetraploid population, and further work with the material has been reported by BLEIER (1950), LAUBE (1950), LÖWENSTEIN (1951) and PLARRE (1954). Also in the Netherlands work has been done with tetraploid rye as described by BREMER and BREMER-REINDERS (1954). Finally, tetraploid rye has interested scientists at several other institutions, *e.g.* ANDERSEN (1956) and MORRISON (1956).

Autotetraploids have been produced in a large number of species with varying success from a practical point of view. According to LEVAN (1948) autopolyploidy promises valuable results in the first hand in *cross-fertilizing* species with a *low chromosome number* and of which man utilizes primarily the *vegetative* parts, *e.g.* root crop and herbage plants. Practically valuable results, however, are not attained merely by the production of a tetraploid population, which is rather to be considered as a new and sometimes valuable raw material for further breeding work.

It is important to know as thoroughly as possible the behaviour and

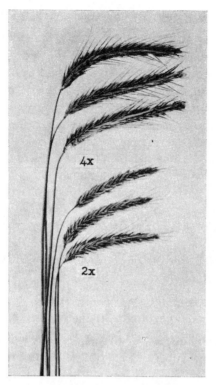

Fig. 1. Some ears of tetraploid and diploid rye of the variety Stål.

reactions of the autotetraploids under the different conditions of breeding, propagation and large scale cultivation. One important problem is the stability of the tetraploid population. Are diploids spontaneously formed and, if so, what are the consequences in practical farming? A related question is: what are the results of intercrossing with, or mechanical admixture of diploids in a tetraploid population? Still another question deals with the frequency of aneuploids and their influence upon the agricultural value of the population.

As already mentioned two varieties of tetraploid rye have been marketed, and the breeding material includes some very promising populations. It may thus be said that rye offers an example of a plant where polyploidy breeding has already given practical results and promises still better ones; in spite of the fact that the grain is the economically most important part of the plant.

[*Editors' Note:* Material has been omitted at this point.]

IV. STUDIES ON THE EFFECT OF VICINISM BETWEEN DIPLOID AND TETRAPLOID RYE

Fairly early it became evident that diploid and tetraploid rye could not be compared in the same yield test, the sterility barrier causing severe deprecation of yield, especially in the tetraploid (MÜNTZING, 1948). It is necessary to compare the tetraploids with wheat in one test, the diploids with the same wheat in another test, sufficiently distant but if possible under the same environmental conditions. The necessity to use indirect comparisons between tetraploids and diploids makes it difficult exactly to estimate the yielding capacity of the former. There is another problem, however, which is perhaps even more important: is it at all possible to grow a tetraploid variety of rye in a region with intense cultivation of diploid rye, where the air must be more or less filled with haploid pollen during flowering time? What is the minimum distance between fields of diploid and of tetraploid rye if a deprecation in yield in both of them is to be avoided?

a. Vicinism experiments at Ugerup in 1950—1951

In order to answer the first of these questions a field of about 1 hectare of Dubbelstål was sown in 1949 on the South Swedish experimental farm Ugerup, situated in a region with intense rye cultivation. The nearest diploid field was only 50 metres distant, all other rye in the region was diploid and there must be a rather dense "pollen cloud" in the air during flowering. Nevertheless the seed setting in the tetraploid field was normal along the borders as well as in the centre and the yield was quite satisfactory.

Next year a field of Dubbelstål was sown in the middle of a large field of diploid rye (Petkus II), the results already being presented in Swedish (HAGBERG, 1953 b). At the same time fields of Dubbelstål, about 2 hectares, and diploid rye were grown at the distance of at least 100 metres from one another.

The degree of fertility was determined in samples of about 50 ears each, taken at random in the large field of Dubbelstål and at certain fixed points in the field of Dubbelstål included in diploid rye. Even in the latter case the sample was a random sample of ears growing at or close to the point decided. Figure 9 gives the results of the fertility analysis, each value being the mean of 25 ears, selected at random from the sample. Along the western border 8 samples were taken immediately on each side of the border, the distance between the points being 3 metres.

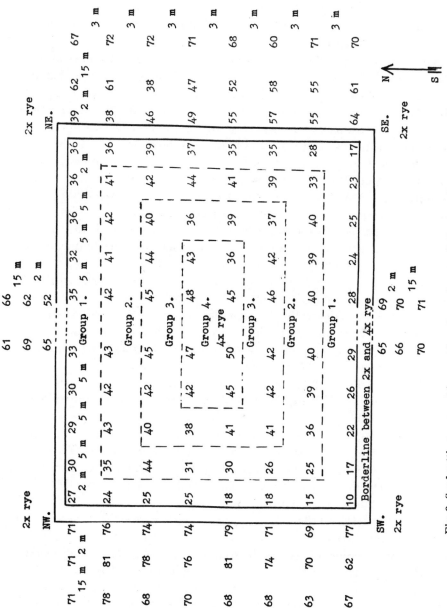

Fig. 9. Seed setting per cent in samples of ears from a field of tetraploid rye, enclosed in a field of diploid rye at Ugerup 1951.

The series was repeated on both sides 2 metres from the border and in the tetraploid again at 7, 12 and 17 metres, in the diploid at 17 metres. A similar series was taken starting from the eastern border, and finally samples were taken in the diploid north and south of the tetraploid.

The seed setting in the comparatively isolated field of Dubbelstål was 60.1 ± 1.01 % while the corresponding diploid field had the unusually low value of 69.0 ± 1.48 %. In the surrounded tetraploid the lowest fertility was found at the southwestern corner with only 10 %. The fertility rose northwards along the western side and eastwards along the southern, with a marked drop at the southeastern corner. The fertility along the northern and eastern sides was again higher with no drop at the northeastern corner. Southwesterly winds were prevailing during flowering. 2—3 metres from the border the seed setting was higher and fairly even except in the southwestern corner. 6—7 metres from the border and further inwards the fertility was fairly even with a mean of 42.3 %. This value is significantly lower than the one found in the "isolated field" which was 60.1 % ($t=13.56$, $P<0.001$).

The samples from the tetraploid field have been primarily grouped into four groups: 1, samples from the border line; 2, samples taken 2—3 metres from the border line; 3, samples taken 6—7 metres from the border; 4, the 8 samples from the centre. There is no significant difference between groups 3 and 4 ($t=1.78$) and in the following comparisons they are pooled. The comparison between group 2 and $3+4$ gives $t=3.04$, $P \sim 0.001$. If 2 is compared only with 3 $t=2.26$ and $P<0.05$. The most striking difference is between group 1 and group 2 with $t=6.35$, $P<0.001$.

Within group 1 the samples from different sides of the field have been compared. The western and southern sides are not significantly different ($t=1.64$); the seed setting along the northern and eastern borders is nearly the same, 32.6 and 32.9, respectively. The western and southern borders are significantly different from the eastern and northern ($t=4.41$, $P<0.001$).

Also within group 2 the sides have been compared. There are no significant differences between the southern, eastern and northern sides (mean values 39, 40 and 42, respectively) but the seed setting on the western side is significantly lower ($t=3.46$, $P<0.01$). The samples along the northern and southern sides are taken 3 metres from the border, those in the east and the west only 2 metres, which partly explains the high values, especially those along the southern side.

The average standard error for a difference between two samples is

TABLE 3. *Chromosome numbers in samples from the trial at Ugerup 1951.*

Sample	2n = 14	21	26	27	28	29	30	Total	% 3x	% aneupl.
SW-corner 4x	0	6		1	33	4		44	14	13
All other 4x	0	10	5	27	265	14	5	326	3	16
NE-corner 2x	237	0			0			237	0	0

2.81, a difference of 7.5 % thus gives P∼0.01. A specially interesting comparison is the one between the corner samples and other samples along the border. Both the southern corners have lower values than the border samples from the middle of the corresponding sides and this "corner effect" is also apparent in the samples next to the corners and in the southwestern corner sample in group 2.

The diploid samples from the western border, which have been "to windward" during flowering, have a significantly higher fertility than those from the eastern border (t=6.49, P<0.001). The same is true for samples taken 2 metres from the border (t=5.08, P<0.001). 17 metres from the tetraploid there is no difference between the eastern and the western side of the diploid.

To the west of the tetraploid there is on the whole no significant difference in fertility between samples taken at the border and at 17 metres into the field. The diploid border samples from the north-eastern corner have a significantly lower fertility than those from the middle of the eastern side, the difference to the southeastern corner is 25 %, t=5.27, P<0.001. The variation between ears is large in the diploids from the eastern border, resulting in high standard errors. It is nevertheless obvious that the variation is continuous from north to south. It might be suspected that the low fertility in some of the diploid samples was due to the occurrence of tetraploids, having been mixed into the diploid by carelessness in sowing. A check of the chromosome number has shown, however, that all the ears in the samples have been diploid, as seen from Table 3.

This table also gives the chromosome numbers in the progeny of the ears sampled in the southwestern corner of the field of tetraploid rye. The fertility of this sample was determined as 10 %. Among seeds formed 14 % were triploids, corresponding to 1.4 % of all flowers. Probably triploid embryos are formed in at least 50—60 % of the flowers and only 2—3 % of these resulted in well developed ripe seed. All data concerning chromosome numbers from the other samples in the tetraploid field have been summarized in the second line of Table 3. In this material 3 % of all seeds studied were triploid. It is thus evident that if a comparatively small field of tetraploids has been subject to an intense cloud of haploid pollen, there will be a certain low percentage of triploids in the progeny. The results from the sample taken from the most exposed corner of the diploid field show that no triploids have been found when the cross has been made with the diploid as the mother plant.

The seed setting of the diploid rye was fairly low at Ugerup in 1951. MÜNTZING (1951, table 12, page 49) presents data for the seed setting at Svalöf in the years 1944—49, and only in 1948 had the diploid Steel rye a lower value than the one obtained for diploid Petkus II in Ugerup 1951. The value for the "isolated" field of Dubbelstål is also somewhat lower than normal, but the difference between the two levels of ploidy is less than usual. In any case, the "pollen cloud" from the numerous large fields of diploid rye around Ugerup has not in 1950 and 1951 had any important effect upon the fertility of the separate fields of tetraploid rye.

The tetraploid field, surrounded by diploid had, however, if the borders are excluded, about 20 per cent units lower fertility than the separate tetraploid field. In this case the pollen cloud from the large surrounding diploid field has been sufficient to cause this very considerable decrease in seed setting in the tetraploid. The result is a decrease in yield. The yield of the tetraploid was 22 decitons/hectare, that of the surrounding diploid 30 dt/ha; the relative yield of the tetraploid is 73 if the diploid=100. The separate field of tetraploid yielded 42 dt/ha, the comparable diploid field 40 dt/ha. In this case the relative value for the tetraploid is 105. ANDERSEN'S (1956) data, from a similar 4x field within a 2x field in 1954, on the whole confirm the results obtained at Ugerup in 1951.

It should be mentioned that one of the neighbours of the experiment farm grew a small field — about 0.25 hectare — of diploid rye close by the "isolated" tetraploid field. This does not seem to have caused any appreciable deprecation of the seed setting in the latter. In this case the tetraploid field was large enough, so that the pollen cloud above it was completely dominated by its own pollen.

The part of the diploid field which was most exposed to pollen from the tetraploid had a considerably decreased seed setting. Tetraploid rye thus may influence the seed setting of diploid rye. This was shown also in the already quoted Table 12 of MÜNTZING (1951). In 1944 and 1945 diploids and tetraploids were included in the same trial. Not only the tetraploids but also the diploids show an abnormally low fertility. Lö-WENSTEIN (1951) has sown alternate plots of diploid and tetraploid rye, 2 metres broad and 40 metres long. There was a marked decrease in fertility in both levels of ploidy, this being most marked along the borders. Similar results have been obtained by OLSSON and RUFELT (1948) in an experiment with white mustard. In another experiment LÖWEN-STEIN also could demonstrate the effect of prevailing winds during flowering; the decrease in seed setting found by him corresponds fairly well with the Ugerup data.

LÖWENSTEIN (l.c.) also showed that strips of tetraploid rye, growing at different distances from large fields of diploid showed an increase in fertility with increased distance from the diploid. At 100—150 metres from the diploid field the seed setting of the tetraploid was normal. He presented some other data, however, which show a somewhat different tendency. The shortest distance between two large fields of diploid — 10 hectares — and tetraploid — 30 hectares — rye was 80 metres. In the tetraploid field, samples taken from the border and up to a distance of 15 metres into the field, showed a decrease in fertility of not less than 15 %. At 15—30 metres into the field the decrease in fertility was about 10 % and it was still 3 % 30—45 metres from the border. In the diploid field the decrease in fertility was about 10 % at the border, whilst further into the field there was a corresponding smaller, but appreciable decrease.

These last results are important and well worth noting. The data from Ugerup has not shown the possible effect of the pollen cloud on large fields to the same extent. The fertility in the German studies were (in both levels of ploidy) considerably higher than we are accustomed to in southern Sweden. Possibly this is one of the reasons for the discrepancies in the results.

b. Vicinism experiments at Svalöf in 1952—1956

In order to be able to give advice to the farmers with regard to "free position" it is necessary to study the effects of vicinism. The most pertinent question is the distance necessary between fields of diploid and tetraploid in order to avoid losses in either. To get further information about this question a series of experiments were carried out at Svalöf in the years 1952—1955. A number of plots of tetraploid rye, each having an area of 200 sq. metres were sown annually. These plots were sited, at

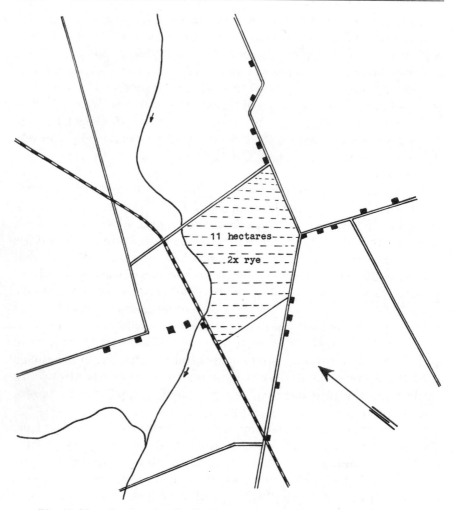

Fig. 10. Map showing the distribution of plots in the trial of 1953—1954.
The filled squares represent the 4x plots.

various distances from a centrally situated large field of diploid rye, and
in such a manner that they surrounded this field on each of its four
sides. In order to avoid as far as possible disturbances to the normal
farming procedures it was necessary to place the plots of tetraploid rye
on the edges of fields which also gave easy access to roads. As an
example of the planning Fig. 10 gives a map of the arrangement in 1953.
During flowering careful observations were made concerning strength
and direction of winds but also in regard to cloudiness, temperature and

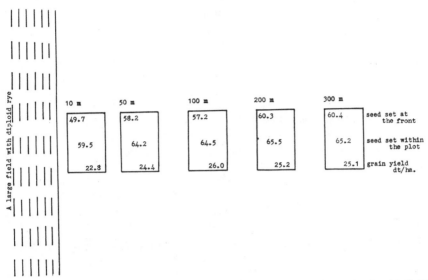

Fig. 11. Five plots of tetraploid rye at different distances from the diploid field, representing three year trials each with the five plots in the four cardinal points.

precipitation. At harvest ear samples were taken, similar to those in Ugerup. Ten samples were taken along the border, of the tetraploid plot, closest to the central diploid field, ten other were taken in the centre of each plot.

Each plot has been harvested and threshed separately and the yield of grain and straw determined. It has been impossible to avoid differences between the plots, even within the same "row". Each year the experiment has covered an area of about 1 square km and it has been impossible to avoid considerable edaphic differences. In spite of this and of a fairly low average yield the data obtained are sufficient as a basis for a discussion of the effects of vicinism.

The most important data are summarized in the diagram, Fig. 11. It presents average seed setting and grain yield in the plots of tetraploid rye at different distances from the large field of diploid. All values given are means of equivalent values from the four "rows" of plots during all three years. Each value, thus, is the mean of 12 primary values; these again, with regard to fertility, are each founded on determinations carried out on 250 ears.

In the plots 10 metres from the diploid field the fertility is 49.7 % along the border nearest the diploid material. In the centre of these plots the fertility is 59.5 %, a surprisingly high value. In the plots at

North

57.3

54.9

52.9

47.4

41.1

West 66.9 65.1 65.2 62.2 48.4 | 2x | 49.5 54.7 54.5 57.9 57.5 East

65.5

66.0

66.6

65.6

67.2

South

Fig. 12. Seed set in the front of the tetraploid plots towards the central field (2x rye) in the trial 1953—54.

50 and 100 metres the border fertility is 57—58 %, in the centre it is 64—65 %, thus reaching normal fertility for tetraploid rye under the conditions prevailing. The plots at 200 and 300 metres have a fertility of about 65 % in the centre, and about 60 % at the border. The low fertility at the border in these cases is probably not due to influence from the diploid field, but rather to the fact that conditions for pollination are always better within a field than at the border. As to yield there is no increase when the distance becomes greater than 100 metres. If the mean yield of the 100, 200 and 300 m plots is counted as 100 the 10 m plots have a relative value of 89.9, the corresponding value for the 50 m plots being 97.2.

The summarized data of Fig. 11 indicates that a distance of 100 metres between a diploid and a tetraploid field should be sufficient, under the conditions prevailing in these experiments to prevent an appreciable decrease in fertility. It is obvious, however, that the prevailing winds during flowering will influence the results. Fig. 12 gives the seed setting along the border of the different plots in the experiment of 1954. During the flowering period of 1954 the prevailing wind was southerly with deviations to southwest and southeast; even without meterological observations this would have been evident from the data. All the plots to the south have a normal fertility, while especially those to the north show a marked decrease; the same is true of the eastern "row" and to a smaller extent also of the western. The effect on the

North

41 (65)

41 (61)

41 (61)

42 (57)

West 41 41 41 39 | 2 x | 38 42 42 42 41 East
(63) (49) (52) (44)   (52) (58) (58) (61) (61)

40 (42)

43 (57)

45 (57)

45 (62)

43 (62)

South

Fig. 13. The percentage of grain of the total yield (straw+grain) in the different tetraploid plots in the trial 1952—53. The average seed set in "front samples" from respective plots are given within brackets.

fertility in the centre of the plots is much less marked as is also the influence on yield.

In a rye population there is a considerable variation in earliness, and the flowering days of different ears vary considerably. In 1955 ears were marked which flowered on the very first day, when the wind was westerly to northwesterly. In the 10 m plot to the east the seed setting among these ears was 34.8 %, while the corresponding ears in the three other 10 m plots had a setting of 63.0, 61.4 and 67.9, respectively. Similar results were obtained from ears flowering during the following day with similar wind conditions. On the fifth day of flowering the wind was from the south to the south-southeast. Ears flowering on this day had a setting of 37.3 % in the northern 10 m plots, 55.3 in the western and 69.7 in the southern plot. Two days later the wind was again westerly; ears flowering this day in the eastern plot had a setting of 53.0 %.

These data show the effect of wind direction more strongly than the annual averages. These only show the effect of prevailing winds, which again is of the greatest importance in regions where winds of a certain direction usually prevail at the time of year when the rye is flowering. In such regions it should be avoided to place a field of tetraploid rye leewards of a diploid one; or at least in such cases the distance between the fields should be considerably more than the 100 metres otherwise suggested as sufficiently safe.

[*Editors' Note:* Material has been omitted at this point.]

VI. SEED CONTROL RULES AND OTHER PRACTICAL RECOMMENDATIONS

The sterility barrier which exists between tetraploid and diploid rye and which is mainly due to an abortion of triploid embryos is a good protection for artificial autotetraploid populations of rye. From the point of view of seed control the risk of contamination by spontaneous crosses with diploid may be considered as negligible. The very few triploid plants which will develop have such a low degree of fertility that they cannot genetically damage a tetraploid population. From this point of view diploid and tetraploid rye may be considered as two well separated species.

A distinctly different question is the influence upon the yield of a field of tetraploid rye of a neighbouring diploid field. Crosses between the two may cause a considerable loss of yield in the part of the tetraploid field closest to the diploid neighbour. According to the results here reported a distance of 100 metres is sufficient in most cases. The prevailing winds during flowering time are of considerable importance and *e.g.* in regions where these are mainly southwesterly it should be avoided to place a tetraploid field to the northeast of a diploid one of any considerable size.

In seed certification there must be a maximum limit to the admixture of diploids permitted. As has been shown the tetraploid population is self cleaning if the frequency of diploids is kept within reasonable limits; the critical value is 32 % or somewhat higher. From a genealogical point of view it would not be dangerous to permit 20 % diploids. On the other hand it must be taken into consideration that the buyer wants a maximum yield. To ascertain this a limit of about 4 % would be sufficient. Since it is rather easy to keep a tetraploid rye free from diploids the permitted admixture of these has in Sweden been fixed at 1 %, but from a biological point of view there does not seem to be anything against an increase of this value as far as to 4 %.

In rye the situation is very favourable because of the high critical value. For comparison it may be mentioned that the corresponding critical value for a mixture of diploids and tetraploids in red and alsike clover is 3—4 % diploids (JULÉN and HAGBERG 1954). If there is a little more than 4 % of diploids in a tetraploid clover population the frequency of diploids will automatically increase in an accelerated tempo until the population is purely diploid. To preserve the tetraploid population it is necessary to keep the diploids below the critical value. Here the possibility to eliminate the small seeds of the diploid by sieving is a very good aid in keeping down the diploid frequency below the critical value.

As a result of the frequently irregular meiosis of the tetraploid there is a fairly high frequency of aneuploid plants, in Dubbelstål about 15 %. Most probably they cause a decrease in the yield of the population, but so far it has not been possible to analyze in detail their influence upon yield.

There is no doubt that on an average aneuploids are weaker than comparable euploids. Their fertility is low, and in the material here reported their seed setting is about one half of that of normal tetraploids. A considerable part of their progeny is probably also aneuploid (*cf.* MÜNTZING's result in *Dactylis*, 1937). It has been shown that hypotetraploids have smaller grains than the euploids and as pointed out earlier it is possible to reduce the frequency of hypotetraploids, by more than one half, by eliminating the small grained fraction of the tetraploid seed. At the same time possible admixtures of diploids will be mainly eliminated, but not the hypertetraploids. On the other hand it has been shown that low fertility is to some extent correlated with large grains and the use of only large-grained seed may result in decrease in fertility, more or less in combination with an increase of aneuploids. As shown in Fig. 16 the correlation between fertility and seed size is not represented by a straight line and many weak plants have both low fertility and small seeds. The elimination by sieving of a small-grained fraction of 20—30 % will probably not damage the population but have a beneficial influence on its composition and yield. Whether the increase in yield is sufficient to pay the increase in seed cost, resulting from the sieving, is at present the subject of further studies.

VII. SOME ASPECTS ON BREEDING TETRAPLOID RYE AND AUTOTETRAPLOIDS IN GENERAL

The importance of autopolyploidy in evolution has been discussed by several authors (*e.g.* MÜNTZING, 1936, CLAUSEN, KECK and HIESEY, 1945, STEBBINS, 1948, 1950, 1956, HAGBERG, 1953, *etc.*). The different points of view on the problem has resulted in different opinions mainly due to the fact that the border line between auto- and allopolyploidy is not always quite sharp. Doubling the chromosomes of a fully homozygous self fertilizer gives the "strict" autopolyploid. When the chromosome number is doubled in a cross bred population containing different caryotypes — variation in chromosome structure, deficiencies, duplications, *etc.*, then this constitutes a step towards allopolyploidy.

A strict autopolyploid is not likely to be more successful than its original diploid. Generally the mere doubling of the chromosome number and the correlated increased cell size is a disadvantage. This has to be compensated by the new, rich possibilities of combinations and mode of inheritance of genes in the autopolyploids. A great genetic variation in a cross bred autopolyploid population is certainly much more promising as raw material for further breeding work. Such a population might be successful in nature; compare some cases of possible autopolyploidy: *e.g. Dactylis glomerata, Hordeum bulbosum, Phleum pratense, Medicago sativa etc.*

In the case of rye an autotetraploid population has not been established in nature without the plant breeders protection of the first tetraploid plants. The natural population would probably never reach the required frequency of 60 per cent tetraploids to prevent them from being eliminated rather quickly. However, as soon as the first autotetraploid population is established this population has a good competitive ability and is self cleaning with regard to competing diploids. The frequency of diploids has to be as high as about 40 per cent before they will become dangerous to the existence of the tetraploid population. The barrier of sterility between diploid and tetraploid rye is of significant importance as is described in the preceding chapters. It is not a question of an incompatibility barrier which is gametic as is common between two different but related species, but a question of zygotic lethality resulting in a decreased fertility and yield of seed. Genetically there is no exchange of hereditary material between a diploid and a competing tetraploid population. As indicated earlier diploid and tetraploid rye may in this respect be considered as two separate species (*cf.* CLAUSEN, KECK and HIESEY, 1945, pp. 150—151).

There are many other reasons for considering artificially produced autopolyploids as new distinct types separated from diploids. It has been shown that tetraploids often require other conditions for optimal development other than those required by the diploids. ELLERSTRÖM (1959 a and b) studied different levels of ploidy in *Phleum pratense* in field trials all over the Scandinavian peninsula. The high ploidy levels gave the best results in areas with high rainfall and/or high latitude. This fact suggests that the testing of tetraploid populations under different environments other than that normal for the corresponding diploids is advisable. RASMUSSON (lecture for the polyploidy section of the Eucarpia, 1958) has given a very good illustration of this fact in tetraploid sugar beets. The experiences from testing a tetraploid Swedish spring rye population in California earlier mentioned is another example. The breeder has to find out which mode of cultivation is optimal for the tetraploids; density of stand, amount of fertilizer added and the different soils and climatic conditions in which the tetraploids are most productive. Day length reaction is another important factor to take into consideration in this respect. Thus, there is a good reason, for the exchange of autotetraploid populations between breeders in different regions, and testing them under widely different conditions. It is also important to bring as much genetic variation as possible into one and the same tetraploid population to create possibilities for complex gene combinations not possible at the diploid level. Selection of heterozygotes of the constitution $A_1A_2A_3A_4$ in such a "melting pot" will be a possibility to avoid homozygosity in the population and to create varieties with a high degree of heterozygosity. Thus, in breeding autotetraploid crops heterosis can be efficiently used where this is not conveniently possible in the corresponding diploids (cf. HAGBERG, 1953 c and 1955).

There are many other problems to be discussed in connection with the breeding of autotetraploids. A few might be briefly mentioned. Due to the mode of inheritance selection is a slower process in autotetraploids than in diploids. In the breeding of diploid cross fertilizers, "synthetics" and/or "F_1-varieties" are produced and released. On the tetraploid level the F_2 generation might have the maximum of heterozygosity. This is dependant on the genetic situation in the inbreds crossed and has to be tested in each case. Cross fertilization being one of the premises for a fairly rapid progress in "polyploidy breeding" made possible by heterozygosity, the self incompatibility system and its effect and function on the polyploid level is very important as is pointed out by several authors, e.g. LEWIS, 1954 and 1956, ATWOOD and BREWBAKER, 1953, and LUNDQVIST, 1957. According to LUNDQVIST the system in rye is functioning better in populations with large genetic variation while in clover the opposite phenomenon seems to occur. In the breeding programme for the tetraploids this has to be taken into consideration.

As to the pairing conditions at meiosis MÜNTZING (1951) found no differences in rye after 7 generations of intense selection for high fertility. Similar results were obtained by MORRISON (1956), while GILLES and RANDOLPH (1951) in maize found a slight reduction of quadrivalent frequency during a period of 10 generations. In rye HILPERT (1957) also found a slight effect of selection for regular distribution of chromosomes at meiosis. The present paper, however, demonstrates, that an increase in fertility is correlated with decrease in seed size and the yield will be the same. If the fertility is decreased as in the plots of 1—4 per cent diploid admixtures the seed size is increased correspondingly and the yield seems to be the same. To increase yield it is necessary to select types which have a higher assimilation capacity and a more "economical" distribution and use of the assimilates giving a higher carbohydrate production per unit area of field. Studies of this kind are now in progress.

A question, which is often discussed, concerns the constancy of the autotetraploid populations. There are cases described where tetraploids show diploid tissues in premeiotic stages and thus diploids in the progeny (*e.g.* GOTTSCHALK, 1958). In a cross fertilizing population such as the autotetraploid rye there is very little chance that diploids eventually formed could in one generation reach the frequency of 30—40 per cent of the individuals in the population required for the diploids to become dominant. Below this frequency of diploids the tetraploid rye is "self cleaning", since single diploids will be sterile. Thus, there is practically no risk of this phenomenon causing the loss of an autotetraploid population of rye.

In the strict autotetraploid populations of the self fertilizers, flax and barley diploids some times occur. These diploids will rapidly increase in the population because of their higher fertility (HAGBERG and ELLERSTRÖM, in manus). It was not possible to decide whether these plants were admixtures or were spontaneously arising. Since 1951 the chromosome number in every plant has been checked in a population of flax. In this very effectively controlled population of flax, diploid plants have never been obtained. However, if diploids are formed in tetraploid populations of self fertilizers there is a great risk of a rapid increase of diploids among the tetraploids.

It is true that autotetraploid populations of rye probably could never arise in nature. However, it is also true, that tetraploid rye plants compete very well with diploids at least under Swedish conditions, and that spontaneously arisen diploid plants would never be able to turn a tetraploid rye population back to diploidy. With the breeders aid tetraploid rye populations can be established and no doubt they constitute a useful and valuable raw material for promising, future breeding work.

Acknowledgements. — This investigation has been supported by the Swedish Agricultural Research Council, and was carried out by the staff at the cytogenetic laboratory of the Swedish Seed Association. The authors are grateful to their colleagues at the laboratory and to the professors MÜNTZING and TEDIN for many helpful suggestions and comments.

SUMMARY

The present paper presents and discusses a collection of relevant data from some investigations on tetraploid rye carried out at the Swedish Seed Association during the last 10 years. The main aim of these investigations was to forestall certain practical difficulties which might be encountered in trying to establish tetraploid rye as a worthwile crop in practical farming.

The frequency of aneuploids has been determined for 9 different seed lots belonging to the tetraploid variety "Dubbelstål". As an average 14.7 % of the seeds are aneuploids, but the frequency was found to vary considerably between different seed lots, the extreme values being 6.8 % and 20.4 %.

The reduction in the seed yield of tetraploid rye caused by the presence of aneuploids is discussed. Figures are given for the reduction of the frequency of aneuploids especially hypoploids, brought about when the small-grained portion of a seed lot is removed.

The seed setting is reduced in tetraploid rye when it is exposed to pollen from diploid rye. This is due to the fact that, in the style of the tetraploid rye, the pollen tube, derived from a haploid pollen grain, grows faster than a pollen tube derived from a diploid pollen grain. Triploid embryos are formed, which show irregular development and, for the most part, disintegrate at an early stage. When diploid plants are heavily exposed to the pollen of tetraploids, the seed setting of the diploids is also reduced due to the abortion of the triploid embryos formed.

The seed setting and yield have been studied in fields of tetraploid rye exposed to pollen from fields of diploid rye. In a small field of tetraploid rye, totally surrounded by diploid rye, the reduction in yield was estimated to be about 30 %. However, fields of tetraploid rye grown in areas, where much diploid rye is grown yielded a normal crop. The haploid "pollen cloud", which must dominate such an area, seems to be of minor importance in this case.

By studying the seed fertility of tetraploid rye grown in plots at various distances from a large field of diploid rye, it was found that a distance of 100 metres, between a diploid and a tetraploid field, is sufficient to avoid the deleterious effect of intercrossing. However, it is pointed out that the minimum distance is influenced by the prevailing winds during the flowering period.

In applying tetraploid rye to practical farming there is always a certain risk of the mechanical admixture of diploids in the tetraploid populations. The effect of such an admixture has been investigated in tetraploid rye to which diploid rye was admixed in frequencies ranging from 1/2 % to 50 % of the total population. A threshold value was found to exist at a frequency of diploids lying somewhere between 32 % and 40 %. At this threshold value the two levels of ploidy are maintaining the original frequency in the progeny. The tetraploid is "self-cleaning" when the frequency of diploids admixed is lower than 32 %. The seed yield of the mixed populations is not reduced, in relation to that of pure tetraploid populations, until the frequency of diploids is higher than 4 %. At this frequency the decrease in seed setting is compensated by a corresponding increase in 1000-grain weight.

In seed certification there must be a minimum limit to the admixture of diploids permissable in tetraploid rye. Since it is rather easy to keep a tetraploid rye free from diploids this limit has already been set at 1 %.

The possibility of improving the yield of tetraploid rye, by sifting off the small sized portion of the seed lot, is discussed. In this way the frequency of aneuploids will be reduced. On the other hand it has been shown that low fertility is to some extent correlated with large grain size. Thus, too severe a sifting may reduce the yield, and give a higher frequency of aneuploids. However, since the relationship between fertility and seed size is not a straight line graph, the removal of a small grained portion, equivalent to 20—30 % of the seed lot, will probably have a beneficial influence on the yield.

Some problems concerning the breeding of autopolyploids are discussed. The reasons for considering artificially produced autopolyploids as new distinct types are pointed out. The fact, that polyploids very often have quite different requirements, compared to those demanded by the corresponding diploids, for optimal growth, makes it advisable for breeders in different regions of the world to exchange polyploid material for testing under widely different conditions.

The importance of bringing as much genetic variation as possible into one and the same tetraploid population, in order to create possibilities for new complex gene combinations, is pointed out. In the breeding of autotetraploid crops heterosis can be efficiently used where this is not conveniently possible in the corresponding diploids.

Since cross fertilization is one of the premises for progress in "polyploidy breeding", the self incompatibility system of the plant has to be taken into consideration. The system in rye functions better, in tetraploid populations, the larger the genetic variation of the tetraploid population is, whilst in tetraploid clover, for example, the situation is reversed.

Selection for a regular distribution of chromosomes at meiosis, resulting in higher seed fertility, has had only a slight effect if any. It is also demonstrated in the present paper that an increase in fertility is correlated with a decrease in seed size, resulting in the same yield of grain. To improve yield it is therefore necessary to select for other characters, such as high assimilation capacity, and a more economical distribution and utilisation of the assimilates within the plant.

In some cases tetraploid populations have been reported to give rise to diploid individuals, through some kind of chromosome reduction in premeiotic stages. In all seed lots of tetraploid rye, so far studied, no diploids have been found. The "self-cleaning" capacity of tetraploid rye serves as a strong protection against the possibility of diploids, eventually formed, reaching the frequency of 30—40 % necessary to enable the diploids to become dominant in the population. In this respect tetraploid rye differs strongly from autotetraploids of self fertilizing species, where there is a great risk of a rapid increase of diploids, because of their higher fertility. Further, tetraploid rye also differs from crops, such as sugar beets, where triploids are developed into mature plants.

Literature cited

ANDERSEN, T. 1956. Forsøg med tetraploid rug 1950—1954. — Tidskr. f. Planteavl 60: 185—197.

ARMSTRONG, J. M., ROBERTSON, R. W. 1956. Studies of colchicine-induced tetraploids of Trifolium hybridum L. 1. Cross and self-fertility and cytological observations. — Canad. J. Agr. Sci. 36: 255—266.

ATWOOD, S. S. and BREWBAKER, J. L. 1953. Incompatibility in autoploid white clover. — Cornell Univ. Memoir 319: 1—47.

BLEIER, H. 1950. Genommutation als neue praktische Zuchtmethode. — D.L.G.-Nachr. f. Pflanzenzucht.

BREMER, G. and BREMER-REINDERS, D. E. 1954. Breeding of tetraploid rye in the Netherlands. I. Methods and cytological investigations. — Euphytica *3*: 49—63.

CHIN, T. C. 1943. Cytology of the autotetraploid rye. — Botanical Gazette *104*: 627—632.

CLAUSEN, J., KECK, D. D. and HIESEY, W. M. 1945. Experimental studies on the nature of species. II. Plant evolution through amphiploidy and autoploidy, with examples from the Madiinae. — Carnegie Inst. of Wash. Publ. *564*: 1—174.

DORSEY, E. 1936. Induced polyploidy in wheat and rye. — Journal of Heredity Vol. 27: 155—160.

ELLERSTRÖM, S. 1949. Cytogenetiska undersökningar på råg (not published thesis).

— 1959 a. The co-variation of chromosome number and environment in *Phleum pratense*. — Hereditas *45*: 461—463.

— 1959 b. Die Konkurrenzfähigkeit polyploider Rassen von *Phleum pratense* unter verschiedenen Milieubedingungen. — In press.

— and HAGBERG, A. 1954. Competition between diploids and tetraploids in mixed rye populations. — Hereditas *40*: 535—537.

GILLES, A. and RANDOLPH, L. F. 1951. Reduction of quadrivalent frequency in autotetraploid maize during a period of 10 years. — Amer. Journ. of Bot. *38*: 12—17.

GOTTSCHALK, W. 1958. Über Abregulierungsvorgänge bei künstlich hergestellten Hochpolyploiden Pflanzen. — Zeitschr. für Vererbungslehre *89*: 204—215.

HAGBERG, A. 1952. Heterosis in F_1 combinations in Galeopsis. I. — Hereditas *38*: 33—82.

— 1953 a. Frekvensen av plantor med avvikande kromosomtal i marknadspartier av tetraploid råg (Dubbelstålråg). — Kungl. Lantbruksakad. Tidskr. *92*: 417—427.

— 1953 b. Kärnansatsen hos Dubbelstålråg vid odling i grannskap av diploid råg. — Sv. Utsädesf. Tidskr. *63*: 63—69.

— 1953 c. Further studies on and discussion of the heterosis phenomenon. — Hereditas *39*: 349—380.

— 1955. Översikt över polyploidiförädlingen i Sverige. — Sv. Utsädesf. Tidskr. *65*: 209—214.

HILPERT, G. 1957. Effect of selection for meiotic behaviour in autotetraploid rye. — Hereditas *43*: 318—322.

HOWARD, H. W. 1939. The size of seed in diploid and autotetraploid *Brassica oleracea* L. — Journ. of Genetics *38*: 325—340.

HÅKANSSON, A. 1953. Endosperm formation after 2x×4x crosses in certain cereals, especially in *Hordeum vulgare*. — Hereditas *39*: 59—64.

— 1956. Seed development of *Brassica oleracea* and *B. rapa* after certain reciprocal pollinations. — Hereditas *42*: 373—396.

— and ELLERSTRÖM, S. 1950. Seed development after reciprocal crosses between diploid and tetraploid rye. — Hereditas *34*: 256—296.

JULÉN, G. and HAGBERG, A. 1954. Några erfarenheter beträffande inblandning av diploider vid fröodling av tetraploida klöverarter. — Svensk Frötidning *11*: 1—4.

JULÉN, U. 1950. Fertility conditions of tetraploid red clover. I. Seed setting of tetraploid red clover in the presence of haploid pollen. — Hereditas *36*: 151—160.

LAUBE, W. 1950. Tetraploider Roggen. — Deutsche landw. Presse 73 h. 7.

LEVAN, A. 1942. The effect of chromosomal variation in sugar beets. — Hereditas *28*: 345—399.

— 1948. The cyto-genetic Department 1931—1947. — Svalöf 1886—1946: 304—323.

LEWIS, D. 1954. Comparative incompatibility in Angiosperms and Fungi. — Advances in Genetics *VI*: 235—285.

— 1956. Incompatibility and plant breeding. — Brookhaven Symp. in Biol. *9*: Genetics in Plant breeding, p. 89—100.

LUNDQVIST, A. 1957. Self-incompatibility in rye. II. Genetic control in the tetraploid. — Hereditas *43*: 467—511.

LÖWENSTEIN, J. Prinz zu 1951. Über die Befruchtungsverhältnisse zwischen diploidem und tetraploidem Roggen. — Zeitschr. f. Pflanzenzüchtung *31*: 104—133.

MORRISON, J. W. 1956. Chromosome behaviour and fertility of tetra Petkus rye. — Canad. Journ. of Agr. Sci. *36*: 157—165.

MÜNTZING, A. 1936. The evolutionary significance of autopolyploidy. — Hereditas *21*: 263—378.

— 1937. The effects of chromosomal variation in *Dactylis*. — Hereditas *23*: 113—235.

— 1943. Aneuploidy and seed shrivelling in tetraploid rye. — Hereditas *29*: 65—75.

— 1948. Några data från förädlingsarbetet med tetraploid råg och rågvete. — Nordisk Jordbrugsforskning p. 499—507.

— 1951. Cyto-genetic properties and practical value of tetraploid rye. — Hereditas *37*: 17—84.

OLSSON, G. and RUFELT, B. 1948. Spontaneous crossing between diploid and tetraploid *Sinapis alba*. — Hereditas *34*: 351—365.

O'MARA, J. G. 1943. Meiosis in autotetraploid *Secale cereale*. — Botanical Gazette *104*: 563—575.

PLARRE, W. 1954. Vergleichende Untersuchungen an diploidem und tetraploidem Roggen (*Secale cereale* L.) unter besonderer Berücksichtigung von Inzuchterscheinungen und Fertilitätsstörungen. — Zeitschr. f. Pflanzenzüchtung *33*: 303—353.

RANDOLPH, L. F. 1941. An evaluation of induced polyploidy as a method of breeding crop plants. — The American Naturalist *LXXV*: 347—363.

SCHILDT, R. and ÅKERBERG, E. 1951. Studier över tetraploid och diploid råg vid Ultunafilialen 1949. — Sv. Utsädesf. Tidskr. *61*: 254—268.

SENGBUSCH, R. v. 1940. Polyploider Roggen. — Züchter *12*: 185—189.

— 1941. Polyploide Kulturpflanzen (Roggen, Hafer, Stoppelrüben, Kohlrüben und Radieschen). — Züchter *13*: 132—134.

STEBBINS, G. L. 1948. Types of polyploids: their classification and significance. — Adv. Genet. *1*: 403—429.

— 1950. Variation and evolution in plants. — N. Y.: Columbia Univ. Press, 643 pp.

— 1956. Artificial polyploidy as a tool in plant breeding. — Brookhaven Symp. in Biol. *9*: Genetics in Plant breeding 37—52.

[*Editors' Note:* The contents list has been omitted.]

12

Reprinted from *Evolution* **13**:311-317 (1959)

NATURAL TRIPLOIDS IN THE ORCHARD GRASS, *DACTYLIS GLOMERATA* L., POLYPLOID COMPLEX AND THEIR SIGNIFICANCE FOR GENE FLOW FROM DIPLOID TO TETRAPLOID LEVELS

DANIEL ZOHARY AND UZI NUR

Department of Botany, The Hebrew University, Jerusalem, Israel

INTRODUCTION

Diploid and tetraploid forms in sexual polyploid groups have been generally regarded as reproductively completely isolated from one another. Whether this is always the case was already questioned by Zohary (1956) and Stebbins and Zohary (1959). In the following work an attempt is made to examine the genetic relationships between diploid and tetraploid populations of the Orchard Grass, *Dactylis glomerata* L., in places where such populations come in contact in Israel. This paper reports on a study of the occurrence of natural triploids in contact areas and on the types of progeny such triploids produce under natural conditions.

Previous studies (Zohary 1956, Stebbins and Zohary 1959) have established the polytypic group of Orchard Grass (*Dactylis glomerata* L.) as a large polyploid complex. This complex contains at least ten distinct diploid (2n = 14) subspecies as well as a large, sexual, tetraploid (2n = 28) superstructure. The diploid forms have each a restricted geographical distribution and they are widely separated from one another. Each diploid is characterized also by comparatively narrow and distinct range of variation. In contrast with the diploids, tetraploid *Dactylis* forms are continuously and widely spread over most of Europe, West Asia and North Africa. The tetraploid level is characterized by a large, continuous range of variation which covers most of the characters found in the various diploids. Cytogenetically all tetraploid forms are interfertile and show typical autopolyploid chromosome behavior.

MATERIALS AND METHODS

Seed collected from the triploid plants was planted in the greenhouse at Jerusalem. Seedlings were later transferred to 8″ pots and grown outside.

For cytological study, material was fixed in 3 : 1 alcohol acetic acid for 24 hours and stored later in 70% alcohol. In plants which flowered, chromosome counts were made in aceto-carmine squashes of microsporocytes. Stages used were anaphase I and diakinesis. At least 8–10 well spread cells were analyzed in each plant. In non-flowering plants chromosome counts were made from shoot meristems. Growing shoots were dissected longitudinally and pretreated for three hours with water solution of paradichlorbenzen and the Feulgen staining and squash method was employed.

To determine the percentage of pollen abortion, pollen grains were stained with 2% aceto-carmine. For each plant examined no fewer than 200 pollen grains were counted.

Seed-set was determined by examination, in each plant, of 50 almost mature spikelets, taken at random from the second and third lower branches of the panicle. Only the lower two florets in each spikelet were examined. A floret was considered fertile when a well developed caryopsis was found in it.

OBSERVATIONS

(1) *Occurrence of Triploids in Contact Areas.* Both diploid (2n = 14) and

200

tetraploid $(2n = 28)$ forms of *Dactylis* are found in Israel (Zohary 1956, Stebbins and Zohary 1959). In the hilly area around the town of Safad, Upper Galilee, these two chromosomal forms are widely and continuously spread. On soft Senonian chalks here, diploid populations have been found to occupy north-facing slopes, while on ridges and on south-facing slopes they are replaced by tetraploid plants (Nur and Zohary 1959). Close contacts between populations are frequent.

A search for natural triploids was conducted in such a place of contact about 200 meters north of Safad. In this location, the contact between the diploid and tetraploid populations was found on a west-facing slope and it consisted of a narrow belt, where diploid and tetraploid plants were found growing together in a mixed population.

This area, about 100 m wide and 150 m long was screened for triploids in the spring of 1957. Triploids in *Dactylis* have already been shown to be highly male-sterile, resulting in non-dehiscence of their anthers (Müntzing 1937). Plants in this area were therefore examined for this morphological trait. Out of about 2000 plants checked, 27 were found to have non-dehiscent anthers.

The next step was cytological examination (meiosis in P.M.C.'s) of the non-dehiscent plants. This was possible in 15 plants. 8 of these were found to be diploids, 4 tetraploids and 3 triploids (No. 1 to No. 3).

A similar check was done in another contact about 300 m east of the previous location. 9 non-dehiscent plants were found, 4 of them triploids (No. 4 to No. 7).

All triploids were quite vigorous and normal in appearance and could not be morphologically distinguished from the adjacent diploids and tetraploids. Careful cytological examination was carried out in the first three triploids. Each was found to have exactly 21 normal chromosomes. In their meiotic behavior these three triploids agree closely with the figures described by Müntzing (1937) in *Dactylis aschersoniana* × *D. glomerata* triploids.

(2) *Seed-set in Triploids.* Two of the triploid plants (No. 1 and No. 2) were marked and left undisturbed in the field. The location of triploid No. 2 was more or less in the middle of the contact zone, while triploid No. 1 grew already in a predominately diploid population. These two triploids were visited again at the end of May 1957 and examined for natural seed-set. Seed-set in the two plants was approximately 5 per cent as compared to values of 25–30 per cent obtained in adjacent diploid and tetraploid plants.

(3) *Progeny of the Triploids.* The seed which was collected from the two triploids was planted in Jerusalem in October 1957. Seed from triploid No. 1 gave rise to 14 seedlings and that of plant No. 2 to 19 seedlings. In both collections, germination was very uneven and already in the seedlings a wide variation in morphology and development was observed. About one third of the progeny showed very weak or retarded growth, which was accompanied in several cases by morphological abnormalities such as very slender shoots, tiny leaves, curling of leaves and tillers. Some of these abnormal seedlings died in early stages of development.

By April 1958 only 23 of the progeny survived. The larger proportion of these were vegetatively normal and well developed plants who came to flower in May and early June. Several other plants looked vigorous vegetatively but produced few or no flowering stalks. Some of the abnormal seedlings with the small, tender leaves and tillers still survived and grew slowly but again did not come to flower. Summary of the observations on the progeny of the triploids is given in tables 1 and 2.

(4) *Cytology of the Progeny of the Triploids.* Chromosome counts made it

TABLE 1. *Progeny from Triploid No. 1*

Plant number	Chromosome number	Morphological characterization	Per cent pollen abortion	Per cent seed-set
DC891–1	28 + 2B	vigorous, many inflorescences	36	32
DC891–2	28	vigorous, many inflorescences	39	44
DC891–3	28 + B	vigorous, few inflorescences	28	30
DC891–4	28	vigorous, few inflorescences	37	32
DC891–5	—	very weak, did not flower	—	—
DC891–6	35 (circa)	weak, only one inflorescence	—	—
DC891–7	28	vigorous, few inflorescences	21	20
DC891–8	28	vigorous with several delicate inflorescences	32	15

possible to classify the progeny of the triploids into the following groups:

(a) Tetraploids. Twelve plants were found to be tetraploids (see tables 1 and 2). Of these, nine plants showed strict tetraploid ($2n = 28$) chromosome number. The other three had, in addition to the normal tetraploid chromosome set, an extra one or two supernumerary chromosomes (fig. 1) and one (DC 892–

15), an extra normal chromosome as well.

Meiosis figures in the tetraploid progeny were very similar to what is observed in regular tetraploid *Dactylis* material (Müntzing 1937, Zohary 1956). Supernumerary chromosomes, when present, resemble closely, in their form (size and staining) and their behavior (precocious movement at M_I, lagging and precocious

TABLE 2. *Progeny from Triploid No. 2*

Plant number	Chromosome number	Morphological characterization	Per cent pollen abortion	Per cent seed-set
DC892–1	28	vigorous, many inflorescences	24	18
DC892–2	28	vigorous, many inflorescences	19	24
DC892–3	35	quite weak vegetatively, broad leaves, only one inflorescence	—	—
DC892–4	35 (circa)	fairly vigorous, few inflorescences	27	26
DC892–5	28	vigorous, fairly many inflorescences	25	4
DC892–6	16	vigorous vegetatively, did not flower	—	—
DC892–7	35	vigorous vegetatively, only one inflorescence	18	—
DC892–8	18	fairly vigorous vegetatively, only one inflorescence	—	—
DC892–9	15	fairly vigorous, only tillers, did not flower	—	—
DC892–10	28	fairly vigorous, few inflorescences	40	17
DC892–11	28	fairly vigorous, many inflorescences	28	28
DC892–12	35	fairly vigorous vegetatively, only one inflorescence	37	—
DC892–13	35	vigorous, leaves and tillers coarse, few inflorescences	15	7
DC892–14	16	weak, did not flower	—	—
DC892–15	29 + B	vigorous, many inflorescences	45	31

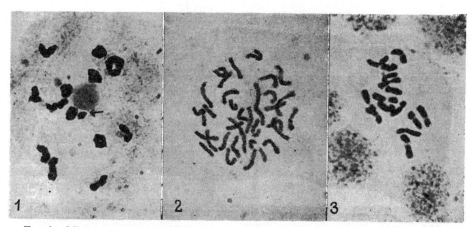

FIG. 1. Microsporocyte in diakinesis in plant DC 891–3 (2n = 28 + B), showing 1 quadrivalent, 12 bivalents and (arrow) a single supernumerary chromosome (× circa 750).

FIG. 2. Metaphase in dividing shoot meristem cell in plant DC 892–7 with 2n = 35 (× circa 750).

FIG. 3. Metaphase in dividing shoot meristem cell in plant DC 892–9 showing 15 chromosomes (× circa 1100).

division at A_I), the B chromosomes which are common in the adjacent diploid population (Zohary and Ashkenazi 1958). The existence of these supernumerary chromosomes was verified also in mitosis of shoot meristems.

(b) Pentaploids. Six plants were classified as pentaploids (see tables 1 and 2). In three of these plants (DC 892–3, DC 892–12 and DC 892–13) good anaphase I figures have been obtained and the 35 elements could be clearly and definitely counted. In another plant (DC 892–4), where meiosis figures were sticky, the chromosome number was estimated to be around 35 (not less than 33). However, it was impossible to verify definitely whether this plant had exactly the pentaploid chromosome number. In plants DC 891–6 and DC 892–7, which almost did not flower, chromosomes were counted in mitosis of shoot meristems (fig. 2).

Morphologically all the six pentaploids were characterized by somewhat coarser leaves and tillers. Only two of them were really vigorous. The development and flowering of the four other plants was much weaker.

(c) Aneuploids. Four aneuploid plants were found among the progeny of the triploids (see table 2), with chromosome numbers of 15, 16 and 18 (fig. 3).

Morphologically all four plants belong to the group of the less vigorous progeny. No one of them flowered.

(d) Plants with Undetermined Chromosome Number. In a single plant, examination of chromosome number was unsuccessful. It was a weak, non-flowering plant (see table 1), in which dividing cells in shoot meristems were very scarce. It is suspected that this plant too, has an aneuploid chromosome number.

DISCUSSION

The most likely explanation for the presence of tetraploid and pentaploid plants among the progeny of the triploids is the assumption that triploids produce an appreciable number of unreduced eggs. When such 3x eggs are fertilized by pollen from adjacent diploid plants, they give rise to tetraploids; when fertilized by pollen from tetraploid plants, they produce pentaploids (see fig. 4). Already Müntzing (1937) assumed production of unreduced eggs to explain occurrence of some pentaploid progeny when crossing triploid *D. glomerata* × *aschersoniana* back to tetraploid *D. glomerata*.

The assumed production of unreduced eggs is strongly supported by another line of evidence—the presence of supernumerary chromosomes in some of the tetraploid progeny. Supernumerary chromosomes have been found to be quite common in diploid *Dactylis* populations in Israel. The diploid population, adjacent to the triploids studied, was extensively investigated (Zohary and Ashkenazi 1958) and found to have a frequency of 0.73 supernumerary chromosomes per plant. In contrast with the diploid material, our cytological survey of *Dactylis* in Israel (Nur and Zohary 1959) revealed no such accessory chromosomes in tetraploid populations. The only exception encountered was a single tetraploid plant near the contact zone itself. Therefore, in the present case, supernumerary chromosomes can be regarded as a cytological marker of the diploid material. They were not found in the two parent triploids. They could only have reached their tetraploid progeny in *haploid* pollen of diploid plants.

Since a large proportion of their viable progeny are tetraploids, triploid *Dactylis* in the contact areas can be regarded as an efficient one-way bridge for gene flow from diploids to tetraploids. If the assumption of unreduced eggs is correct, the resulting tetraploids receive two chromosome sets from the donor diploid population (see scheme in fig. 4, a). The normal development and fertility observed in the polyploid progeny (see tables 1 and 2) is another point that should be stressed. Thus the newly created tetraploids should survive and compete well with ordinary tetraploids.

The present data indicate, therefore, that the reproduction barriers between diploid and tetraploid populations in *Dactylis* are not as complete as was previously assumed in cases of polyploids and that one-way gene flow from diploids to tetraploids does exist. To these indications one has to add still another theoretical possibility of a similar gene flow, in contact areas, brought by occasional formation of unreduced gametes in the diploid plants themselves and their union with 2n gametes produced by tetraploid plants. However, this possibility has yet to be demonstrated. Another point worth mentioning is the full accordance of these data with the varia-

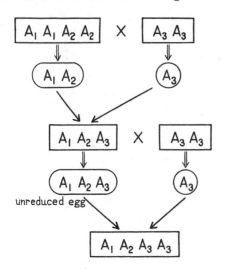 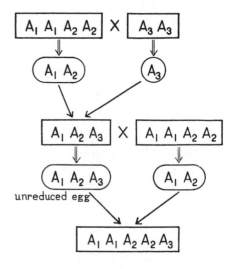

a. TETRAPLOID PROGENY   b. PENTAPLOID PROGENY

FIG. 4. Schematic representation of the production by triploids of *a.* tetraploid progeny; *b.* pentaploid progeny.

tion patterns found in the diploid and tetraploid populations in areas where they come in contact. In many such cases (Zohary 1956) tetraploid morphology largely overlaps the diploids and separation of the two chromosomal levels from one another on a morphological basis is impossible.

These indications of incomplete genetic barriers between the two chromosomal levels in *Dactylis* are important for general considerations of the mode of evolution of tetraploid superstructures. Two main evolutionary models have been suggested to account for the formation of large, polyploid superstructures such as found in *Dactylis glomerata*.

The first model explains the origin of the superstructure by strict autopolyploidy of its diploid progenitors, followed by fusion on the tetraploid level. This mode of origin seems highly unlikely on the basis of available information (for details see Stebbins 1947, 1950).

The other mode of origin was suggested by Stebbins (1950) who assumed the formation of the tetraploid superstructure as a result of various intervarietal crosses between distinct diploid forms, followed by chromosome doubling in the hybrids and fusion of such initial tetraploid populations into a continuous superstructure.

A serious difficulty in the last model is the need to assume a large number of contacts between the diploids contributing to the polyploid superstructure. Theoretically every diploid "pillar" should have come in contact and hybridized with at least one other diploid. Geographically and historically in some cases this is difficult to visualize.

If the reproductive barriers between the diploid and tetraploid levels are incomplete, the large continuous superstructure needs not necessarily be the result of fusion of a large number of tetraploid forms, each of independent origin. It could have evolved from a small number of initial intervarietal tetraploid combinations which spread to new regions

and incorporated genetic material from additional diploid forms via triploids. The work of Heiser (1949, 1951) in *Helianthus annuus* has demonstrated that introgression from related species can greatly enrich the adaptive range of a given species and facilitate its rapid spread. Since the reproductive barrier between diploid and tetraploid levels in *Dactylis* is likewise not absolute, a similar process could have operated here and greatly assisted in the spread of the initial intervarietal polyploids. The advantage of such a model is that one does not need to postulate a large number of direct contacts between the diploid forms themselves to account for the production of all tetraploid combinations necessary to cover the entire range of variation met in the superstructure.

Another point worth mentioning is that introgression via triploids could operate only in cases of autopolyploid chromosome behavior. It could not work in an allopolyploid genetic system. If found of general occurrence in evolutionary successful autopolyploid complexes, it should system.

In conclusion, the present information strongly suggests that the model of the polyploid complex, as presented by Babcock and Stebbins (1938) and Stebbins (1950) is yet a simple one, and that in *Dactylis glomerata,* and possibly also in other sexual polyploid groups, the whole complex should be regarded as genetically and taxonomically a much more closely knit unit than has been previously assumed.

SUMMARY

Natural triploids occur in contact areas between diploid and tetraploid populations of *Dactylis glomerata* in Israel and such triploids set seed in their natural habitat.

The triploids apparently produce a large proportion of unreduced eggs. Fertilization of such eggs by haploid pollen (from diploid plants) and diploid pollen (from tetraploid plants) results m the formation of tetraploids and penta-ploids respectively.

By producing a large proportion of vigorous tetraploid progeny, triploids can serve as an efficient bridge for one-way gene flow from diploid level to tetraploid level. The evolutionary significance of such type of introgression for the construction of tetraploid superstructure in polyploid complexes is pointed out.

Acknowledgments

The authors are indebted to Ford Foundation for a research grant which partly supported this study. We also wish to express our thanks to Miss A. Horovitz for growing the plants, and to Mr. David Apirion for assistance in examination of pollen abortion and seed-set.

Literature Cited

Anderson, E. 1953. Introgressive hybridization. Biol. Rev., **28**: 280–307.

Babcock, E. B. and G. L. Stebbins Jr. 1938. The American species of *Crepis:* their relationships and distribution as affected by polyploidy and apomixis. Carnegie Inst. Washington, Publ. No. 504, 200 pp.

Heiser, C. B. 1949. Study in the evolution of the sunflower species *Helianthus annuus* and *H. Bolanderi*. Univ. Calif. Publ. Bot., **23**: 157–208.

——. 1951. Hybridization in the annual sunflowers: *Helianthus annuus* × *H. debilis* var. *cucumerifolius*. Evolution, **5**: 42–51.

Müntzing, A. 1937. The effects of chromosomal variation in *Dactylis*. Hereditas, **23**: 113–235.

Nur, U. and D. Zohary. 1959. Distribution patterns of diploid and tetraploid forms of *Dactylis glomerata* L. in Israel. Bull. Res. Counc., Israel, Sect. D: Botany, **7D**: 13–22.

Stebbins, G. L. Jr. 1947. Types of polyploids: their classification and significance. Advances in Genetics, **1**: 403–429.

——. 1950. Variation and evolution in plants. New York, Columbia Univ. Press, 641 pp.

—— and D. Zohary. 1959. Cytogenetic and evolutionary studies in the genus *Dactylis*. I. The morphology, distribution and interrelationships of the diploid subspecies. Univ. Calif. Publ. Bot. (in press).

Zohary, D. 1956. Cytogenetic studies in the polyploid complex of *Dactylis glomerata* L. Ph.D. thesis. Univ. of California, Berkeley.

—— and I. Ashkenazi. 1958. Different frequencies of supernumerary chromosomes in diploid populations of *Dactylis glomerata* in Israel. Nature, **182**: 477–478.

13

Copyright ©1974 by the British Genetical Society
Reprinted from *Heredity* **32**:171–181 (1974)

INTROGRESSION FROM TETRAPLOID DURUM WHEAT TO DIPLOID *AEGILOPS LONGISSIMA* AND *AEGILOPS SPELTOIDES*

ALIZA VARDI

Agricultural Research Organization, Volcani Center, Bet Dagan, Israel

Received 14.iii.73

SUMMARY

Tetraploid to diploid introgression by means of interspecific triploid hybrids was followed in two species combinations:

(i) *T. durum* × *Ae. longissima ABSˡ* combination.

(ii) *T. durum* × *Ae. speltoides ABS* combination.

Triploids set rare seeds when exposed to massive back pollination by their diploid parents.

Third-hybrid generation progenies of both combinations were diploid $2n = 14$ or almost diploid $2n = 15$ or 16, with high fertility, and bore a close morphological resemblance to their respective diploid parent.

1. INTRODUCTION

IN previous papers (Vardi and Zohary, 1967; Vardi, 1970) evidence was presented to indicate that, in wheats, genetic variation can be enriched through introgression from diploid to polyploid species. Furthermore, it was demonstrated that interspecific triploid hybrids serve as an effective bridge in such gene transfer.

The present study assesses the reverse possibility, namely, gene flow from tetraploid to diploid entities. An indication that such a process can actually take place has been shown in the work of Kihara (1954), who backcrossed the triploid hybrid *Aegilops caudata* × *Aegilops cylindrica* to its diploid parent and obtained several seeds. Our aim was to examine the possibility of this type of transfer on a larger scale by using two related diploid species, *Aegilops longissima* and *Aegilops speltoides*. These two species differ in their effect on meiotic pairing when crossed with tetraploid or hexaploid wheat (Riley, 1960).

2. METHODS

Meiosis in P.M.C.s served both for the determination of chromosome numbers and for the study of chromosome pairing. Anthers were fixed in 3:1 alcohol-acetic acid for 24 hours, stored in 70 per cent alcohol and stained in acetocarmine.

Since chromosome numbers varied from plant to plant, pairing was expressed not only in values of " chiasmata per cell ". The values of " association per chromosome " were also used, *i.e.* twice the number of chiasmata per cell divided by the number of chromosomes in the examined plant. In the parental lines, the F_1 triploid, the F_2, and the majority of F_3 plants, this value was based on examination of 30 randomly picked metaphases.

Contribution from the Volcani Center, Agricultural Research Organization, Bet Dagan, Israel. 1973 Series, No. 120-E.

Pollen fertility was determined by dissecting mature anthers soaked in 4 per cent acetocarmine and scoring about 500 pollen grains per plant. Grains were considered normal when they were rounded and well stained.

Seed fertility was determined by examination of two lower florets in the spikelet. A floret was considered fertile if a well-developed kernel was found in it. In the more fertile plants a sample of 100 florets (*i.e.* 50 spikelets) was employed. In semi-sterile plants, and particularly in the triploids themselves, seed set was determined by examination of all available spikes.

3. Experimental procedures and results

The following two interspecific triploid combinations were produced and utilised:

    (i) *ABS^l* combination:
        Triticum durum (AABB) × *Aegilops longissima (S^lS^l)*

    (ii) *ABS* combination:
        Triticum durum (AABB) × *Aegilops speltoides (SS)*.

As shown by Sarkar and Stebbins (1956) and Riley *et al.* (1958), genome S is closely related to genome B.

These two combinations are identical to *ABS^l* (low pairing) and *ABS* (high pairing) triploids employed previously (Vardi and Zohary, 1967; Vardi, 1970) in the study of introgression from diploid to tetraploid wheat. Each F_1 triploid hybrid combination was planted intermixed with its diploid parental lines and isolated from other *Triticum* or *Aegilops* species. The functional male-sterile F_1 hybrids were thus exposed to backcross pollination by their diploid parent. This open-pollination was supplemented by some artificial back-pollination. Only a few backcross seeds were obtained from the highly sterile triploids. Second-generation hybrid derivatives raised from these seeds were interplanted with their parents in order to promote seed production. *Ae. longissima* was used as pollen donor for *durum* × *longissima* derivatives, and both *T. durum* and *Ae. speltoides* were used to provide pollen for *durum* × *speltoides* derivatives. Because some second-generation plants were semi-fertile, the seeds produced may possibly have been a mixture of selfing and second backcross products. Selected families of third-generation hybrid derivatives were also grown. Fig. 1 illustrates the experimental design employed.

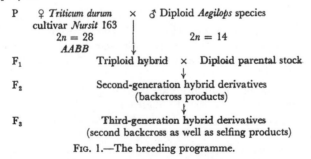

FIG. 1.—The breeding programme.

The results obtained can be summarised as follows:

(i) *Introgression from* T. durum *to* Ae. longissima

(a) *F_1 triploid ABS^l hybrids*

Four ABS^l triploid plants were used. The present triploids were vegetatively vigorous and meiosis was characterised by an almost total lack of pairing (table 1), the number of chiasmata, 1·20 per cell, being very low. This behaviour has been previously described in plants representing a similar genome combination (Vardi and Zohary, 1967). Anthers in all four plants failed to dehisce and contained about 1 per cent of stainable pollen. However, back-pollination to the diploid *Ae. longissima* resulted in an occasional set of well-developed seeds.

(b) *Second-generation hybrid derivatives*

Fifteen backcross products were examined. Morphologically, most plants were intermediate between the F_1 triploid hybrid and typical *Ae. longissima*. However, some F_2 plants bore a close resemblance to their *longissima* parental line.

Chromosome numbers in the second hybrid generation varied from $2n = 22$ to $2n = 28$ (table 1). Eleven plants were tetraploid or almost tetraploid and contained 26-28 chromosomes. Twelve plants displayed five or six bivalents in meiosis (table 1). Furthermore, some F_2 individuals had an occasional trivalent or quadrivalent. Pollen and seed fertility in the second generation was low and several plants failed to set seed (table 1). It is, however, of interest to note the relatively high seed fertility of plant no. 6859-57 ($2n = 23$). Although no pollen counts were made in this plant, the dehiscence of its anthers indicated low pollen abortion. This plant was used as seed parent to most of the F_3 progeny analysed.

The cytological data thus corroborate our previous findings (Vardi and Zohary, 1967) that unreduced or almost unreduced female gametes are, in the main, the functional gametes produced by ABS^l triploids. In other words, the data on chromosome number and meiotic pairing indicate that the majority of F_2 plants obtained from $ABS^l \times S^lS^l$ backcrosses were roughly of an ABS^lS^l constitution. The occasional formation of trivalents and quadrivalents in the F_2 plants is also noteworthy. This indicates the occurrence of chromosomal exchange during meiosis in the triploid, presumably between homoeologous chromosomes.

(c) *Third-generation hybrid derivatives*

A sample of 29 plants, derived from plant 6859-57 ($2n = 23$) and an additional single plant obtained from plant 6859-51 ($2n = 28$), were examined. All F_3 plants resembled *Ae. longissima* very closely.

As seen from table 3, all third-generation derivatives had lower chromosome numbers than plants of the second generation. Twenty-seven plants were diploid ($2n = 14$) and three were almost diploid ($2n = 15$). Chromosome pairing in the $2n = 14$ plants was normal, *i.e.* there were seven bivalents with 1·86-1·98 chiasmata per chromosome. Plants with a $2n = 15$ number, that produced a single trivalent, may have been trisomic for the additional chromosome. However, $2n = 15$ plants that displayed seven bivalents plus a single univalent, probably contained an additional chromosome of the *A* or *B* genome. In contrast to the second generation, high restoration of pollen fertility (table 5) and marked recovery in seed set (table 6) were achieved.

209

TABLE 1

Backcross of durum × longissima ABS¹ triploid to its diploid parent: cytology and fertility of F₁ hybrid and second-generation derivatives

Accession No.	Chromosome No. (2n)	No. of cells examined	Chromosome association in metaphase I				Associations per chromosome	Fertility	
			Univalents	Bivalents	Trivalents	Quadrivalents		Per cent normal pollen	Per cent seed-set
F₁ Triploid hybrid									
T. durum × Ae. longissima (genomes ABS¹)									
6859-1	21	46	18·63 (11-21)	1·15 (0-5)	0·02 (0-1)	—	0·11	1·40	4·10
F₂ Second hybrid generation									
6859-3	28	30	16·00 (12-21)	5·60 (2-8)	2·26 (0-2)	—	0·54 ± 0·12	2·40	0
6859-17	28	30	13·93 (11-18)	5·80 (3-8)	0·73 (0-3)	0·06 (0-1)	0·66 ± 0·12	7·60	0·35
6859-26	26	30	10·16 (5-14)	6·50 (3-10)	6·83 (0-1)	0·13 (0-1)	0·93 ± 0·14	0	0
6859-28	27	30	13·00 (7-17)	5·66 (2-10)	0·80 (0-2)	0·06 (0-1)	0·74 ± 0·18	4·18	0
6859-29	28	20	17·75 (14-24)	5·00 (2-8)	0·05 (0-1)	—	0·59 ± 0·33	8·38	0
6859-35	26	30	11·00 (5-16)	6·46 (2-9)	0·60 (0-3)	0·06 (0-1)	0·87 ± 0·02	3·00	0
6859-40	22	23	17·30 (14-20)	2·34 (1-4)	—	—	0·25 ± 0·09	—	—
	—	—					—	0	0·23
6859-42	28	30	10·80 (6-14)	6·90 (4-10)	1·13 (0-3)	—	0·89 ± 0·12	4·55	0
6859-43	25	30	20·16 (17-23)	2·16 (0-5)	0·16 (0-1)	—	0·23 ± 0·10	0	1·25
6859-51	28	30	13·13 (9-18)	6·73 (5-9)	0·46 (0-1)	—	0·81 ± 0·15	3·00	0·54
6859-52	27	30	12·70 (8-16)	6·26 (4-9)	0·50 (0-2)	0·06 (0-1)	0·73 ± 0·14	0·80	0·38
6859-53	26	30	12·86 (9-16)	5·90 (3-8)	0·40 (0-2)	0·03 (0-1)	0·78 ± 0·13	3·60	0·22
6859-55	28	30	12·10 (8-17)	6·63 (4-10)	0·83 (0-2)	0·03 (0-1)	0·84 ± 0·16	3·00	0
6859-57	23	11	10·09 (5-13)	6·45 (5-7)	—	—	1·08 ± 0·17	*	16·70

* Relatively high pollen fertility was indicated by anther dehiscence in this plant.

TABLE 2

Backcross of durum × speltoides ABS triploid to its diploid parent: cytology and fertility of F₁ Hybrid and second-generation derivatives

Accession No.	Chromosome No. (2n)	No. of cells examined	Chromosome association in metaphase I				Associations per chromosome	Fertility	
			Univalents	Bivalents	Trivalents	Quadrivalents		Per cent normal pollen	Per cent seed-set
F₁ Triploid hybrid									
T. durum × Ae. speltoides (genomes ABS)	21	44	5·07 (1-9)	4·62 (0-5)	2·50 (0-5)	—	1·03	2·20	0·31
F₂ Second hybrid generation									
6860-2	21	30	4·76 (2-9)	4·56 (2-7)	2·33 (0-4)	0·03 (0-1)	1·22±0·13	5·40	0·11
6860-6	21	30	5·06 (2-8)	4·43 (2-7)	2·13 (0-2)	0·16 (0-2)	1·12±0·12	25·05	0·74
6860-7	21	30	9·40 (6-13)	5·00 (3-7)	0·53 (0-2)	—	0·75±0·13	0·60	0
6860-8	21	30	5·53 (3-8)	5·43 (3-7)	1·50 (0-4)	—	1·18±0·11	26·20	21·42
6860-9	21	30	6·13 (3-10)	5·83 (4-8)	1·66 (0-3)	—	1·13±0·14	19·00	0·53
6860-10	21	11	4·50 (2-7)	5·66 (3-8)	1·50 (0-3)	0·16 (0-1)	1·22±0·18	24·10	0·70
6860-12	22	30	6·71 (4-12)	5·90 (4-8)	0·93 (0-3)	0·13 (0-1)	1·02±0·16	23·40	0·44

In summary, the third generation of hybrid derivatives of the ABS^l combination showed an almost complete stabilisation at the diploid level. This rapid establishment of the diploid chromosome number appears to be due to a striking elimination of redundant chromosomes during meiosis of F_2 plants. Presumably the 14 chromosomes in third-generation derivatives represent more or less the complete genome of *Ae. longissima*.

(ii) *Introgression from* T. durum *to* Ae. speltoides

(a) *F_1 triploid ABS hybrid*

Three *ABS* triploid plants were employed. As in previously studied *ABS* triploids (Vardi, 1970), meiosis was characterised by four to seven

TABLE 3

Frequency distribution of chromosome numbers in third hybrid generation

Triploid combination and accession numbers	Chromosome No. in F_2 parent	No. of F_3 plants examined	14	15	16	...	21	22	23	24	25	26	27	28	29
(i) *T. durum × Ae. longissima*															
6859-51	28	1	1	—	—		—	—	—	—	—	—	—	—	—
6859-57	23	29	27	2	—		—	—	—	—	—	—	—	—	—
(ii) *T. durum × Ae. speltoides*															
6860-6	21	12	1	2	—		4	2	—	1	1	—	—	—	1
6860-8	21	53	39	13	1		—	—	—	—	—	—	—	—	—

bivalents and the frequent occurrence of trivalents (as many as five trivalents in some microsporocytes). The relatively high frequency of trivalents is probably due to a suppression of the 5B diploidisation effect by *Ae. speltoides* chromosomes (Riley and Chapman, 1964).

Also in the present experiment, *durum × speltoides* triploids were functionally completely male-sterile and their anthers did not dehisce. Yet, massive exposure to *Ae. speltoides* pollen resulted in the formation of seven well-developed backcross seeds.

(b) *Second-generation hybrid derivatives*

All seven second-generation plants resembled their F_1 triploid parents in general morphology, although some variation between plants was obtained. Six of the seven plants had a triploid chromosome number ($2n = 21$), while one had a number of $2n = 22$ (table 2). Pairing of chromosome in the F_2 plants was very similar to that observed in the original *durum × speltoides* F_1 triploids. But, significantly, in four of the seven F_2 plants, one to two quadrivalents were also detected (table 2). Another conspicuous difference between the F_1 and second-generation hybrid derivatives was the partial restoration of pollen fertility (table 2). In some F_2 plants anthers dehisced partially. One of these male-fertile plants, 6860-8, also showed a remarkable recovery of seed fertility (table 2).

Although the triploid chromosome number was maintained in the individuals of the second generation, the cytological data for F_1 triploids

TABLE 4

Chromosome pairing in third hybrid generation: frequency distribution of " association per chromosome " values

Triploid combination and accession numbers	Chromosome No. in F₂ parent	No. of F₃ plants examined	Association per chromosome														
			0.51-0.60	0.61-0.70	0.71-0.80	0.81-0.90	0.91-1.00	1.01-1.10	1.11-1.20	1.21-1.30	1.31-1.40	1.41-1.50	1.51-1.60	1.61-1.70	1.71-1.80	1.81-1.90	1.91-2.00
(i) *T. durum* × *Ae. longissima*																	
6859-51	28	1	—	—	—	—	—	—	—	—	—	—	—	—	1	—	—
6859-57	23	28	—	—	—	—	—	—	—	—	—	—	—	—	1	7	20
(ii) *T. durum* × *Ae. speltoides*																	
6860-6	21	10	—	3	2	1	—	1	1	—	1	1	—	—	—	1	—
6860-8	21	45	—	—	—	—	—	—	—	—	—	—	7	9	14	14	—

suggest that the genomic constitution of the F_2 derivatives of these triploids is the outcome of some chromosome reshuffling between the original parental genomes. This means that second-generation progenies are already the products of homoeologous recombination which occurred during meiosis of the original F_1 triploids. Furthermore, the present results are in full agreement with previous data on *ABS* triploids (Vardi, 1970). Here, as in the

TABLE 5

Frequency distribution of pollen-fertility values scored in third hybrid generation

Triploid combination and accession numbers	Chromosome No. in F_2 parent	No. of F_3 plants examined	Pollen-fertility classes (in per cent)							
			0–5	6–15	16–30	31–45	46–60	61–75	76–90	91–100
(i) *T. durum* × *Ae. longissima*										
6859-51	28	1	—	—	—	—	—	—	1	—
6859-57	23	28	—	—	—	—	—	—	5	23
(ii) *T. durum* × *Ae. speltoides*										
6860-6	21	8	2	5	1	—	—	—	—	—
6860-8	21	37	—	—	—	1	1	6	14	15

TABLE 6

Frequency distribution of seed-set values scored in third hybrid generation

Triploid combination and accession numbers	Chromosome No. in F_2 parent	No. of F_3 plants examined	Seed-set classes (in per cent)							
			0–5	6–15	16–30	31–45	46–60	61–75	76–90	91–100
(i) *T. durum* × *Ae. longissima*										
6859-57	23	27	—	3	1	3	7	6	5	2
(ii) *T. durum* × *Ae. speltoides*										
6860-6	21	10	9	1	—	—	—	—	—	—
6860-8	21	47	—	2	2	4	10	14	14	1

previously examined triploids, it is obvious that the main type of functional gamete produced by *durum* × *speltoides* triploids has roughly an *AB* constitution.

(c) *Third-generation hybrid derivatives*

Two families of the third generation were grown and analysed. Twelve plants were derived from seed parent 6860-6 and 57 plants from the most fertile of the F_2 plants, 6860-8. Chromosome numbers in the family derived from plant 6860-6 varied from 14 to 29 (table 3). This was expected, since the seed parent 6860-6 was explosed to both *T. durum* and *Ae. speltoides* pollen and was itself partially male fertile. Nine of the derivatives had a high number of chromosomes (21-29) and can be considered products of crosses to $n = 14$ pollen. Morphologically, they resembled the triploid hybrid more than they did *Ae. speltoides* and were marked by high sterility (tables 5, 6).

Three plants had chromosome numbers of 14-15. They showed six or seven bivalents in meiosis plus an additional trivalent or univalent. Morphologically, they resembled *Ae. speltoides*. These plants can be considered as products of crosses to $n = 7$ or $n = 8$ pollen.

In contrast to the family derived from F_2 plant 6860-6, in which a wide range of chromosome numbers was segregated, the family derived from F_2 plant 6860-8 contained mainly diploid $2n = 14$, or almost diploid $2n = 15$ or 16, individuals (table 3). Plants of this family must be considered to have arisen from crosses to $n = 7$ or $n = 8$ pollen. Since anthers in the 6860-8 parent dehisced, most of the progeny were presumably products of self-pollination. Chromosome pairing in $2n = 14$ plants was considerable but not completely regular ($11 \cdot 07$-$13 \cdot 03$ chiasmata per cell). Most of the $2n = 15$ plants were apparently trisomic and a single trivalent was observed frequently. Pollen fertility and seed set were normal or almost normal (tables 5 and 6). All the progenies of 6860-8 resembled *Ae. speltoides* rather closely in general growth habit, morphology and ear shape.

In summary: as in the previous combination ABS^l, the third-generation hybrid derivatives ABS combination also showed an almost complete stabilisation at the diploid level. This was unexpected since pairing behaviour of triploid F_2 plants of the ABS combination still resembled that of the triploid F_1. The high pairing observed in F_1 and F_2 ABS plants indicated that chromosome exchange was taking place and thus the *speltoides*-like diploid ($2n = 14$) individuals obtained in F_3 were presumably introgressants.

4. Discussion

The data obtained suggest that tetraploid to diploid introgression can occur in wheats. As in diploid to tetraploid introgression (Vardi and Zohary, 1967; Vardi, 1970), the occasional gametes produced in the triploid F_1 hybrid, which are genomically more or less balanced, are necessary for this process. When such ABS^l gametes are produced in *T. durum* × *Ae. longissima* triploids and roughly AB gametes in *T. durum* × *Ae. speltoides* triploids, products of the backcross to the respective diploid parent will be mainly tetraploid, or almost tetraploid, ABS^lS^l plants in the first combination and more or less triploid ABS plants in the second combination. Rare trivalents in F_2 samples from the ABS^l combination and quadrivalents in the samples of the ABS combination indicate that intergenomic chromosome exchange has occurred during the meiosis of the F_1 triploid hybrids (tables 1 and 2). It is assumed that these exchanges take place between homoeologous chromosomes. One of the main subsequent problems in tetraploid to diploid introgression is an excess of chromosomes which have to be eliminated before stabilisation at the diploid level can occur.

(i) In the *durum* × *longissima* combination, complete or almost complete elimination of A and B genomes is achieved already in F_3. Stabilisation at the diploid level is indicated by the full or almost full fertility of F_3 plants and their close morphological resemblance to *Ae. longissima*. In an F_2 plant, 6857-59 ($2n = 23$), which gave rise to viable diploid F_3 products, cytological data (table 1) indicate that there were only relatively minor donations, if any, from the A and B genomes of wheat. As seen from table 1, this F_2 plant produced only bivalents and univalents.

215

Another *durum* × *longissima* F$_2$ plant, 6859-51 ($2n = 28$), showed an average of 0·46 trivalent per cell in M_I (table 1). The single third-generation derivative obtained from F$_2$ plant 6859-51 differed morphologically from the 6859-57 family. This segregant bore close resemblance to a local single-awned or asymmetrically awned form of *Ae. longissima*, which occurs sporadically on both sides of the Jordan Rift Valley (Judean Hills, Moab, Gilead). This wild *Ae. longissima* form differs from the typical *Ae. longissima* described by Eig (1929) in its more delicate spike morphology and in its single or asymmetrical awns which carry two lateral teeth on their adaxial side. It is noteworthy that some of the sites of this *longissima* morph are areas in which *Ae. longissima* and *T. dicoccoides* form mixed populations. This wild morph may represent an established product of an introgression process similar to that witnessed under experimental conditions in the 6859-51 line.

(ii) In the *durum* × *speltoides* combination there is a parallel rapid elimination of chromosomes and the main products in the third hybrid generation are diploid ($2n = 14$) or almost diploid ($2n = 15$ or 16). Here the F$_3$ plants resemble typical *Ae. speltoides* morphologically. This indicates a relatively small donation from the *A* genome.

Homoeologous exchanges in the *ABS* F$_1$ triploid may be a contributing factor to the relatively low fertility in the second hybrid generation. This is without doubt one of the reasons why the various F$_2$ plants ($2n = 21$ and $2n = 22$) showed variation in chromosome pairing and fertility (table 2). Because of these homoeologous exchanges in F$_1$ *ABS* triploids, it is impossible to determine the genomic constitution of F$_2$ plants through cytological observations, as is possible in most *ABS*^{l} F$_2$ plants. Without information on the genomic make-up of F$_2$ plants it is difficult to account for diploidy in the third hybrid generation. Diploid F$_3$ plants can be the result of either genomic segregation or lethal duplication which leads, during meiosis of the F$_2$ seed parent, to an elimination of chromosomes. Such processes may have led to diploids in the third-generation plants of the 6860-8 family.

That the *speltoides*-like 6860-8 F$_3$ plants bear a modified genome and differ from true *speltoides* plants is indicated by their lower chiasma frequency (11·00-13·03 chiasmata per cell as against 13·65 chiasmata per cell in *Ae. speltoides*) and by occasional univalents. Similar relatively low pairing was also observed in F$_4$ hybrids obtained from crosses between *speltoides*-like F$_3$ ($2n = 14$) individuals of families 6860-6 and 6860-8 and within family 6860-8. On the other hand, when the same *speltoides*-like F$_3$ parent plants were crossed with true *Ae. speltoides*, pairing in the resulting hybrids was fully normal. In 23 such hybrids, chiasma frequency per cell varied from 12·67 to 13·87. Thus, pairing was lower in the crosses between two recombined genomes (*speltoides*-like × *speltoides*-like) than in crosses between a recombined and a pure *Ae. speltoides* genome (*speltoides*-like × *Ae. speltoides*). Yet, in both series of crosses, 5B suppressor alleles were present, as was verified by crossing *speltoides*-like F$_3$ plants with *T. aestivum* (unpublished data). In other words, the *speltoides*-like individuals are evidently introgressants, but the pure *Ae. speltoides* genome cancels the pairing irregularities caused by introgression. However, in nature any such processes of introgression would result in crosses between true *Ae. speltoides* and *speltoides*-like introgressants, both being outcrossed. Within a few generations derivatives of such crosses would achieve normal pairing and be cytologically indistinguishable from *Ae. speltoides*.

The data presented confirm that interspecific triploid hybrids in the wheat (*Triticum-Aegilops*) group do act as bridges for tetraploid to diploid gene transfer. When such transfers, or transfers in the reverse direction, take place in nature the recombined products become almost indistinguishable morphologically and cytologically from their diploid or tetraploid parents within only a few generations. Such introgression enriches natural genetic variation. Thus, if the *B* genome of wheat is a modified genome, as suggested by Kimber and Athwal (1972), it is possible that the two diploid species used in this experiment contributed to the *B* genome. Tetraploid to diploid introgression may also explain the appearance of local forms, such as the *Ae. longissima* form described here, and the comparatively large morphological variation in diploids that are related to polyploids. Examples of such polyploid-diploid species pairs are *Ae. triuncialis* and *Ae. umbellulata*, *T. dicoccoides* and *T. boeoticum*, and *Ae. cylindrica* or *Ae. crassa* and *Ae. squarrosa*. Thus, polyploid to diploid introgression can enrich the very diploid which gave rise to the polyploid.

Acknowledgments.—The author is grateful to Drs A. Horovitz and M. Feldman for critical reading and for many helpful suggestions in the preparation of the manuscript.

5. References

EIG, A. 1929. Monographisch-kritische Uebersicht der Gattung *Aegilops*. *Rep. Spec. Nov. Reg. Veget. Beih.*, *55*, 79-81.

KIHARA, H. 1954. Consideration on the evolution and distribution of *Aegilops* species based on the analyser-method. *Cytologia*, *19*, 336-357.

KIMBER, G., AND ATHWAL, R. S. 1972. A reassessment of the course of evolution of wheat. *Proc. Natn. Acad. Sci. U.S.A.*, *69*, 912-915.

RILEY, R. 1960. The diploidisation of polyploid wheat. *Heredity*, *15*, 407-429.

RILEY, R., AND CHAPMAN, V. 1964. Cytological determination of the homoeology of *Triticum aestivum*. *Nature*, *203*, 156-158.

RILEY, R., UNRAU, I., AND CHAPMAN, V. 1958. Evidence on the origin of the B genome of wheat. *J. Hered.*, *49*, 91-99.

SARKAR, P., AND STEBBINS, G. L. 1956. Morphological evidence concerning the origin of the B genome of wheat. *Am. J. Bot.*, *43*, 297-304.

VARDI, A. 1970. Introgression from diploid *Aegilops speltoides* to tetraploid durum wheat. *Heredity*, *25*, 85-91.

VARDI, A., AND ZOHARY, D. 1967. Introgression in wheat via triploid hybrids. *Heredity*, *22*, 541-560.

14

Reprinted from pages 403–419 of *The Genetics of Colonizing Species*, H. G. Baker and G. Ledyard Stebbins, eds., Academic, New York, 1965

Colonizer Species in the Wheat Group

DANIEL ZOHARY*

DEPARTMENT OF BOTANY, THE HEBREW UNIVERSITY OF JERUSALEM, ISRAEL

The Old World belt of Mediterranean agriculture (southwestern Asia and the Mediterranean basin) furnishes one of the most conspicuous examples of widespread destruction of indigenous vegetation cover and its replacement by open, man-made formations. Since the neolithic agricultural revolution some 10,000 years ago, the prehistoric forest and steppe-forest formations in this belt have been largely destroyed by human activity (fire, axe, agriculture, and overgrazing by domestic stocks). They have been replaced by open shrub formations, steppe-like vegetation, and secondary segetal man-dependent cover. The herbaceous and shrubby plants which predominate in the landscape today have either enormously expanded their distribution in the last few millennia or are entirely newly evolved "weeds." Among the successful annual colonizers of these opened-up territories are several wild members of the wheat group.

The wild *Aegilops* and *Triticum* species thus offer us a critical case for our present attempt to clarify the genetic systems and evolutionary mechanisms involved in colonization of newly opened, man-made habitats. The "weedy" *Aegilops* species are still found side-by-side with ecologically specialized and geographically much more limited species. Most important, the wheat group is almost unique among weedy plants, as to the extent to which it has been studied cytogenetically. It has been also well investigated taxonomically and ecologically. Thus the basic elements of information necessary for evolutionary interpretation are available here and a synthesis can be attempted.

The aim of this paper is to review the main features of some two

* The author's study of the wheat group is aided by Grant FG-Is-129 from the United States Department of Agriculture.

dozen species of *Aegilops* and *Triticum,* contrasting weeds with nonweeds, and to outline the various elements of the genetic system which facilitated rapid evolution and massive colonization in this group of Mediterranean annual grasses.

SPECIES AND SPECIES GROUPS

Twenty-two wild species are recognized in *Aegilops* and *Triticum* (see Table I). Ten are diploid ($2n = 14$) and 12 are polyploid—mostly tetraploids ($2n = 28$), but also hexaploids ($2n = 42$). All polyploids show allopolyploid chromosome behavior in meiosis. All species, with the exception of diploid *Ae. speltoides,* are facultative selfers, i.e., are predominantly self-pollinated.

The cytogenetic affinities between the various species have been intensively studied [see reviews by Kihara (1954) and Sears (1948, 1959)]. Diploids can be arranged (see Table I) in, at least, six genomic groups which are also fully separated from one another by sterility barriers. The genomes of the polyploids have been analyzed and a genomic formulation is available for each of them (Kihara, 1954; Kihara *et al.*, 1959). As pointed out by Zohary and Feldman (1962), polyploids readily fall into three species clusters (see Table I), each characterized by the presence of a common or pivotal genome. Species grouped in a given cluster differ from one another by virtue of their additional or differential genome (or sometimes two genomes in hexaploids). Kihara and his associates established, by means of genome analysis, that the pivotal genome in each polyploid cluster is, more or less, identical with a chromosome set of a given diploid analyzer. In contrast, the differential genome in each species was found to be usually modified, i.e., only partially homologous with the chromosomes of any known diploid. Thus, most polyploid *Aegilops* and *Triticum* species show a peculiar situation: the presence of a modified genome side by side with an unaltered one. Moreover, the unchanged genome is a common one—to a whole species group.

DIPLOIDS VERSUS POLYPLOIDS

A comparison between diploids and polyploids in *Aegilops* and *Triticum* reveals sharp contrasts between the two chromosomal levels. These are summarized in the following paragraphs.

DIPLOIDS

On the diploid level one deals with taxonomically "good species." In other words, diploids present us with easily definable units which are separated from one another by clear morphological discontinuities. Each

TABLE I

Species and Species Groups in *Aegilops* and *Triticum*[a]

Natural units	Species	Genome type
Diploid genomic groups		
Genome B (= S)	*Ae. bicornis* (Forsk.) Jaub. et Sp.	S[b]
	Ae. sharonensis Eig	S[l]
	Ae. longissima Schweinf. et Musch.	S[l]
	Ae. speltoides Tausch	S
Genome D	*Ae. squarrosa* L.	D
Genome C	*Ae. caudata* L.	C
Genome M	*Ae. comosa* Sibth. et Sm.	M
	Ae. uniaristata Vis.	M^u
Genome C^u	*Ae. umbellulata* Zhuk.	C^u
Genome A	*T. boeoticum* Boiss.	A
	T. monococcum L.[b]	A
Polyploid complexes		
Genome D species cluster	*Ae. crassa* Boiss. 4x	DM^{cr}
	Ae. crassa Boiss. 6x	DD^{2}M^{cr}
	Ae. juvenalis (Thell). Eig.	DC^{u}M^j
	Ae. ventricosa Tausch	DM^v
	Ae. cylindrica Host	DC
Genome C^u species cluster	*Ae. triuncialis* L.	C^{u}C
	Ae. columnaris Zhuk.	C^{u}M^c
	Ae. biuncialis Vis.	C^{u}M^b
	Ae. triaristata Willd. 4x	C^{u}M^t
	Ae. triaristata Willd. 6x	C^{u}M^{t}M^{t2}
	Ae. ovata L.	C^{u}M^o
	Ae. variabilis Eig	C^{u}S^v
	Ae. kotschyi Boiss.	C^{u}S^v
Genome A species cluster	*T. dicoccoides* Koern.	AB
	T. timopheevi Zhuk.	AB (= AG)
	T. dicoccum Schübl.[b]	AB
	T. durum Desf.[b]	AB
	T. aestivum L.[b]	ABD

[a] Genomic formulation after Kihara (1954) and Kihara *et al.* (1959).
[b] Cultivated wheats.

diploid species manifests specific growth habits and characteristic spike morphology. In comparison to polyploids, the range of variation within any diploid species is limited, sometimes even very narrow.

Morphologically, a most conspicuous evolutionary divergence among the diploids is found in their seed-dispersal devices. As in many other annual groups specialization in seed-dispersal apparatus is of prime importance and clearly reflects the major trends of adaptive radiation among these grasses. Different and contrasting dispersal units are met with among diploids. As a matter of fact each diploid genomic group presents us with a distinct dispersal apparatus. For comparison of the various devices see Fig. 1.

In parallel with their morphological divergence, diploids are also easily divided on a basis of genetic affinities into at least six distinct genomic groups (see Table I). These groups are efficiently isolated from one another reproductively; hybrids between them are absolutely sterile.

Diploids are, as a rule, restricted in their distribution and, in sharp contrast with the polyploids, are almost exclusively restricted to the eastern Mediterranean basin. Table II presents details concerning their distribution. About half of the diploid species are very specific or specialized in their ecological (edaphic and climatic) requirements and occupy very small areas. Other diploids show somewhat wider eco-geographical amplitudes (which are, however, not as wide as in polyploids!) and this is often correlated with weedy or segetal tendencies. But only three diploids have attained a relatively wide distribution in the Middle East and were able to achieve mass colonization of segetal or man-made secondary habitats (compare data set in Table II). The relatively successful colonizers among the diploids are *Ae. squarrosa* (the bearer of genome D), *Ae. umbellulata* (genome C^u), and *Triticum boeoticum* (genome A). As will be discussed presently these three diploids are also the donors of the common or pivotal genomes to the three polyploid clusters.

POLYPLOIDS

The numerous polyploid *Aegilops* species present us with an entirely different situation. Compared with the diploids, tetraploids as well as hexaploids are extraordinary variable. The wide ranges of variation are usually coupled with blurred specific boundaries. In other words, species delimitation on the polyploid level is often difficult and arbitrary, and series of intermediate forms interconnect the major morphological types. Significantly such blurred boundaries are the rule among sympatric *Aegilops* species sharing a common genome. Polyploid cluster C^u and polyploid cluster D (see Table I) should thus both be considered as aggregates or complexes of forms. As will be shown later, this lack of morphological discontinuity reflects the occurrence in nature of loose genetic connections between polyploids sharing a common genome.

Another major difference between diploids and polyploids is the pronounced "weediness" of the latter. The polyploid *Aegilops* species do not show the marked ecological specificity which is so characteristic of the majority of their diploid relatives. Instead, polyploid *Aegilops* have a wide ecogeographical amplitude and a distinctly weedy nature. The

Fig. 1. Seed dispersal units in diploid *Aegilops* and *Triticum*. From left to right S = *Ae. longissima, Ae. sharonensis, Ae. bicornis,* and the two morphs of *Ae. speltoides;* D = *Ae. squarrosa;* A = *T. boeoticum;* Cᵘ = *Ae. umbellulata;* M = *Ae. uniaristata* (left) and *Ae. comosa* (center and right); C = *Ae. caudata.*

TABLE II

GEOGRAPHICAL AND ECOLOGICAL CHARACTERIZATION OF DIPLOID SPECIES OF
Aegilops AND (WILD) *Triticum*

Species and genome type	Distribution and habitats
Ae. bicornis (S^b)	Relatively limited distribution: southern Israel, Lower Egypt, Cyrenaica. Restricted to xeric sandy soils
Ae. sharonensis (S¹)	Very limited distribution: endemic to the sandy soils of the Israeli Mediterranean coastal plain
Ae. longissima (S¹)	Relatively limited distribution: Israel, Jordan, southern Syria. Occupies sandy loams in the Mediterranean coastal plain and, in addition, a variety of steppe-like habitats, mainly in the sagebrush formation
Ae. speltoides (S)	Medium size range: Common in the "fertile crescent" belt (northern Israel, Syria, southern Turkey, northern Iraq, western Iran), more sporadically spread over the Anatolian plateau. Occupies open steppe-like herbaceous formations and cleared-up maquis, also common in alluvial plains and edges of cultivation
Ae. squarrosa (D)	Widely spread over central Asia from northern Iraq to Iran, Transcaucasia, Transcaspia, Afghanistan, and Pakistan. Occupies a wide array of habitats from xeric sagebrush steppes to opened up temperate forests. Common also as weed in cultivation
Ae. caudata (C)	Medium size range: spread over Greece, Turkey, northern Syria, and northern Iraq. Occupies steppe and steppe-like herbaceous formations, as well as opened maquis. Also at edges of cultivation
Ae. comosa (M)	Relatively limited distribution: restricted to Mediterranean formations in Greece, the Aegean Islands, and western Turkey. Inhabits mainly dwarf shrub and maquis formations, as well as cleared-up areas and edges of cultivation
Ae. uniaristata (M^u)	Very limited distribution: scattered Mediterranean maquis and dwarf shrub formations in Greece and the Marmara Sea area, and apparently more common in the Adriatic zone of Yugoslavia
Ae. umbellulata (C^u)	Widely spread over western Asia: northern Iraq, most of Turkey, northern Syria, northern and western Iran, Transcaucasia. Occupies a wide range of habitats, e.g., Irano-Anatolian inner steppes, herbaceous steppe-like formations in the "fertile crescent," opened-up Mediterranean maquis in western Turkey, also common at edges of cultivation and roadsides
Tr. boeoticum (A)	Widely spread over western Asia and the southern Balkans: from Greece to Turkey, Syria, northern Iraq, northern and western Iran, Transcaucasia. A component of the open, herbaceous park forest and steppe-like formations in the "fertile crescent" belt, inhabits opened-up areas and edges of cultivation throughout the area

polyploid *Aegilops* colonize almost exclusively opened-up secondary formations and highly disturbed or segetal habitats. The majority of the polyploids (compare diploids and polyploids in Maps 1 and 2), have spread over such man-made habitats far beyond the restricted eastern Mediterranean territory of the diploids. Today they colonize massively the whole Mediterranean agricultural belt—from the Atlantic shore in the west to Central Asia in the east. Among the most successful colonizers are *Ae. triuncialis, Ae. biuncialis, Ae. triaristata, Ae. ovata,* and *Ae. cylindrica.* Two of these (*Ae. triuncialis* and *Ae. cylindrica*) have also established themselves in North America.

In their mode of seed dispersal the polyploids do not show the wide contrasts and the numerous trends characteristic of the diploid level. Variation in spike morphology is wide—but in basic structure the twelve polyploid *Aegilops* and *Triticum* are all centered around only three trends of seed-dispersal models. Seven *Aegilops* species, grouped in the largest and most widespread cluster (C^u cluster in Table I), have a many awned "umbrella"-type of dispersal unit. Thus, they follow the evolutionary trend set by diploid *Ae. umbellulata* (compare spikes in Fig. 2). The four other *Aegilops* polyploids, the members of the D cluster, clearly utilize the "barrel"-type dissemination device. They all follow the trend which has evolved in *Ae. squarrosa* (compare spikes in Fig. 3). The third and smallest polyploid cluster, that of the wild wheats (*T. dicoccoides* and *T. timopheevi*), is characterized by an arrowlike disarticulating spikelet—a dispersal device elaborated by the wild diploid wheat *T. boeoticum.*

To sum up, each of the three polyploid species clusters is characterized not only by a common, pivotal genome but also by a common basic trend in the seed-dispersal mechanism. Moreover, the dissemination "themes" employed by the polyploids are those evolved, and previously tested, by the three pivotal diploid species, i.e., the contributors of the common genomes to the polyploid clusters. It is noteworthy that the three "themes" utilized on the polyploid level are evolutionarily the three most successful devices among diploids. As already mentioned, *Ae. umbellulata* (genome C^u), *Ae. squarrosa* (genome D), and *T. boeoticum* (genome A), have achieved the widest distribution and are the most successful colonizers among diploid species.

SPATIAL RELATIONSHIPS AND STRUCTURE OF NATURAL POPULATIONS

That the "weedy" polyploid *Aegilops* species largely overlap in their geographical distribution was already clear to the two monographers of

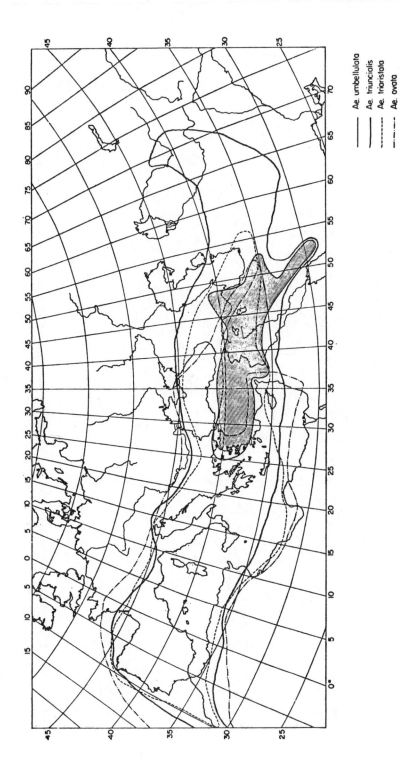

MAP 1. Distribution areas of some polyploid members of the C^u genome cluster in Aegilops and their pivotal diploid species *Ae. umbellulata*.

Ae. umbellulata
Ae. triuncialis
Ae. triaristata
Ae. ovata

225

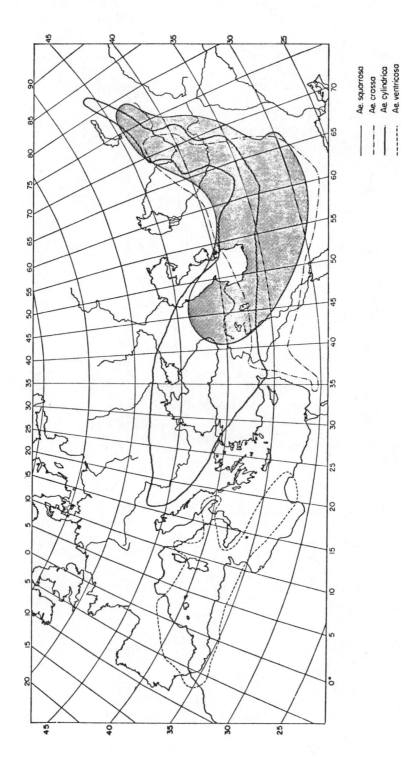

MAP 2. Distribution areas of some polyploid members of the D genome cluster in *Aegilops* and their pivotal diploid species *Ae. squarrosa.*

Ae. squarrosa
Ae. crassa
Ae. cylindrica
Ae. ventricosa

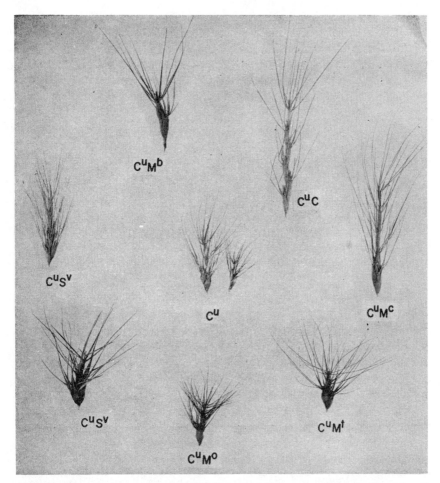

Fɪɢ. 2. The many awned "umbrella"-type seed-dispersal unit as displayed by the seven polyploid members of the C^u cluster and their pivotal diploid species *Ae. umbellulata*. C^u = *Ae. umbellulata*; C^{u}C = *Ae. triuncialis*; C^{u}M^c = *Ae. columnaris*; C^{u}M^t = *Ae. triaristata*; C^{u}M^o = *Ae. ovata*; C^{u}S^v = *Ae. variabilis* (lower unit) and *Ae. kotschyi* (upper unit).

the genus (Eig, 1929, 1936; Zhukovsky, 1928). This largely overlapping distribution is also well illustrated in the examples of the areas of poly-ploids plotted in Maps 1 and 2. Thus, in the center of this genus (the Middle East countries) one finds a concentration of the majority of the polyploid species in the same floristic areas, often together with their diploid relatives as well. What was not appreciated till recently is that these colonizer polyploids are not only geographically sympatric; they

commonly colonize the very same sites! This is particularly apparent
in cases of heavily interfered with habitats such as degraded maquis and
park forests, opened-up or depleted steppes, abandoned fields, and edges
of cultivation. Such habitats, in the geographic center of *Aegilops*
(Greece, Turkey, northern Iraq, Syria, and Israel), frequently harbor
3–5 or even larger number of species in mixed stands. This cohabitation

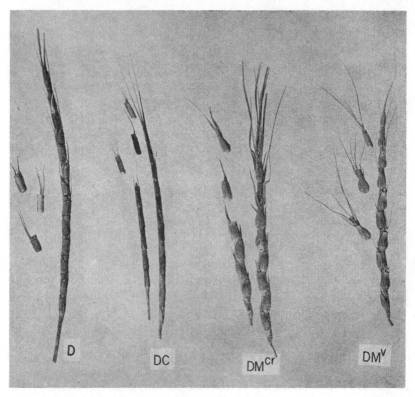

Fig. 3. The "barrel"-type seed-dispersal unit as displated by three polyploid
members of D cluster and their pivotal diploid species *Ae. squarrosa*. D = *Ae.
squarrosa;* DC = *Ae. cylindrica;* DM^cr = *Ae. crassa;* DM^v = *Ae. ventricosa.*

is particularly clear in the case of the polyploid members of the C^u
cluster. Mixed stands of various combinations between these seven
tetraploids are so common in the Middle East countries as to be regarded
the rule rather than the exception (for examples see Zohary, 1963).
Moreover, in Turkey, northern Syria, and northern Iraq, polyploids of
the C^u cluster grow together with their pivotal diploid *Ae. umbellulata*
and other diploid species.

Natural polyploid *Aegilops* populations in the Mediterranean basin are also conspicuous in their intraspecific polytypic nature. As a rule, a given polyploid is represented in any given station not by a single form but by many lines. Populations are usually found to be composed of a wide array of morphologically easily recognizable types.

NATURAL HYBRIDIZATION

Another striking contrast between diploids and polyploids is the reproductive isolation of the diploid genomic groups from one another as compared to loose genetic connections between polyploids. Genetic links and gene flow through occasional spontaneous hybridization and subsequent introgression are apparently a general rule in the case of the seven polyploid species sharing the pivotal C^u genome. There are good indications that natural hybridization is not at all rare also in case of other polyploids. In examining mixed stands in Israel, Turkey, and Greece, we repeatedly encountered hybrids and hybrid products in some twelve different species combinations. Diagram 1 presents the various connections between the polyploids of the C^u cluster. Significantly, hybridization products are relatively rare in stable habitats; they are much more

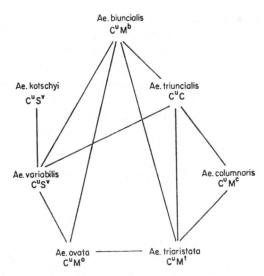

DIAGRAM 1. Natural hybridization between the seven polyploid members of the C^u cluster of *Aegilops* (sect. *Pleionathera* sensu Eig. 1929). Schematic representation of genetic connections between tetraploids encountered in Israel, Turkey, and Greece. Compiled from Zohary and Feldman (1962), Feldman (1963), and field notes of trips to Turkey and Greece in 1959 and 1962.

frequent in sites recently disturbed by man. In such places one occasionally comes across whole series of intermediates and recombinants bridging two (or sometimes a larger number) of the coexisting polyploid species. In some favorable stations, the frequency of pronouncedly intermediate and introgressed individuals (some yet semisterile) can reach 1–5%.

THE PROCESS OF INTROGRESSION

We were able to follow in some detail the genetic connections between the three weedy *Aegilops* tetraploids which are widespread in Israel, namely, *Ae. variabilis* (genome formulation $C^u C^u S^v S^v$), *Ae. ovata* ($C^u C^u M^o M^o$), and *Ae. biuncialis* ($C^u C^u M^b M^b$). These three grasses regularly build mixed populations in Israel and in recently disturbed habitats show characteristic introgressive hybridization variation patterns (for details see Zohary and Feldman, 1962; Feldman, 1963). Sporadic interspecific F_1 hybrids between these three tetraploids, sharing the C^u genome, were repeatedly detected in mixed stands. They are functionally male-sterile since their pollen is largely aborted and the anthers do not dehisce. These interspecific hybrids, however, are heavily exposed to open-pollination by parental pollen in nature, and significantly they do set some (backcross) seed. Progeny raised from such hybrids (Feldman, 1963; B. Pazi, unpublished data, 1963) segregated widely and showed remarkable restoration of fertility in the second and the third hybrid generations. Some of the second hybrid generation plants already were both vigorous and almost fully fertile. These data indicate that introgression between the Israeli polyploid *Aegilops* species is a neat and quick process. Its critical early stages are buffered by a common genome (C^u) and further balanced by subsequent backcrossing. When pollen fertility is restored, introgressed types are quickly fixed by self-pollination.

INTRASPECIFIC CHROMOSOMAL VARIATION

As already pointed out (Zohary and Feldman, 1962), hybridization between amphidiploids should be expected to produce intraspecific chromosomal variation—parallel to morphological variation. Moreover, since species involved share a common pivotal genome such variation should be concentrated in the modified genome of each polyploid species.

There are several reports on intraspecific chromosomal variation in several polyploid *Aegilops* species (for examples, consult Zohary and Feldman, 1962). In our laboratory, Feldman (1963) recently studied intraspecific differences between six Israeli collections of tetraploid *Ae.*

variabilis. Examination of F_1 hybrids between these lines revealed the presence of several translocations in this species (up to two in a given F_1 combination) as well as additional chromosomal differences reflected by some reduction in pairing and other irregularities in meiosis. Feldman further crossed the *variabilis* lines to a standard *Ae. ovata* and standard *Ae. longissima.* Hybrids with *ovata* showed similar chromosomal association, whereas hybrids with *Ae. longissima* differed one from another —indicating the concentration of the intraspecific differences in the modified genome S^v.

THE GENETIC SYSTEM OF THE
COLONIZER AEGILOPS SPECIES

The data presented seems to indicate that the colonizer polyploids in *Aegilops* cannot be considered as simple units—each with its independent amphidiploid history. Instead, the aggressive polyploids are genetically linked to form species clusters. In other words, they build up compound, loosely interconnected polyploid superstructures. It is this system of compound amphidiploidy in combination with a mating system of predominant self-pollination that have apparently enabled the polyploids to build up genetic variation rapidly and evolve successfully as aggressive weedy annuals.

The origin of polyploids is envisaged as follows: tetraploids within each cluster are considered to be the derivatives of only a restricted number of initial amphidiploid combinations, most likely between the same diploids which exist today. But initial amphidiploids which shared a common genome hybridized with one another, their common genome serving as a buffer in the process of hybridization. As a result numerous recombinations were possible—but mostly between the unshared chromosome sets—resulting in their differential modification and the establishment of a modified or differential genome side by side with an unaltered set. The existing polyploid forms within each cluster represent the new combinations which have been favored by natural selection. Their modified genomes do not necessarily contain chromosomes or chromosome segments linearly derived from an original diploid parent. Instead, such modified genomes represent new chromosomal recombinations derived from two, three, or even larger numbers of original diploid genomes.

The apparent and recent evolutionary success of the weedy polyploid *Aegilops* is not attributed to polyploidy as such but to the fact that polyploidy made possible the establishment of a large, common gene pool. While diploids are completely isolated from one another, several initial amphidiploids could establish genetic connections. Thus, on the polyploid level, the genetic material of the six divergent diploid genomic groups was brought together and could be recombined and remolded. Schematic representation of this mode of formation is given in Diagram 2.

The genetic system established is admirably suited for rapid colonization. Polyploids in each cluster have a rich potential of genetic variation (a pool which on the diploid level is split among at least six independent groups!). Flexibility is assured by formation of mixed stands and by occasional hybridization. On the other hand, plants are predominantly self-pollinated. Successful combinations are thus rapidly fixed.

Another feature that emerges from the present data is that species buildup on the polyploid level was far from random; instead it followed the three most successful adaptive trends previously explored, tested, and

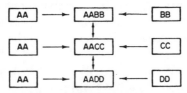

DIAGRAM 2. Schematic representation of the mode of formation of a polyploid species cluster. Diploid AA is the donor of the common genome; it forms initial amphidiploids with diploids BB, CC, and DD. Hybridization between the initial amphidiploids results in recombination and modification of B, C, and D genomes, whereas the common A genome remains constant and buffers the hybridization process.

established on the diploid level (by *Ae. squarrosa, Ae. umbellulata,* and *T. boeoticum*). In other words, the evolutionary most successful or "weedy" types on the diploid level were, so to speak, preadaptive, and they contributed the pivotal genomes or the main evolutionary "themes" for the polyploid clusters. The other diploid genomic groups provided material for chromosomal recombination, i.e., the material for variations on these three basic "themes."

Finally one should also bear in mind that the buffering effect of the common pivotal genome in the process of hybridization between polyploids is not necessarily the only means for establishing loose connections on the polyploid level here. As already stressed by Stebbins (1956), polyploids as such are better balanced, in comparison to diploids, to withstand the drastic effects of incorporation of alien genetic material. This is certainly the case in the wheat group where there is ample experimental evidence to support this notion. We have several indications that, except for connections between polyploids sharing a common genome, other types of interspecific connections exist in nature as well and might also have contributed to the common pool available on the polyploid level. We have found spontaneous hybridization in several cases of *Aegilops* species not sharing a common genome. There are also indications for one-way introgression from diploids to tetraploids—via natural triploid hybrids.

CONCLUDING REMARKS

The annual colonizer species of *Aegilops* present us with a specific combination of elements: compound amphidiploidy, buffering pivotal genomes, and predominance of self-pollination. I wonder how widespread such a genetic system is among allopolyploid plant groups? Similar connections, based on common genomes, might be found to occur also in some other polyploid complexes; I suspect their existence in *Agropyrum*. But, as mentioned in the last section, even among allopolyploid plants such a device is only one among several possibilities that can facilitate introgression and the building up of a large pool of genetic variation.

The following characteristics, however, of the *Aegilops* polyploid cluster might have a wider significance in considerations of the evolution of colonizer species:

(1) The occurrence in the species cluster of a common trend of adaptive specialization (the "theme") which is in fact a preadaptation of the pivotal diploid.

(2) The operation of a dual system. There is a conservative gene complex in control of the evolutionary "theme" (in the pivotal genome), side-by-side with a pool of recombinable material (in the modified or differential genomes).

A dual system—where the genes controlling the preadaptive "theme" are held together by one part of the chromosomal complement and where wide "variation on the theme" is provided by a second part of the chromosomal complement—may be not restricted to the *Aegilops* colonizers. It is, undoubtedly, an arrangement designed for "canalization" of the evolving population in its preadaptive, basic "theme," guarding it from haphazard variation. Unrestricted recombination might easily throw the population off balance and off the adaptive peak. This is particularly a danger when fusion of widely divergent units is involved.

Whether dual systems are really widespread in colonizer species is yet only a matter of speculation. Localized chiasmata in diploid organisms might actually reflect the same trend. There is one group of colonizer species where such duality is well shown, namely, the North American *Oenothera*. The translocation systems which characterize the weedy *Euoenothera* forms have the same effect. They build a conservative gene complex (the uniform chromosomal arms) side-by-side with a chromosomal system for genetic variation (the differential segments).

SUMMARY

The "weedy" *Aegilops* polyploids present us with a genetic system of compound amphidiploidy. Polyploids within a cluster constitute a group of closely interconnected units. The main "adaptive theme" is set up by the common, pivotal genome. Genetic variation is provided by hybridization and recombination in the other genomes, i.e., the fusion of genetic material from several genomes which are already completely separated on the diploid level. The system established is well suited for rapid colonization. Polyploids in each cluster have a rich potential of genetic variation. Flexibility is assured by formation of mixed stands and by occasional hybridization. On the other hand, plants are predominantly self-pollinated. Successful combinations are thus rapidly fixed.

REFERENCES

Eig, A. (1929). Monographisch-kritische Uebersicht der Gattung *Aegilops*. *Rept. Spec. Nov. Reg. Veget. Beih.* 55, 1–228.

Eig, A. (1936). *Aegilops.* In "Die Pflanzenareale" (E. Hannig and H. Winkler, eds.), 4 Reihe, Heft 4, pp. 43–50, maps 38–41. Fischer, Jena.

Feldman, M. (1963). Evolutionary studies in the *Aegilops–Triticum* group with special emphasis on causes of variability in the polyploid species of section *Pleionathera*. Ph.D. Thesis, The Hebrew University, Jerusalem (in Hebrew).

Kihara, H. (1954). Considerations on the evolution and distribution of *Aegilops* species based on the analyzer-method. *Cytologia (Tokyo)* 19, 336–357.

Kihara, H., Yamashita, H., and Tanaka, M. (1959). Genomes of 6 species of *Aegilops.* *Wheat Inform. Serv.* 8, 3–5.

Sears, E. R. (1948). The cytology and genetics of wheats and their relatives. *Advan. Genet.* 2, 239–270.

Sears, E. R. (1959). Weizen I: The systematics, cytology and genetics of wheats. In "Handbuch der Pflanzenzuechtung" (H. Kappert and W. Rudorf, eds.) Vol. II, pp. 164–187. Parey, Berlin.

Stebbins, G. L. (1956). Artificial polyploidy as a tool in plant breeding. *Brookhaven Symp. Biol.* 9, 37–52.

Zohary, D. (1963). The evolution of genomes in *Aegilops* and *Triticum*. *Intern. Wheat Genet. Symp., 2nd., Lund, 1963* (in press as suppl. to *Hereditas*).

Zohary, D., and Feldman, M. (1962). Hybridization between amphidiploids and the evolution of polyploids in the wheat (*Aegilops–Triticum*) group. *Evolution* 16, 44–61.

Zhukovsky, P. M. (1928). A critical-systematical survey of the species of the genus *Aegilops* L. (Russian with English summary). *Bull. Appl. Bot., Genet. Plant Breeding* 18, 417–609.

[*Editors' Note:* The Discussion has been omitted.]

Part V

GENE EXPRESSION AND PHYSIOLOGY

Editors' Comments
on Papers 15 Through 18

This section contains papers concerned with the genic effects of polyploidy. Of the numerous articles on this topic, many are concerned with genic diploidization, the process whereby genes lose their function due to mutation. However, this problem can be addressed only experimentally because whether a gene has lost its function or is inhibited in its expression is a moot question in almost all studies published to date. Analyses of newly synthesized polyploids whose diploid progenitors' genotypes are known are necessary to answer the diploidization question.

Paper 15 is a genic analysis of the only known unambiguous example of the natural origin of segmental allotetraploid species. The paper fully documents the earlier findings by Ownbey (Paper 6), and it gives some evidence that the polyploid with the greatest heterozygosity is not necessarily the most successful. The occurrence of novel heteromeric enzymes not found in either parent is clearly demonstrated in the polyploids.

Levy and Levin (Paper 16) have not presented new kinds of information, but they clearly and concisely review literature on gene inactivation to that time. The paper is important because it clearly shows partial suppression of some metabolic pathways, demonstrates novel compounds not found in either putative parent, and points to gene suppression in polyploids as a probable means of gene inactivation. A later paper by Levy (1976) on artificial tetraploids verifies the findings from natural hybrids.

Paper 17 is of considerable interest because it examines the qualitative and quantitative differences in multicellular gametophyte and sporophyte generations. The study demonstrates clear differences between tetraploid gametophytes and tetraploid sporophytes and quantitative differences between diploid and tetraploid sporophytes.

The effect of polyploidy on cell and organ size have long been known. Hall (Paper 18) presents evidence indicating that the increase in radius of a root meristem is correlated with a change in tolerance to soil temperature because of changes in critical oxygen pressure. This could be one reason why there is an ecological difference in distribution of tetraploids and diploids in some species.

REFERENCE

Levy, M., 1976, Altered Glycoflavone Expression in Induced Autotetraploids of *Phlox drummondii, Biochem. Syst. Ecol.* **4:**249–254.

15

Reprinted from *Evolution* **30**:818–830 (1976)

GENETIC AND BIOCHEMICAL CONSEQUENCES OF POLYPLOIDY IN *TRAGOPOGON*[1]

M. L. Roose and L. D. Gottlieb

Department of Genetics, University of California, Davis 95616

Most studies of the evolution of polyploid plant species have emphasized phylogenetic issues, for example, identification of diploid progenitors and clarification of polyploid complexes. They have largely utilized evidence from comparative morphology, karyotypes and cytogenetic analysis of interploidal hybrids, as well as biochemical profiles of certain classes of compounds such as flavonoids and seed proteins. Although these studies helped greatly to elucidate the mode of origin and ancestry of many polyploid species, they were not concerned with explaining one of the most intriguing features of polyploidy which is that, in many plant genera, the polyploids are more widely distributed over more habitats than their diploid progenitors. This is a problem of the first rank because at least one-third of the Angiosperms and a higher proportion of the ferns are polyploid.

Several recent hypotheses have proposed that the wide capabilities of allopolyploids (we use allotetraploids as an example) are a direct biochemical consequence of their possession of two divergent diploid genomes which provides them with a multiplicity of enzymes relative to both diploid parents as well as a high proportion of novel enzymes (Fincham, 1969; Barber, 1970; Manwell and Baker, 1970). Enzyme multiplicity may extend the range of environments in which normal development can take place and, thereby, might account for the frequently wider distribution of polyploids. This may be true even if the tetraploid as a species is less poly-

morphic than its diploid parents. Tetraploid individuals contain a second copy of each gene and, therefore the population can maintain a higher proportion of heterozygous individuals than can a diploid population, and individual tetraploid plants have the potential to express more enzymes than diploid ones.

If the duplicated genes of tetraploid species specify different polypeptide subunits of multimeric enzymes, novel heteromers or "hybrid" enzymes are produced. When the diploid parents differ in their allelic contents at these genes, as is often the case, the tetraploid derivative expresses novel heteromers that are never produced in either parental species. Such enzymes may have distinctive properties (Schwartz and Laughner, 1969; Fincham, 1972; Schwartz, 1973; Scandalios, 1974). The possession of several allelic forms of the same enzyme may permit tetraploid plants to maintain a sufficient flux through metabolic pathways in the event of changed reaction conditions associated with varying environments (Johnson, 1974). Increased number of genes may also result in increase in concentration and activity of their enzyme products, thereby increasing the flow through rate-limiting reactions which in turn may provide additional substrate for other biosynthetic pathways. Thus, in somatic cell cultures of *Datura*, enzyme activity in trisomic lines was proportional to the number of coding structural genes in the genome (Carlson, 1972). Most studies of tetraploid plant species have reported that they express all or nearly all of the enzymes specified by the different alleles inherited from their diploid parents, for example, wheat (Mitra and Bhatia, 1971), cotton (Cherry et al., 1972),

---

[1] This paper is dedicated to the memory of Professor Marion Ownbey who discovered and described the tetraploid species of *Tragopogon*.

tobacco (Smith et al., 1970; Reddy and Garber, 1971; Sheen, 1972), safflower (Efron et al., 1973), beans (Garber, 1974), and *Stephanomeria elata* (Gottlieb, 1973a). Intergenomic variation has been detected in the structural genes encoding each of nine genetically independent systems studied in hexaploid wheat (Hart, 1969, 1970, 1973, 1975, and personal communication). This contradicts the earlier conclusion of Brewer, Sing and Sears (1969), based on analysis of 12 enzyme systems, that the level of multiplicity implied by the divergent diploid patterns has not been maintained in hexaploid wheat.

In many cases, allotetraploid species may be fixed heterozygotes because all individuals express the same multiple enzyme phenotype. This comes about because chromosome doubling gives rise in the tetraploid to two pairs of identical homologues that preferentially pair with each other at meiosis. Alleles of each of the duplicated genes go to the same pole so that each gamete receives one copy of each. Following fertilization, a heterozygous phenotype is reconstituted in the tetraploid even though each gene is homozygous.

The only unambiguous examples of the recent natural origin of allotetraploid plant species are two tetraploid species of *Tragopogon*, *T. mirus* and *T. miscellus*, which were discovered and elegantly described by Ownbey (1950). The original populations of both species described by Ownbey are nearly all still extant, and *T. miscellus* has become one of the most common weeds in vacant lots in and around Spokane, Washington, and to the east. These species provide crucial evidence about the initial genetic and biochemical consequences of allotetraploidy because they originated during the present century and it is likely that their genomes have undergone little if any modification since their origin. Therefore, we undertook an extensive electrophoretic survey of enzyme variation in both species and a number of populations of their diploid parents. The basic evidence sought, in this initial study,

was the proportion of the tetraploid genomes that was fixed in a heterozygous state as a result of the combination of different alleles inherited from their diploid parents.

MATERIALS AND METHODS

The three diploid species, *Tragopogon dubius*, *T. porrifolius* and *T. pratensis* are widespread in the Old World and have become widely naturalized in the United States. In southeastern Washington and adjacent Idaho they first appeared during the present century and have successfully invaded waste places, roadsides, fields, and pastures. The species are biennial, germinating and forming leaf rosettes one year and flowering stems the following year. They are self-compatible and capable of self-pollination. Their populations are generally small, numbering fewer than 500 individuals and often only several hundred individuals. The three species are morphologically sharply delimited without overlap in many features (Ownbey, 1950) and can also be distinguished cytologically (Ownbey and McCollum, 1954). Their populations are frequently sympatric and wherever this occurs, readily detected highly sterile F_1 hybrids are formed (Ownbey, 1950). The two tetraploid species, *T. mirus* and *T. miscellus*, originated in the vicinity of Pullman, Washington, and Moscow, Idaho, two small towns about ten miles apart, and are now known from about 20 localities in the region. *Tragopogan mirus* is also found in Flagstaff, Arizona (Brown and Schaack, 1972), and *T. miscellus* in Gardiner, Montana, and Sheridan, Wyoming (Ownbey, pers. comm.). Ownbey (1950) clearly documented their parentage using evidence from morphology, karyotypes, and genetic analysis: *T. dubius* × *T. porrifolius* gave rise to *T. mirus*, and *T. dubius* × *T. pratensis* to *T. miscellus*. With the generous help of the late Dr. Ownbey, seeds from nearly all of the populations of both tetraploid species and a number of populations of the three diploid species were collected in June and

239

TABLE 1. *Locations and sample sizes of populations of species of* Tragopogon *studied for electrophoretic variation in enzymes.*

T. porrifolius	N
Corvallis, Benton Co., Oregon	29
Pullman, Whitman Co., Washington	27
Palouse, Whitman Co., Washington	28

T. mirus	
Pullman, Whitman Co., Washington (3 sites)	62
Palouse, Whitman Co., Washington	23
Garfield, Whitman Co., Washington	13
Tekoa, Whitman Co., Washington	28
Rosalia, Whitman Co., Washington	14
Flagstaff, Coconino Co., Arizona	13

T. dubius	
Tacoma, Pierce Co., Washington	36
Portland, Multnomah Co., Oregon	30
Moscow, Latah Co., Idaho	19
Burney, Shasta Co., California	19
Oak Creek Campground, Coconino Co., Arizona	19
Rosalia, Whitman Co., Washington	18

T. miscellus	
Moscow, Latah Co., Idaho	35
Pullman, Whitman Co., Washington	15
Spangle, Spokane Co., Washington	14
Spokane, Spokane Co., Washington (2 sites)	41
Stateline, Kootenai, Co., Idaho	15

T. pratensis	
Garfield, Whitman Co., Washington	14
Moscow, Latah Co., Idaho	30
Seattle, King Co., Washington	23

July, 1974, for electrophoretic studies. Collection sites and sample sizes are listed in Table 1.

In general, each population was sampled by walking across it and collecting seeds from a single head of each of approximately 30 to 40 plants selected at random. The seeds were germinated and the seedlings grown in the greenhouse for approximately 12 weeks until they were assayed. One seedling individual per mother plant in nature was examined. Germination and survival were better than 95%. Thirteen different enzyme systems were analyzed by horizontal starch gel electrophoresis: esterase (*EST*), leucine amino peptidase (*LAP*), acid phosphatase (*APH*), peroxidase (*PER*), glutamate oxaloacetate transaminase (*GOT*), glutamate dehydrogenase (*GDH*), malate dehydrogenase (*MDH*), phosphoglucoisomerase (*PGI*), alcohol dehydrogenase (*ADH*), glucose-6-phosphate dehydrogenase (*G6PD*), malic enzyme (*ME*), superoxide dismutase (*SOD*), and phosphoglucomutase (*PGM*).

Crude extracts were obtained by grinding one or two mature rosette leaves (0.3 g) in 0.75 ml 0.1% 2-mercaptoethanol, buffered with 0.1 M Tris-HCl pH 7.0. All extraction procedures were done in ice baths. The homogenates were applied to paper wicks and electrophoresis conducted as described previously (Gottlieb, 1973*b*). However, a Poulik gel buffer system (Yang, 1971) was used for *GDH*, *PGI*, *G6PD*, and *SOD*. For *MDH*, the gel buffer was 0.005 M histidine, buffered to pH 8.0, with an electrode buffer of 0.41 M citrate pH 8.0.

Enzyme assay conditions were as follows: *GOT*, *PGI*, *GDH*, and *PER* were not modified from Gottlieb (1973*c*). *ADH*: 4 ml 95% ethanol, 40 mg NAD, 30 mg NBT, 2 mg PMS, 95 ml 0.1 M Tris-HCl pH 8.0; *EST*: 30 mg alpha-naphthyl acetate, 60 mg beta-naphthyl acetate, otherwise not modified from Gottlieb (1973*b*); *APH*: 150 mg Na-alpha-naphthyl acid phosphate, 100 ml 0.2 M sodium acetate pH 5.0, presoaked 30 min. in 0.4 M sodium acetate pH 5.0 at 4 C before changing to assay solution, otherwise not modified from Gottlieb (1973*c*); *LAP*: presoaked 30 min. in 0.5 M Tris-maleate pH 6.5, otherwise not modified from Gottlieb (1973*c*); *G6PD*: 75 mg Na-glucose-6-phosphate, 10 mg NADP, 25 mg NBT, 2 mg PMS, 100 ml 0.1 M Tris-HCl pH 8.0; *ME*: 280 mg L-malic acid, 20 mg NADP, 30 mg NBT, 5 mg PMS, 1 ml 0.1 M MgCl₂, 100 ml 0.1 M Tris-HCl pH 8.6; *MDH*: 40 mg NAD, 30 mg NBT, otherwise not modified from Gottlieb (1973*b*); *PGM*: 75 mg disodium alpha-D-glucose-1-phosphate, 5 ml 0.00017 M dipotassium alpha-D-glucose-1, 6-diphosphate, 5 ml 0.1 M MgCl₂, 40 units glucose-6-phosphate dehydrogenase, 5 mg NADP, 10 mg MTT, 2 mg PMS, 90 ml 0.02 M Tris-HCl pH 8.0; *SOD*: incubate *G6PD* in light.

In general all enzyme systems were examined in each individual. However, *ADH* was assayed independently because it was obtained from crude extracts of whole germinated seedlings (age: 72–96 hrs.), crushed in two drops of grinding buffer. Not all individuals were run for *MDH* and *ME*. Enzyme patterns reported are sharp discrete bands repeatable on different leaves from the same individual. All individuals which varied from the common pattern were reexamined to confirm the difference. Plants from different populations and species were run side by side on the same gel, in several combinations, to compare enzyme mobilities. Enzymes are identified by their migration distance in mm from the origin when the brown borate front (bromphenol blue on histidine gels) migrated 80 mm; assignment of *ADH* is based on a 90 mm front. *LAP* enzymes were clearly resolved in *T. dubius* and both tetraploid species by doubling the duration of electrophoresis. In these buffer systems, all enzymes migrated to the anode except *PER* and *ADH-3* which migrated to the cathode.

The enzymes were assigned to genes on the basis of criteria described previously (Gottlieb, 1973*b*). In addition to the surveys of populations, progeny tests were made of seedlings grown from seeds collected from single open-pollinated plants in nature. *ADH* was examined in a large number of progenies of all the species except *T. pratensis* which showed no variation. Segregations in *Est-4*, *Lap-1*, and *GDH* were also studied in a few progenies. In certain cases, chromosome counts were made of mitotic cells in root tips and the same individual was also assayed for its enzyme phenotype. In the complex enzyme systems, i.e., those specified by more than one gene (*EST*, *LAP*, *ADH*, *GOT*), the enzymes from the different diploid species were assigned to what are considered homologous genes on the basis of similari-

ties of their electrophoretic mobility and, in esterase, substrate preference was also used. Five of the systems (*PGM*, *PGI*, *MDH*, *APH*, and *SOD*) showed additional regions of activity that did not resolve well and these are not reported.

RESULTS

Genetic variation in the diploid species.— The 13 enzyme systems appear to be specified by a minimum of 21 genes: *Est-1*, *Est-2*, *Est-3*, *Est-4*, *Lap-1*, *Lap-2*, *Aph*, *Per*, *Got-1*, *Got-2*, *Got-3*, *Gdh*, *Mdh*, *Pgi*, *Adh-1*, *Adh-2*, *Pgm*, *Adh-3*, *G6pd*, *Me*, *Sod*. Very little or no variation was observed within the three diploid species. None of the genes was polymorphic in *T. pratensis*, only two genes (*Adh-2* and *Est-2*) were polymorphic in *T. porrifolius*, and the most widespread species, *T. dubius*, was polymorphic at four genes (*Est-4*, *G6pd*, *Got-3*, *Adh-1*) (Table 2). Genes were considered monomorphic when all individuals in a species displayed a single enzyme band with the same electrophoretic mobility. Only a few heterozygous individuals were observed in the survey of the populations: in *T. porrifolius*, a single individual was heterozygous at *Adh-2* in two populations; and in *T. dubius*, a single individual was heterozygous at *Est-4*. Additional heterozygous individuals were found in several of the progeny tests which suggests that more would have been detected in the surveys with larger sample sizes.

All the individuals studied in *T. porrifolius* and *T. pratensis* displayed three *GOT* enzymes with the same mobilities and these are presumed to be specified by three genes. In *T. dubius*, all populations except Oak Creek expressed only the two more anodal of these enzymes. The Oak Creek populations was polymorphic for the presence and absence of the third enzyme. For the present purposes, we consider that the presence of the enzyme is specified by an "active" allele of *Got-3* and its absence by a "null" allele. Thus, the other

TABLE 2. *Frequencies of alleles at 12 genes that differentiate* Tragopogon porrifolius, T. dubius, *and* T. pratensis. *The alleles are identified by the migration from the origin in mm of the enzymes.* N *is the average number of individuals scored for each enzyme system.*

Locus/ allele	T. porrifolius			T. dubius						T. pratensis
	Cor	Pal	Pul	Tac	Por	Mos	Bur	Oak	Ros	3 pop.
N	29	28	27	36	30	19	19	19	18	67
Est-2										
54				1.0	1.0	1.0	1.0	1.0	1.0	
57										1.0
59	1.0	1.0	.89							
62			.11							
Est-3										
44				1.0	1.0	1.0	1.0	1.0	1.0	1.0
53	1.0	1.0	1.0							
Est-4										
35				1.0	.27	.65	1.0	1.0	1.0	1.0
40	1.0	1.0	1.0		.73	.35				
Lap-1										
30	1.0	1.0	1.0							
32				1.0	1.0	1.0	1.0	1.0	1.0	
35										1.0
Lap-2										
25				1.0	1.0	1.0	1.0	1.0	1.0	
27										1.0
29	1.0	1.0	1.0							
Aph										
49										1.0
52	1.0	1.0	1.0							
55				1.0	1.0	1.0	1.0	1.0	1.0	
Gdh										
16	1.0	1.0	1.0							
21				1.0	1.0	1.0	1.0	1.0	1.0	1.0
G6pd										
29	1.0	1.0	1.0		.03		.05			
31				1.0	.97	1.0	.95	1.0	1.0	1.0
Adh-1*										
30				.57	.23	.67	.05	.45	.80	
35	1.0	1.0	1.0							1.0
40				.43	.77	.33	.95	.55	.20	
Adh-2										
18										1.0
25	.34	.98	.11							
30	.66	.02	.89	1.0	1.0	1.0	1.0	1.0		
null									1.0	
Adh-3										
5										1.0
15	1.0	1.0	1.0	1.0	1.0	1.0	1.0	1.0	1.0	
Got-3										
25	1.0	1.0	1.0					.20		1.0
null				1.0	1.0	1.0	1.0	.80	1.0	

* Allele frequencies in *T. dubius* are based on the assumption that when allele 40 is present it is homozygous, except in Rosalia (see text).

242

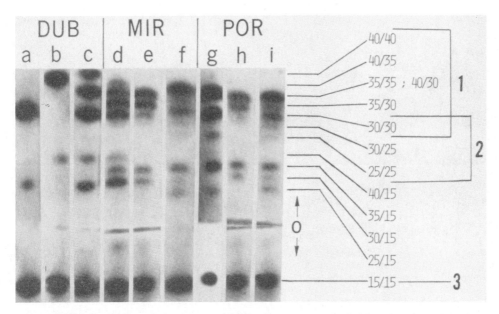

Fig. 1. Electrophoretic phenotoypes of alcohol dehydrogenases in the diploid species *Tragopogon dubius* and *T. porrifolius* and their tetraploid derivative *T. mirus*. The isozymes are specified by three genes, designated *Adh-1*, *Adh-2* and *Adh-3*; the subunit composition of each isozyme is indicated on the right. The origin is indicated by O, and the arrows show the directions of migration.

populations of *T. dubius* are considered to be homozygous for the null allele. An alternative hypothesis that *T. dubius* has only two *GOT* genes and that the additional variant in the Oak Creek population represents a polymorphism at *Got-2* was rejected because 1) no heteromeric enzyme with intermediate mobility was observed (*GOT* is dimeric in maize (MacDonald and Brewbaker, 1972), wheat (Hart, 1975), and *Stephanomeria* (Gottlieb, 1973a), and regularly expresses a third variant with intermediate mobility in heterozygous individuals): and 2) progeny tests did not show segregation for the difference.

The *ADH* phenotypes of the three diploid species are illustrated in Figures 1, 2, and 3. The genetic control of the enzymes has been tentatively inferred (pending completion of formal genetic analysis) on the basis of evidence from progeny tests of open-pollinated plants from the field collections, the additivity of the *ADH* profiles in the tetraploid species, and dif-

ferences in enzyme expression in individuals of different age and environmental condition. These lines of evidence suggest that the *ADH* enzymes are specified by three genes (designated *Adh-1*, *Adh-2*, and *Adh-3*) which code subunits that associate to form both intra- and intergenic enzymes similar to the *ADH*s in maize (Freeling and Schwartz, 1973) and sunflower (Torres, 1974). All three genes are expressed in roots of one-to-seven-day-old seedlings (Fig. 3). The *Adh-1* and *Adh-2* enzymes migrate to the anode whereas the *Adh-3* enzyme migrates to the cathode. Intergenic heteromers specified jointly by *Adh-3* with both *Adh-1* and *Adh-2* migrate anodally and are located approximately halfway between the intragenic enzymes. Older rosette plants do not express any *ADH* unless they are flooded for at least three days. Following such induction, the enzymes specified only by *Adh-1* and *Adh-2* are present in roots and they have electrophoretic patterns

243

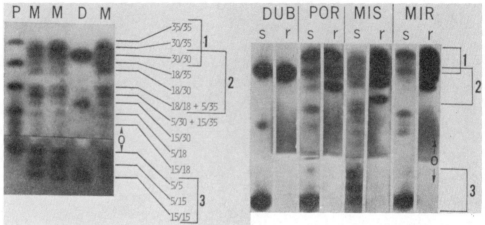

FIG. 2. Electrophoretic phenotypes of alcohol dehydrogenases in the diploid species *Tragopogon dubius* and *T. pratensis* and their tetraploid derivative *T. miscellus*. The isozymes are specified by three genes, designated *Adh-1*, *Adh-2* and *Adh-3*; the subunit composition of each isozyme is indicated on the right. The origin is indicated by O, and the arrows show the directions of migration.

FIG. 3. Electrophoretic phenotypes of alcohol dehydrogenases in seedlings (S) and roots of induced rosette plants (R) in *Tragopogon dubius*, *T. porrifolius*, *T. miscellus* and *T. mirus*.

similar to those in the young seedlings (Fig. 3).

The *ADH* phenotype of *T. pratensis* was uniform in all individuals and consisted of five enzyme bands in seedlings but only the three anodal ones in older rosette plants (Fig. 2). Three different *ADH* phenotypes were found in *T. porrifolius*. Two of them (phenotypes G and H) occur when *Adh-2* is homozygous for allele 25 and 30, respectively (Fig. 1). Individuals heterozygous for both alleles at *Adh-2* have nine enzyme bands in seedlings (phenotype I) with only the five anodal bands present in older plants (Fig. 1). Three *ADH* phenotypes were also observed in *T. dubius*. Phenotype C is comparable to phenotypes G and H of *T. porrifolius* in number of enzymes (Fig. 1). The three banded phenotype A in seedlings (older individuals have only the single anodal band) presumably resulted from *Adh-1* and *Adh-2* specifying electrophoretically indistinguishable enzymes (Fig. 1). The Rosalia population of *T. dubius* displayed phenotype B in addition to phenotypes A

and C (Fig. 1). Single field-collected individuals from this population produced progeny individuals with all three phenotypes, suggesting that the mother plants were heterozygous at *Adh-1* for different alleles. This interpretation requires that *Adh-2* in Rosalia was homozygous for a null allele that has not yet been detected in other conspecific populations.

Genetic differentiation among diploid species.—The three diploid species of *Tragopogon* were monomorphic for the same allele at nine of the 21 genes examined: *Est-1*, *Got-1*, *Got-2*, *Per*, *Mdh*, *Pgi*, *Pgm*, *Me*, and *Sod*. They were strongly differentiated at the other genes. *Tragopogon porrifolius* was the most distinct species since it did not share alleles with either of the other diploid species at six genes, and was monomorphic for an allele at two genes which were rare in *T. dubius* and absent in *T. pratensis* (Table 2). *Tragopogon dubius* and *T. pratensis* were fixed for alternate alleles at seven genes (Table 2). Nearly half of the genes distinguished *T. pratensis* and *T. porrifolius* (Table 2). None of the genes was polymorphic in the three species for the same alleles. This substantial allelic divergence among the three diploid species (33% to 48% of the genes fixed for

TABLE 3. *Electrophoretic phenotypes in the tetraploid species of* Tragopogon *which illustrate additivity of enzymes specified by alleles inherited from their diploid parents:* T. mirus (T. dubius × T. porrifolius) *and* T. miscellus (T. dubius × T. pratensis). *A dash indicates that an enzyme phenotype in the tetraploid was not additive (i.e., the electrophoretic mobility of the enzyme in the two diploid parents was the same). The enzymes are identified by their migration from the origin in mm.*

Enzyme	Enzyme Phenotypes				
	porrifolius	→ mirus ←	dubius	→ miscellus ←	pratensis
EST-2	59	54/59	54	54/57	57
EST-3	53	44/53	44	–	–
EST-4	40	35/40	35	–	–
LAP-1	30	30/32	32	32/35	35
LAP-2	29	25/29	25	25/27	27
APH	52	52/55	55	49/55	49
GDH	16	16/21	21	–	–
G6PD	29	29/31	31	–	–
ADH-1	35	30/35;35/40	30;40	30/35	35
ADH-2	–	–	30	18/30	18
ADH-3	–	–	15	5/15	5

alternate alleles depending on the pair of species compared) is highly concordant with the sharp differentiation among them in many morphological characters (Ownbey, 1950).

Enzyme multiplicity in the tetraploid species.—Since the diploid species possess different alleles at a very high proportion of their genes, their tetraploid derivatives receive both of them and, therefore, the tetraploid genomes produce substantially heterozygous phenotypes. Thus, all populations of *T. mirus* were fixed or nearly so in a heterozygous state for eight to ten of the 21 genes examined, depending on the population, and *T. miscellus* was fixed in the heterozygous state for seven genes, including five in common with *T. mirus* (Table 3). For those genes at which both diploid parents had the same single variant, the tetraploid derivative also displayed only the same single enzyme form.

Three of the enzymes that showed additive profiles are multimeric (*G6PD*, *GDH*, and *ADH*) so that the tetraploid species expressed novel heteromeric enzymes with intermediate mobilities in addition to the homomeric ones characteristic of each diploid parent. Thus, in *T. mirus*, a single novel heteromer of the dimeric *G6PD* and at least three novel heteromers of the multimeric *GDH* were

observed. The tetraploids displayed a remarkable number of *ADH*s since these enzymes can be composed of polypeptide subunits specified by different alleles of the same gene as well as by different genes. The *ADH* phenotype of *T. miscellus* contained 13 distinguishable isozyme bands. This is because this tetraploid inherited alleles from *T. pratensis* specifying enzymes at 5 mm (cathodal), 18 mm and 35 mm from the origin, and alleles from *T. dubius* specifying enzymes at 15 mm (cathodal) and 30 mm. Heterodimeric enzymes form in all possible combinations, namely, 5/15, 5/18, 5/30, 5/35, 15/18, 15/30, 15/35, 18/30, 18/35, and 30/35. The electrophoretic phenotype includes these 10 enzymes plus the five homodimeric ones (5/5, 15/15, 18/18, 30/30 and 35/35); however, only 13 are observed because of overlaps between 18/18 and 5/35 and between 5/30 and 15/35 (Fig. 2). Thus, each of the *ADH* isozymes of *T. miscellus* can be fully accounted for by simple additivity of the polypeptide subunits specified by alleles from its two diploid parents.

The three different *ADH* phenotypes of *T. mirus* can also be interpreted as the result of the addition of subunits coded by *ADH* genes inherited from its diploid parents. Thus, phenotype D of *T. mirus*

245

TABLE 4. *Electrophoretic variation in enzymes in populations of* T. miscellus.

Enzyme Phenotype	Pullman	Spangle	4 other pop.
N	15	14	86
LAP-1			
32/35	1.0	.93	1.0
35		.07	
ADH-1			
30	.07		
30/35	.93	1.0	1.0
ADH-2			
18/30	.93	1.0	1.0
30	.07		
ADH-3			
5/15	.93	1.0	1.0
15	.07		

TABLE 5. *Electrophoretic variation in enzymes in populations of* T. mirus.

Enzyme Phenotype	Pul	Pal	Tek	Ros	4 other pop.
N	33	20	28	14	54
EST-4					
35	.03				
35/40	.84		1.0	1.0	1.0
40	.13	1.0			
APH					
52		.04			
52/55	1.0	.87	1.0	1.0	1.0
55		.09			
GDH					
16/21	1.0	.96	1.0	1.0	1.0
21		.04			
ADH-1					
30	.05		.07		
30/35	.95		.93	1.0	1.0
35/40		.93			
40		.07			
ADH-2					
25/30			.93	1.0	
30	1.0	1.0	.07		1.0

results from the addition of phenotype C of *T. dubius* and phenotype H of *T. porrifolius*; phenotype E of *T. mirus* results from the addition of phenotype A of *T. dubius* and phenotype H of *T. porrifolius*; and phenotype F of *T. mirus* adds phenotype A of *T. dubius* and phenotype G of *T. porrifolius* (Fig. 1). Phenotype E in *T. mirus* cannot be distinguished from phenotype H of *T. porrifolius* because *T. dubius* contributed alleles to *T. mirus* that specify enzymes having mobilities identical to those specified by an allele from *T. porrifolius*.

With only a few exceptions, nearly all of the genes in all of the populations of each tetraploid species were monomorphic for the same single allele. In *T. miscellus*, a single individual in the Pullman population had a unique *ADH* phenotype and one plant in the Spangle population had a unique *LAP-1* phenotype (Table 4). In both cases, the individuals had the non-additive phenotype of *T. dubius*, one of its diploid parents. Therefore, additional seedlings that had the same mother plant in the field were examined both for their enzyme phenotype and for their chromosome number in order to confirm their identification. For *LAP-1*, eight additional seedlings were also found to be homozygous

for allele 35, and the mitotic chromosome number of the two sampled individuals was 2n=24, the expected tetraploid number. For *ADH*, ten additional seedlings were studied. These segregated both the standard *T. miscellus* phenotype and phenotype A of *T. dubius*. Two individuals of each were sampled and they had the tetraploid chromosome number. Thus, in both cases, it was shown that rare individuals of *T. miscellus* possessed a non-additive phenotype for enzymes that in conspecific individuals show perfect additivity.

Very similar results were obtained in *T. mirus* which displayed more genetic variation than *T. miscellus*. The Palouse population had a unique phenotype among those studied at *EST-4* and *ADH-1*, and the Tekoa and Rosalia populations had distinctive phenotypes at *ADH-2* (Table 5). Like *T. miscellus*, a few plants of *T. mirus* displayed the phenotype of only one diploid parent for enzymes (*ADH, GDH, APH*, and *EST-4*) that showed additivity in all other conspecific individuals examined (Table 5). Chromosome counts confirmed the tetraploid chromosome number of several of the individuals with non-additive phenotypes for *ADH* and *GDH*.

DISCUSSION

The primary genetic consequence of the combination of the genomes of the diploid species of *Tragopogon* to form the tetraploid species, *T. mirus* and *T. miscellus*, is that the tetraploid species have a fixed heterozygous multi-enzyme phenotype specified by 43% and 33%, respectively, of the 21 duplicated genes examined in each of them. In contrast, only six individuals in the three diploid species were heterozygous at even a single gene. The substantial fixed phenotypic heterozygosity in the tetraploids results from their inheriting different alleles from their diploid progenitors which, in turn, reflects the complete divergence between the diploids in allelic constitution at these genes. With very few exceptions, heterozygous phenotypes were displayed by all individuals of each tetraploid species and, unlike heterozygosity at the diploid level, these phenotypes do not segregate at meiosis. Every one of the enzymes detected in each tetraploid was fully accounted for by simple additivity of polypeptide subunits specified by the alleles of its respective diploid parents. The tetraploids also expressed novel heteromeric enzymes of the multimeric *ADH*, *GDH* and *G6PD* which were not produced in either one of their diploid parents. The proportion of fixed heterozygosity in the tetraploids may actually be higher than presently identified because the electrophoretic assay only utilizes a few criteria to distinguish enzymes. Other probes, e.g., heat treatment and urea denaturation, might reveal additional fixed heterozygosity.

Many populations of both tetraploid species contained one or several individuals which did not express the characteristic additive phenotype for certain enzymes, otherwise observed in all other conspecific individuals. The general mode of origin of allotetraploid populations from single interspecific F_1 hybrid individuals suggests that descendents would be highly uniform, at least during the time period shortly following their origin and in the absence of additional hybridization. However, if even a small amount of pairing occurs between the chromosomes contributed by the two diploid parents, recombination followed by self-fertilization may lead to the formation of individuals homozygous for the chromosome segments of one of the parents. Ownbey (1950) reported the occurrence of multivalent chromosome associations in the tetraploids and it is a likely hypothesis that the occasional individuals observed in the present study that did not express additivity for single enzymes reflect past events of recombination between homoeologous chromosomes in such multivalents. Continued multivalent formation in an inbreeding polyploid will eventually eliminate enzyme multiplicity at loci where multiplicity is not favored by selection. To what extent this occurs is not clear.

The marked genetic divergence among the three diploid species at about 40% of the genes examined is fully consistent with their sharp morphological differences as well as the near total sterility of their interspecific F_1 hybrids. Ownbey (1950) described their differences firmly: "The genetic hiati are broad, sharp, and absolute, and there simply is no biological intergradation between the entities." It is of interest that two-dimensional chromatography of certain flavonoids failed to distinguish a single component that was species specific, although intra-specific variation was substantial (Brehm and Ownbey, 1965). No other group of plants has been studied as extensively both for electrophoretic variation in enzymes and flavonoid patterns and, therefore, it is not yet possible to know whether the results reported here are general. But this does seem plausible since changes in the amino acid sequences of a large number of polypeptides are more likely to reflect early

247

stages in genetic divergence than are changes in secondary metabolites such as flavonoids which are products of enzyme-catalyzed biosyntheses.

The observed patterns of enzyme additivity and variation in both tetraploid species fully confirm their ancestry as documented by Ownbey (1950). In addition, the results reveal the remarkable sensitivity of electrophoresis for discriminating populations. Thus, Ownbey (1950) and Ownbey and McCollum (1953, 1954) presented morphological and cytological evidence that the Pullman, Palouse and Tekoa populations of *T. mirus* had independent phylogenetic origins from different pairs of parents of the same two diploid species. The electrophoretic evidence is consistent with this hypothesis. The Palouse population appears to contain only the *Est-4⁴⁰* allele, provided by its *T. dubius* parent, while other populations are fixed or nearly so for *Est-4³⁵/⁴⁰*, with a different *T. dubius* parent providing their *Est-4³⁵* allele. The Palouse population is also unique in possessing *Adh-1⁴⁰*, although the allele is common in *T. dubius*. The Tekoa population (along with the recently discovered Rosalia population) has *Adh-2²⁵* of its *T. porrifolius* parent while all the other populations of *T. mirus* are monomorphic for *Adh-2³⁰*. The remaining populations of *T. mirus* form a distinctive third group possessing *Est-4³⁵/⁴⁰*, *Adh-1³⁰/³⁵*, and *Adh-2³⁰*. Flavonoid patterns distinguished the Garfield population of *T. mirus* from the Pullman population (Brehm and Ownbey, 1965). Differences between populations of *T. miscellus* in a cytoplasmic factor governing ligule length also suggested that this species had at least two independent origins (Ownbey and McCollum, 1953); the electrophoretic results furnished no information on this hypothesis.

The present range of the tetraploids is very small in comparison to that of the diploids, but further expansion seems probable. The ecological success of the tetraploids is thus far apparent in the persis-

tence of their populations since Ownbey's original discovery of them in 1949, and in the dramatic increase in the range of *T. miscellus*, now one of the most common weeds in and around Spokane, Washington. This success presumably reflects, in some part, their substantial enzyme multiplicity. The present electrophoretic results set the stage for specific *ad hoc* studies of the biochemical and physiological consequences of this multiplicity. Such studies in other species suggest strongly that heterozygous enzyme phenotypes are likely to increase biochemical versatility in a number of ways: novel properties of heteromeric enzymes (Fincham, 1972; Schwartz, 1973); increased levels of enzyme activities (Berger, 1974; Murray and Williams, 1973; Carlson, 1972); production of novel metabolites (Levy and Levin, 1974); specialization of enzyme function leading to tissue specificity and new regulatory control patterns (Bender and Ohno, 1968; Ohno, 1970); increased amounts of diploid constituents (Rick and Butler, 1956); and maintenance of sufficient flux through metabolic pathways under changed reaction conditions (Johnson, 1974).

Based on numbers of individuals and extent of geographic distribution, *T. miscellus* appears, at the present time, to be more successful than *T. mirus*. However, the electrophoretic evidence suggests that the genome of *T. mirus* is more heterozygous than that of *T. miscellus*, reflecting the greater genetic divergence between its parents. This observation reinforces an important a priori assumption that the initial success of a tetraploid species is not a simple function of its amount of enzyme multiplicity. Many other factors are relevant to the likelihood of establishment including chromosome pairing behavior which affects fertility, developmental interactions between its two genomes, and numerous aspects of its environment. The biochemical hypotheses do serve to focus attention on measurable factors that influence evolutionary success and thereby they also avoid the vague and general speculations that frequently have emerged from correlations of ploidy level with latitude, elevation and plant community.

Summary

The tetraploid species of *Tragopogon*, *T. mirus* and *T. miscellus*, are the only unambiguous examples of the recent natural origin of allotetraploid plant species. Both species originated during the present century and it is likely that their genomes have undergone little or no modification since their origin: *T. dubius* × *T. porrifolius* gave rise to *T. mirus*, and *T. dubius* × *T. pratensis* to *T. miscellus*. Thus, the two tetraploid species provide critical evidence regarding the initial genetic and biochemical consequences of allotetraploidy.

Electrophoretic analysis of variation in 13 enzyme systems specified by a minimum of 21 genes revealed that the three diploid species are monomorphic or nearly so for different alleles at about 40% of their genes, a result fully concordant with their sharp morphological separation. The combination of these divergent diploid genomes to form the tetraploid species has provided them with substantial phenotypic heterozygosity. Thus, *T. mirus* has a fixed heterozygous multi-enzyme phenotype specified by 43% of its 21 duplicated genes examined and *T. miscellus* is similarly heterozygous at 33% of its genes. In contrast, very few individuals in the diploid species were heterozygous at even a single gene.

Every one of the enzymes detected in each tetraploid species was fully accounted for by simple additivity of polypeptide subunits specified by alleles inherited from its diploid parents. The tetraploids also express novel heteromeric enzymes not produced in either one of their parents. The observed patterns of additivity in both tetraploid species fully confirm their ancestry as documented by Ownbey (1950). Their substantial enzyme multiplicity may extend the range of environments in which normal development can take place and thereby may contribute to their apparent ecological success (*T. miscellus* is one of the most common weeds in and around Spokane, Washington).

Acknowledgments

We gratefully acknowledge the generous help of the late Professor Marion Ownbey who took us to the sites of the tetraploid Tragopogons and carefully explained the intricacies of their variation patterns and relationships. We also thank R. K. Brown for collecting seeds for us of *Tragopogon mirus* from Flagstaff, Arizona, and of *T. dubius* from Oak Creek, Arizona. B. G. Brehm, Gary Hart, and G. J. Brewer provided useful comments on the manuscript. This research was supported by NSF grant GB 39873 to the junior author.

Literature Cited

Barber, H. N. 1970. Hybridization and the evolution of plants. Taxon 19:154–160.

Bender, K., and S. Ohno. 1968. Duplication of the autosomally inherited 6-phosphogluconate dehydrogenase gene locus in the tetraploid species of cyprinid fish. Biochem. Genet. 2:101–107.

Berger, E. 1974. Esterases of Drosophila. II. Biochemical studies of esterase-5 in *D. pseudoobscura*. Genetics 78:1157–1172.

Brehm, B., and M. Ownbey. 1965. Variation in chromatographic patterns in the Tragopogon dubius-pratensis-porrifolius complex (Compositae. Amer. J. Bot. 52:811–818.

Brewer, G. J., C. F. Sing, and E. R. Sears. 1969. Studies of isozyme patterns in nullisomic-tetrasomic combinations of hexaploid wheat. Proc. Nat. Acad. Sci. 64:1224–1229.

Brown, R. K., and C. G. Schaack. 1972. Two new species of Tragopogon for Arizona. Madrono 21:304.

Carlson, P. 1972. Locating genetic loci with amphiploids. Molec. Gen. 114:273–280.

Cherry, J. P., F. Katterman, and J. Endrizzi. 1972. Seed esterases, leucine aminopeptidases and catalases of species of the genus *Gossypium*. Theor. App. Gen. 42:218–226.

Efron, Y., M. Peleg, and A. Ahsri. 1973. Alcohol dehydrogenase allozymes in the safflower genus *Carthamus* L. Biochem. Genet. 9:299–308.

Fincham, J. 1969. Symposium talk, 11th International Botanical Congress, Seattle, Wash.

——. 1972. Heterozygous advantage as a likely general basis for enzyme polymorphisms. Heredity 28:387–391.

FREELING, M., AND D. SCHWARTZ. 1973. Genetic relationships between the multiple alcohol dehydrogenases of maize. Biochem. Genet. 8: 27–36.

GARBER, E. 1974. Enzymes as taxonomic and genetic tools in Phaseolus and Aspergillus. Israel J. Med. Sci. 10:268–277.

GOTTLIEB, L. D. 1973a. Genetic control of glutamate oxaloacetate transaminase in the diploid plant Stephanomeria exigua and its allotetraploid derivative. Biochem. Genet. 9: 97–107.

——. 1973b. Enzyme differentiation and phylogeny in Clarkia franciscana, C. rubicunda and C. amoena. Evolution 27:205–214.

——. 1973c. Genetic differentiation, sympatric speciation and the origin of a diploid species of Stephanomeria. Amer. J. Bot. 60:545–553.

HART, G. 1969. Genetic control of alcohol dehydrogenase isozymes in Triticum dicoccum. Biochem. Genet. 3:617–625.

——. 1970. Evidence for triplicate genes for alcohol dehydrogenase in hexaploid wheat. Proc. Nat. Acad. Sci. 66:1136–1141.

——. 1973. Homoeologous gene evolution in hexaploid wheat. Proc. 4th Intern. Wheat Genet. Symp. p. 805–810.

——. 1975. Glutamate oxaloacetate transaminase isozymes of Triticum: evidence for multiple systems of triplicate structural genes in hexaploid wheat. p. 637–657. In Markert (ed.) Isozymes III: Developmental Biology. Academic Press, N. Y.

JOHNSON, G. 1974. Enzyme polymorphism and metabolism. Science 184:28–37.

LEVY, M., AND D. LEVIN. 1974. Novel flavonoids and reticulate evolution in the Phlox pilosa-P. Drummondii complex. Amer. J. Bot. 61:156–167.

MACDONALD, T., AND J. BREWBAKER. 1972. Isoenzyme polymorphism in flowering plants. VIII. Genetic control and dimeric nature of transaminase hybrid maize isoenzymes. J. Hered. 63:11.

MANWELL, C., AND C. M. BAKER. 1970. Molecular biology and the origin of species. Univ. Wash. Press, Seattle.

MITRA, R., AND C. BHATIA. 1971. Isoenzymes and polyploidy. 1. Qualitative and quantitative isoenzyme studies in the Triticinae. Genet. Res., Camb. 18:57–69.

MURRAY, B., AND C. WILLIAMS. 1973. Polyploidy and flavonoid synthesis in Briza media L. Nature 243:87–88.

OHNO, S. 1970. Evolution by gene duplication. Springer-Verlag, N. Y.

OWNBEY, M. 1950. Natural hybridization and amphiploidy in the genus Tragopogon. Amer. J. Bot. 37:487–499.

OWNBEY, M., AND G. McCOLLUM. 1953. Cytoplasmic inheritance and reciprocal amphiploidy in Tragopogon. Amer. J. Bot. 40:788–796.

——. 1954. The chromosomes of Tragopogon. Rhodora 56:7–21.

REDDY, M., AND E. GARBER. 1971. Genetic studies of variant enzymes. III. Comparative electrophoretic studies of esterases and peroxidases for species, hybrids and amphiploids in the genus Nicotiana. Bot. Gaz. 132:158.

RICK, C., AND L. BUTLER. 1956. Cytogenetics of the tomato. Advance. Genet. 8:267–382.

SCANDALIOS, J. 1974. Isozymes in development and differentiation. Ann. Rev. Plant Physiol. 25:225–258.

SCHWARTZ, D. 1973. Single gene heterosis for alcohol dehydrogenase in maize: the nature of the subunit interaction. Theor. App. Gen. 43:117.

SCHWARTZ, D., AND W. LAUGHNER. 1969. A molecular basis for heterosis. Science 166:626.

SHEEN, S. 1972. Isozymic evidence bearing on the origin of Nicotiana tabacum. Evolution 26:143–154.

SMITH, H. H., D. HAMILL, E. WEAVER, AND K. THOMPSON. 1970. Multiple molecular forms of peroxidases and esterases among Nicotiana species and amphiploids. J. Hered. 61:203–212.

TORRES, A. 1974. An intergenic alcohol dehydrogenase isozyme in sunflowers. Biochem. Genet. 11:301–308.

YANG, S. 1971. Appendix. Studies in Gen. VI. Univ. Texas Publ. 7103:85–90.

16

Reprinted from *Natl. Acad. Sci. (USA) Proc.* **68**:1627–1630 (1971)

THE ORIGIN OF NOVEL FLAVONOIDS IN *PHLOX* ALLOTETRAPLOIDS

M. Levy and D. A. Levin

ABSTRACT   The flavonoids of two diploid species of *Phlox* and two of their allotetraploid derivatives were characterized. The tetraploids accumulate five flavonoids not observed in the parental species. Generally, novel compounds are less highly glycosidated than, but otherwise similar to, parental compounds that they partially or totally replace. We propose that hybridity and polyploidy have repressed or suppressed the activity of certain ancestral genes responsible for the production of glycosidating enzymes. Accordingly, the novel compounds are merely the accumulated or modified precursors of parental compounds.

The development of a specific morphological, physiological, or biochemical expression is dependent upon appropriate genes, the external environment, and the genetic environment. An alteration in the latter may disrupt the processes that permit normal gene activity, so that a novel expression emerges. Interspecific hybridization is one means of subjecting co-adapted gene complexes to a manifestly different genetic milieu. The disruptive nature of one or more generations of hybridization on the developmental feedback and regulatory devices in plants is seen in aberrant and unstable development (1), cleistogamy (2), tumor formation (3), variegation (3), breakdown of self-incompatibility (4), imbalance in phenolic aglycone production and glycosidation (5), elimination of organ-specific differences in phenolic compounds (6), and the formation of multiple nucleoli during meiosis (7). An increase in ploidal level also alters the genetic milieu, as seen in the repression of genes coding for ribosomal RNA or enzymes (8). Redundancy of less than the entire genome may have a similar effect (9).

The disturbance of regulatory mechanisms due to a change in genetic environment could cause allopolyploids to experience disruptions in the biosynthesis of secondary constituents such as flavonoids, which have been useful in documenting reticulate evolution. Although such disruption has not been demonstrated, a recent chromatographic survey of flavonoids in *Phlox* suggests that it may have occurred (10). The present investigation was undertaken to characterize the flavonoids of two diploid phloxes and their common allopolyploid derivatives, determine the presence of novel compounds in the latter, and decide whether such compounds might owe their origin to modifications of the ancestral genetic milieu. The species in question are *P. pilosa* L. (2n), *P. drummondii* Hook. (2n), *P. villosissima* (Gray) Whitehouse (4n), and *P. aspera* E. Nels. (4n). Consideration of morphology, ecogeography, karyology, and seed protein chemistry indicates that both tetraploids were derived from the aforementioned diploids (11). The flavonoid data are compatible with this view.

MATERIALS AND METHODS

Twenty-five flowering plants of each species were individually chromatographed to confirm previously described chromatographic patterns and determine which non-anthocyanin flavonoids were reliable genomic markers. Single compounds were isolated and purified by ascending two-dimensional paper chromatography. The first dimension was developed in butanol–acetic acid–water 4:1:2, and the second in acetic acid–water 1:10. 300–700 chromatograms were used to obtain an adequate amount of material for analysis (about 3 mg). In order to determine the oxygenation pattern of each compound, we recorded absorption spectra in spectral methanol alone, and with sodium methoxide, neutral and acidic aluminum chloride, sodium acetate, and boric acid, before and after acidic hydrolysis of oxygen-linked sugars. The procedure is described by Mabry *et al.* (12). Compounds were hydrolyzed with 7% sulfuric acid. The hydrolyzable sugars were identified by paper chromatography according to the method of Pridham (13). The C-glycosidic moieties of some compounds have not been completely characterized. However, the equivalence of compounds has been judged by cochromatography.

RESULTS

All the *Phlox* compounds proved to be glycoflavones, i.e., sugars were attached to the flavone nucleus by carbon linkages rather than oxygen linkages. All of the compounds were glycoflavone derivatives of either apigenin (4',5,7,-trihydroxyflavone) or luteolin (3',4',5,7-tetrahydroxyflavone). The tentative determinations of the glycoflavones are presented in Table 1. *O*-Rhamnosyl-6-*C*-xylosyl derivatives of luteolin and apigenin have been identified in a *P. drummondii* cultivar by UV, NMR, and mass spectroscopy (14). The presence of these compounds in the native *P. drummondii* was supported by cochromatography with the known xylosyl derivatives.

Fifteen compounds were isolated from the four species. *Phlox drummondii* contained eight flavonoids, four of which occurred exclusively in that species. *Phlox pilosa* contained four flavonoids, one of which was species-specific. *Phlox aspera* and *P. villosissima* accumulate compounds of their diploid progenitors, but not all of them. The former species also accumulates three non-parental compounds, and the latter species four non-parental components (Table 1). Some of the non-parental compounds were shared by both tetraploids, others were not, so that when both tetraploids are considered we find five non-parental (herein referred to as novel) compounds. Four of the five are *C*-monoglycosides; the fifth is an *O-C*-diglycoside. The parental complement contains only one *C*-

TABLE 1. *Flavonoids of diploid and tetraploid Phlox*

Apigenin derivatives	Species*	Luteolin derivatives	Species*
Vicenin (6,8-di-C-glycoside)	D,P,A,V	Lucenin (6,8-di-C-glycoside)	D,P,A,V
6-C-xylosyl-O-rhamnoside	D,A,V	6-C-xylosyl-O-rhamnoside	D
6-C-xylosyl-O-glucoside	A†,V†	6-C-glycoside$_1$-O-rhamnoside	D,A,V
6-C-glycosyl$_4$-O-rhamnoside‡	D	6-C-glycoside$_1$	A†,V†
		8-C-glycoside$_1$	A†
6-C-glycosyl$_4$-O-glycoside	D	6-C-glycosyl$_2$-di-O-xyloside§	P
6-C-glycoside$_3$	V†	6-C-glycosyl$_3$	V†
		6-C-glycosyl$_4$-O-rhamnoside	D
		6-C-glycoside$_5$	P,A

* Species designations: D = *P. drummondii*; P = *P. pilosa*;
A = *P. aspera*; V = *P. villosissima*.

† Novel compound.

‡ Numerical subscripts refer to type of sugar. Glycosides with the same subscript are judged equivalent.

§ May simply be 6-C-glycosyl$_2$-O-xyloside.

monoglycoside (Table 1). The novel compounds have not been detected in minor concentrations or in rare variants in either *P. pilosa* or *P. drummondii*.

DISCUSSION

The presence of novel glycoflavones has been demonstrated in the allotetraploids *P. aspera* and *P. villosissima*. All of these compounds are closely related in biogenesis to compounds found in their progenitors. The principal difference between the novel and parental compounds lies in the number of oxygen-linked sugars attached to the basic glycoflavone. The novel compounds usually are of less complex structure, and may replace certain parental compounds. The situation in *Phlox* is in contrast to the additive profiles typical of allopolyploids in other genera (15). The question thus arises as to the origin of the novel products.

As a prelude to a specific hypothesis, let us consider the biosynthesis and genetics of flavonoids. Flavonoids are constructed from a condensation of phenylpropanoid and polyacetate-derived C_6 moieties (16). *In vivo* and *in vitro* studies have demonstrated that *C*-linked sugars are added to the developing flavonoid before the ring is formed, and that *O*-linked sugars are added after ring formation, as are other substituents (17). Moreover, it is clear that flavonoids are assembled along a stepwise biosynthetic pathway which may branch and which may be blocked at various points. The genetic control of glycosidation also has been demonstrated. One of the best-known studies involves *Streptocarpus hybrida*, which has five genes controlling anthocyanidin glycoside synthesis (18). Four of the genes independently mediate specific steps in the biosynthetic pathway. The simple genetic control of steps in flavonoid biosynthesis seems to be the rule rather than the exception (19).

It seems reasonable to assume that flavonoid glycosidation in *Phlox* occurs in well-defined steps, most of which are mediated by single genes. If we accept this assumption, we may construct putative biosynthetic pathways showing the biogenetic relationships of the novel and parental compounds.

As seen in Fig. 1, novel compounds may be only a step removed from parental compounds. *Phlox villosissima* and *P. aspera* accumulate relatively large quantities of luteolin 6-C-glycoside$_1$ and luteolin 6-C-glycosyl$_1$-O-rhamnoside, the former being absent and the latter being present in *P. drummondii* (Fig. 1, I). *Phlox villosissima* accumulates luteolin 6-C-glycoside$_2$, whereas it is absent from *P. pilosa*, which contains luteolin 6-C-glycosyl$_2$-O-diglycoside (Fig. 1, II). *Phlox villosissima* also accumulates apigenin 6-C-glycoside$_3$, whereas *P. drummondii* contains apigenin 6-C-glycosyl$_3$-O-glycoside (Fig. 1, III). In these cases, perhaps the enzymes necessary for the terminal step in biosynthesis are not functioning or are doing so at undetectable levels.

Other novel compounds are not simply precursors of parental compounds. Both allotetraploids accumulate apigenin 6-C-xylosyl-O-glucoside and apigenin 6-C-xylosyl-O-rhamnoside. *Phlox drummondii* accumulates the latter but not the former (Fig. 1, IV). Ostensibly, a new enzyme is present in the tetraploids which is capable of effecting an O-sugar linkage with xylose. The tetraploids have a branched biosynthetic pathway with apigenin 6-C-xyloside serving as a common precursor. The allotetraploids and *P. drummondii* contain luteolin 6-C-glycosyl$_1$-O-rhamnoside, although it is present in low concentration in the former. On the other hand, *P. aspera* and *P. villosissima* are accumulating relatively large quantities of the putative precursor of this glycoside; and *P. aspera* opened a new pathway leading from a precursor to luteolin 8-C-glycoside$_1$ (Fig. 1, V).

The partial or complete blocks in flavonoid biosynthesis inferred thus far have been late in the biosynthetic pathway. Blocks ostensibly exist at earlier stages of flavonoid biosynthesis as well, for parental compounds are not invariably replaced by less complex ones in the allotetraploids. Two compounds not having close relatives in the latter are luteolin 6-C-xylosyl-O-rhamnoside and luteolin 6-C-glycosyl$_4$-O-rhamnoside, both of which are produced by *P. drummondii* (Fig. 1, VI and VII).

We propose that the complete or partial blocks in the aforementioned pathways may be due to repression or suppression of gene action brought about by the interaction of disparate genomes and polyploidy. Suppression is evident in the nearly complete failure of some biosynthetic steps to be accomplished and in the accumulation of the immediate precursor. Repression can only be inferred. We cannot discount the possibility that genes were lost prior to or after chromosome doubling. However, the chance that genes were lost prior to chromosome doubling actually is remote, for F_1 hybrids between *P. pilosa* and *P. drummondii* are sterile (10). Accordingly, the first steps in the evolution of the tetraploids must not have included reassortment of genome components at the diploid level.

The new biosynthetic capabilities of the tetraploids (Fig. 1, IV and V) may be due to derepression of parental genes, the reconstruction of ancestral biosynthetic pathways through the complementation of related genomes, or the evolution of new biosynthetic pathways. If the first alternative is applicable, the genes in question must have been contributed by *P. drummondii* because its biosynthetic pathways have substrates (apigenin 6-C-xyloside and luteolin 6-C-glycoside$_1$) upon which the activated genes may act. *Phlox pilosa* lacks these substrates.

There is considerable evidence in the literature demonstrat-

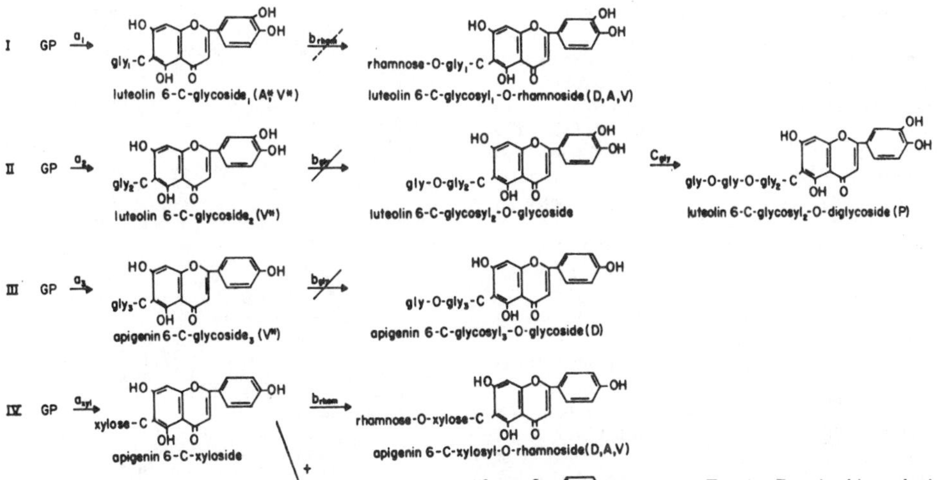

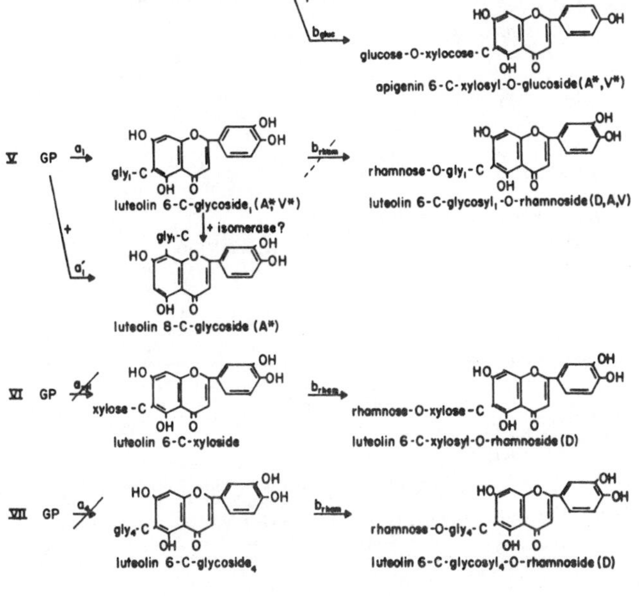

Fig. 1. Putative biosynthetic pathways (I–VII). Dashed diagonal lines refer to gene suppression, solid diagonal lines to gene repression, and + to gene derepression. Glycoflavone precursors (GP) are 6-*C*-glycosidated by "*a*" enzymes and *C*-monoglycosides are *O*-glycosidated by "*b*" enzymes. Species designations in parentheses are the same as in Table 1, and indicate in which taxa products are accumulated. Asterisks indicate that the products are novel. The novel isomeric change in V "a_1'" may be due to either an isomerase or an 8-*C*-glycosidase, "a_1'".

ing the disruptive effect of hybridity and polyploidy on genetic regulatory mechanisms. Consider first the effect of hybridity on such mechanisms. Interspecific hybrids in *Lilium* fail to glycosylate ferulic acid in a normal fashion, 60% of the ferulic acid in a hybrid seed being free, as compared to 6% in intraspecific seed (5). Interspecific hybrids in *Nicotiana* may manifest positive or negative chemical heterosis for phenol content and for the associated polyphenoloxidase and peroxidase systems (20). An example of regulatory gene suppression comes from *Dicentra*, where hybrids accumulate flavonoids less highly glycosylated than the parental compounds, while the parents show only a trace of the precursor (21). Suppression also is suggested in *Phaseolus*, where major constituents of species hybrids occur as trace components in one species (22). The disruption of regulatory mechanisms in hybrids is not unique to plants. The inhibition of the autosomally inherited gene for liver alcohol dehydrogenase has been noted in chicken–quail hybrids (23). An asynchronous activation of parental alleles at tissue-specific gene loci occurs in trout hybrids during early stages of development (24), and for some enzymes the maternal isozyme pattern may persist through these stages (25). The presence of only maternal gene activity during early development has also been reported in *Rana* hybrids (26). Interactions between genomes affecting the gene activity are evident in somatic cell hybrids as well (27).

Consider next the effect of polyploidy on gene activity. In plants, the alterations of physiological and biochemical properties following chromosome doubling have long been known (28). The effect of such doubling on specific genes is only now coming to light. The repression of redundant loci is suggested in *Triticum*, where the number of isozymes in hexaploid wheat is not a summation of the number of isozymes of the parental genomes; only one of 12 enzyme systems studied was polymorphic (29). In a similar vein, only the nucleolar organizers of one genome are operative in allotetraploids of several genera (7). Evidence has been presented for the duplication of alcohol dehydrogenase genes in maize, one of which is normally repressed but can be derepressed under stress (8). The biosynthesis of flavonoids has been upset in autotetraploids of raspberries and *Phlox subulata* by the presence of novel compounds (30, and unpublished observations). The compounds have not been characterized, but may well be precursors that are accumulated as a result of gene suppression or repression. Turning to animals, we note that there is repression of an immunohemoglobin in malignant plasma cells of the mouse, the amount of immunohemoglobin produced being an inverse function of ploidal level in cell clones (31).

Schwartz (9) proposes that the repression of redundant loci may be the rule rather than the exception. Ohno (32) contends that all regulated structural genes in a newly arisen tetraploid may be subjected to super-repression, which may be escaped by the simultaneous functional diversification of redundant regulator and structural genes. He maintains that the concordant duplication of a regulatory gene and a structural gene upsets the balance between the two in favor of the former. This view is in accord with that of Britten and Davidson (33), who state the "the likelihood of utilization of new DNA for regulation is far greater than the likelihood of invention of a new and useful amino acid sequence...". Following these arguments, one must conclude that a polyploid contains a very large number of redundant and (or) silent genes.

In conclusion, there is ample evidence that alterations in the genetic environment accomplished via hybridization or polyploidy may repress or suppress gene activity. Moreover, it is clear that flavonoid glycosidation occurs in a stepwise fashion, with most steps being under single gene control. Accordingly, it is feasible to relate the origin of the novel flavonoids in *Phlox* to such altered gene activity in a novel genetic environment.

The authors express sincere thanks to Drs. V. Grant, C. L. Markert, T. Mabry, A. W. Galston, and N. Giles for their critical reading of the manuscript. We are indebted to T. Mabry for providing the xylosyl derivatives used in cochromatography. This investigation was supported in part by the Society of the Sigma Xi and by grants GB-6743 and GB-17987 from the National Science Foundation.

1. Reviewed by Stebbins, G. L., *Advan. Genet.*, **9**, 147 (1958); Levin, D. A., *Amer. Natur.*, **104**, 343 (1970).
2. Mather, K., and A. Vines, *Heredity*, **5**, 196 (1951); Khoshoo, T. N., R. C. Mehra, and K. Bose, *Theor. Appl. Genet.*, **33**, 133 (1969).
3. Smith, H. H., *Advan. Genet.*, **14**, 1 (1968).
4. Martin, F. W., *Genetics*, **60**, 101 (1968).
5. Asen, S., and S. L. Emsweller, *Phytochemistry*, **1**, 1969 (1962).
6. Alston, R. E., and J. Simmons, *Nature*, **195**, 825 (1962); Alston, R. E., H. Rosler, K. Naifeh, and T. J. Mabry, *Proc. Nat. Acad. Sci. USA*, **54**, 1458 (1965).
7. Walker, S., *Ann. Mo. Bot. Gard.*, **56**, 261 (1969); Levin, D. A., and M. Levy, *Brittonia*, **23**, in press.
8. Navasin, M., *Cytologia*, **5**, 169 (1934); Day, A., *Aliso*, **6**, 25 (1965); Levin, D. A., *Evolution*, **22**, 612 (1968); Sing, C. F., and G. J. Brewer, *Genetics*, **61**, 391 (1969).
9. Schwartz, D., *Proc. Nat. Acad. Sci. USA*, **56**, 1431 (1966).
10. Levin, D. A., *Evolution*, **22**, 612 (1968).
11. Levin, D. A., and B. A. Schaal, *Amer. J. Bot.*, **57**, 977 (1970).
12. Mabry, T. J., K. R. Markham, and M. B. Thomas, *The Systematic Identification of Flavonoids* (Springer-Verlag, New York, 1970).
13. Pridham, J. B., *Anal. Chem.*, **28**, 1967 (1956).
14. Mabry, T., H. Yoshioka, S. Sutherland, S. Woodland, W. Rahman, M. Ilyas, J. Usmani, R. H. Rizui, and J. Chopen, *Phytochemistry*, in press.
15. Summarized by Alston, R. E., in *Evolutionary Biology*, ed. T. H. Dobzhansky, M. K. Hecht, and W. C. Steere (Appleton-Century-Crofts, New York, 1967), Vol. 1.
16. Harborne, J. B., *Comparative Biochemistry of the Flavonoids* (Academic Press, New York, 1967); Grisebach, H., in *Recent Advances in Phytochemistry*, ed. T. J. Mabry, R. E. Alston and V. C. Runeckles (Appleton-Century-Crofts, New York, 1968), Vol. 1.
17. Barber, G. A., *Biochemistry*, **1**, 463 (1962); Barber, G. A., *Arch. Biochem. Biophys.*, **7**, 204 (1962); Wallace, J. W., T. J. Mabry, and R. E. Alston, *Phytochemistry*, **8**, 93 (1969).
18. Lawrence, W. J. C., and V. C. Sturgess, *Heredity*, **11**, 303 (1957); Harborne, J. B., *Phytochemistry*, **2**, 85 (1963).
19. Alston, R. E., in *Biochemistry of Phenolic Compounds*, ed. J. B. Harborne (Academic Press, New York, 1964).
20. Sheen, S. J., *Theor. Appl. Genet.*, **40**, 45 (1970).
21. Fahselt, D., and M. Ownbey, *Amer. J. Bot.*, **55**, 334 (1968).
22. Schwartze, P., *Planta*, **54**, 152 (1959).
23. Castro-Sierra, E., and S. Ohno, *Biochem. Genet.*, **1**, 323 (1968).
24. Hitzeroth, H., J. Klose, S. Ohno, and U. Wolf, *Biochem. Genet.*, **1**, 287 (1968).
25. Klose, J., H. Hitzeroth, H. Ritter, E. Schmidt, and U. Wolf, *Biochem. Genet.*, **3**, 91 (1969).
26. Wright, D. A., and F. H. Moyer, *J. Exp. Zool.*, **163**, 215 (1966).
27. Davidson, R., in *Heterospecific Genome Interaction*, ed. V. Defendi (Wistar Institute, Philadelphia, 1969).
28. Noggle, G. R., *Lloydia*, **10**, 19 (1947).
29. Sing, C. F., and G. J. Brewer, *Genetics*, **61**, 391 (1969).
30. Haskell, G., *Heredity*, **23**, 129 (1968).
31. Cohn, M., *Cold Spring Harbor Symp. Quant. Biol.*, **32**, 211 (1967).
32. Ohno, S., *Evolution by Gene Duplication* (Springer-Verlag, New York, 1970).
33. Britten, R. S., and E. H. Davidson, *Science*, **165**, 349 (1969).

17

Copyright ©1974 by Plenum Publishing Corporation
Reprinted from *Biochem. Genet.* **12**:429-440 (1974)

Polyploidy and Gene Dosage Effects on Peroxidase Activity in Ferns

A. E. DeMaggio[1] and J. Lambrukos[2]

Received 9 July 1974—Final 13 Aug. 1974

Total soluble proteins, peroxidase, and peroxidase isozymes were examined in polyploid series of fern gametophytes and sporophytes. A distinctive pattern of protein bands was associated with gametophytes and sporophytes and the pattern did not vary within each phenotype with increases in the genome. Peroxidase activity per cell increased in direct proportion to increases in the genome and was determined to be gene dosage related. Slight differences in the patterns of peroxidase isozyme bands were associated with increases in the chromosome complement in both series of plants, but major variations were found between gametophyte and sporophyte. Quantitative analysis of peroxidase activity in each band revealed both increases and decreases in individual isozymes as ploidy increased. These findings suggest the involvement of regulatory mechanisms controlling isozyme activity.

KEY WORDS: gene dosage; peroxidase; polyploidy; fern.

INTRODUCTION

Gene duplication by polyploidy is recognized as an effective mechanism for increasing the number of both structural and regulatory cistrons in the genome. For many organisms, bacteria, plants, and animals, a direct correlation exists between gene dosage and the amount or activity of a particular protein. While exceptions to the gene dosage–protein relationship are known (Becak and

---

[1] Department of Biological Sciences, Dartmouth College, Hanover, New Hampshire.
[2] Dartmouth Medical School, Hanover, New Hampshire.

Pueyo, 1971), it has been suggested (Carlson, 1972) that the presence of additional genes increases the overall rate of transcription and results in increases in activity. Other hypotheses to account for increased protein in cells of higher ploidy have been proposed (Ohno, 1970). Aside from speculation on the role of polyploidy in evolution, there is little evidence for a direct role of increased genetic material during development and differentiation.

To study morphological and biochemical changes accompanying gene duplication, we have utilized polyploid series of fern gametophytes ($1n$, $2n$, and $4n$) and sporophytes ($2n$ and $4n$). These plants possess conspicuous alternation of generations, and gametophytes and sporophytes are phenotypically different. It is therefore possible to investigate changes which occur through duplication of the genome in phenotypically similar forms, either gametophytes or sporophytes. Also, it is possible to compare changes taking place in phenotypically different but genotypically identical forms, for example, $2n$ or $4n$ gametophytes vs. $2n$ or $4n$ sporophytes.

In previous studies (DeMaggio et al., 1971), we determined that in the polyploid fern series cell and nuclear volumes increase in relationship to increases in gene dosage. Also, studies of cell cycle kinetics (Hannaford and DeMaggio, 1970) demonstrated that the duration of DNA synthesis is similar for cells of different ploidies. These results suggest that, in addition to increases in structural features of the cell, an increase in the number of genes is accompanied by proportional increases in enzymes associated with DNA synthesis. The present study was undertaken to establish a direct relationship between gene dosage and protein in this polyploid series.

Photosynthetic and respiratory rates do not appear to be gene dosage related for these plants (DeMaggio and Stetler, 1971); therefore, enzymes participating in primary metabolic pathways or enzymes primarily localized within these organelles were not considered. Peroxidase was chosen for this investigation since it has been studied extensively in plants, is predominantly found in the soluble fraction of the cell, and has multiple molecular forms which may be associated with disease resistance and plant growth.

MATERIALS AND METHODS

The polyploid gametophytes and sporophytes of *Todea barbara* (L.) Moore were grown under sterile conditions as previously described (DeMaggio, 1961). Plants were removed from culture tubes 71–74 days after the last subculture, separated, and weighed. Plants used for extraction and separation of total protein were treated as described by Caponetti et al. (1972). For peroxidase isolation and isozyme analysis, the procedure employed was essentially that of Seevers et al. (1971). Total peroxidase activity of the extracts was determined using p-phenylenediamine as substrate (Luck, 1963). Peroxidase activity was calculated by plotting absorbance vs. time for experimental and blank cuvettes. Horseradish peroxidase served as standard, and the reaction rate was expressed as purpurogallin units of peroxidase activity. Protein determinations were performed according to the procedure outlined by Goldthwaite and Bogorad (1971).

The electrophoresis system of Seevers *et al.* (1971) was used. The volume of sample loaded on each gel never exceeded 25 μl, and volumes were adjusted to provide equal amounts of peroxidase activity. Gels were stained at room temperature in 0.332 M sodium acetate, pH 5.0, containing 1.3 mM benzidine and fresh 1.3 mM hydrogen peroxide for 6 hr or overnight. Gels were scanned with a Gilford model 224 linear scanner at 340 nm. R_f values were calculated by assigning the value of 1.00 to the fastest-moving isozyme band. Quantitative analysis of isozyme bands was accomplished by calculating the area under each peak on the scan using a compensating polar planimeter and representing this value as a percentage of the total scan area.

RESULTS

Total Soluble Proteins

Separation and examination of total soluble proteins from gametophytes and sporophytes revealed several distinct, measurable, major protein bands (Fig. 1). In the gametophyte series $1n$, $2n$, and $4n$, plants displayed similar banding patterns. Quantitative analysis of gels loaded with equivalent amounts of protein indicated that slight variation in intensity existed in one or two bands. However, the majority of separated bands were of equal intensity in $1n$, $2n$, and $4n$ plants.

The pattern of proteins separated from sporophytes was almost identical in $2n$ and $4n$ plants. Some minor differences in band intensity between the two genotypes were noted, but in general the separated proteins appeared quantitatively and qualitatively similar in $2n$ and $4n$ sporophytes.

A comparison of protein banding pattern between gametophytes and sporophytes indicated marked variations in the protein spectra of these morphologically dissimilar plants. While the methods employed permitted separation of only a few of the many soluble proteins, the data do suggest that the protein pattern observed is correlated with the plant's phenotypic expression rather than with its genotype.

Total Peroxidase Activity

For measurements of peroxidase activity and analyses of isozyme patterns,

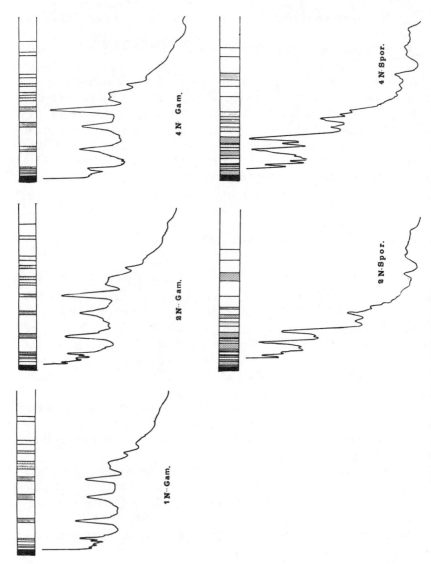

Fig. 1. Banding patterns of total soluble proteins in gametophytes and sporophytes.

Table I. Total Peroxidase Activity

Genome constitution	Purpurogallin units per gram fresh weight	Purpurogallin units per cell
Gametophyte		
$1n$	3.3×10^{-2}	2.63×10^{-8}
$4n$	6.6×10^{-2}	9.09×10^{-8}
Sporophyte		
$2n$	6.3×10^{-2}	—
$4n$	5.1×10^{-2}	—

two gametophyte genotypes ($1n$ and $4n$) and two sporophyte genotypes ($2n$ and $4n$) were compared. Total peroxidase activity was determined for each group of plants and the results were expressed on the basis of fresh weight of plant material and, in the case of gametophytes, as peroxidase activity per cell (Table I).

In gametophytes, peroxidase activity per gram fresh weight doubled with a fourfold increase in genetic material. In the sporophytes, little change in peroxidase activity was observed with a doubling of chromosomes. We earlier determined that an increase in cell size accompanies duplication of the genome in these plants. It follows that every gram of tissue is represented by fewer but larger cells. Therefore, as the genome increased in both gametophytes and sporophytes the activity of peroxidase per cell also would be expected to increase. Only in the gametophyte series, however, was it possible to calculate enzyme activity per cell. Because these plants are structurally simple and their cells homogeneous, we were able to determine that in haploids 1 mg of tissue contained approximately 1267 cells and in tetraploids 1 mg of tissue contained 726 cells. Using these figures, we estimate that in gametophytes enzyme activity per cell increased 3.5-fold with a fourfold increase in ploidy. These results indicate that total peroxidase activity increased with increases in the genome and suggest that total enzyme activity is gene dosage related.

Qualitative and Quantitative Analysis of Isozyme Patterns

Gels were loaded with sufficient extract to provide equal amounts of peroxidase activity for each gel. Stained gels were examined visually and the R_f and relative intensity of each band was recorded. Gels were then scanned and the visual bands were correlated with recording traces for complete analyses. Comparisons were made between distinct, separable bands. When two bands were closely associated and not easily resolved, they were treated as one.

Better separation of isozymes was achieved with $4n$ than with $1n$ gameto-

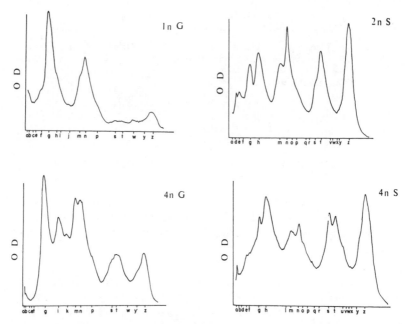

Fig. 2. Gel scans of peroxidase isozymes in gametophytes and sporophytes.

phytes. Comparison of the banding patterns indicated that, despite the difference in ploidy, plants of both genotypes possess similar isozymes. The gel scans in Fig. 2 show that the major differences in isozyme patterns between 1n and 4n plants are due primarily to changes in activity of individual bands. In haploid tissue, bands e and f were prominent and bands m and n stained with unequal intensity, while in tetraploid tissue bands e and f were barely detectable and bands m and n appeared equal in staining intensity. Among the other differences, bands i, m, s, t, and z represented isozymes present in greater concentrations in 4n than in 1n tissue. Band j in 1n tissue and band k in 4n tissue are not unique to their respective plants. Although the bands migrated at different rates and in repeated analyses always differed from one another in position on the gel, the closeness of their R_f values leads us to believe that they represent the same isozyme or isozymes.

Quantitative measurements of peroxidase activity in each band (Table II) established that bands i, s, and t were approximately three times more active in the 4n than in the 1n series. In other bands (h, m, y, and z), smaller but significant increases were also noted. Increased peroxidase activity was not restricted to isozymes of the 4n series. In the 1n gametophytes, large increases in activity were noted when isozymes c, e, f, and g were compared to similar isozymes in the 4n material.

Table II. Peroxidase Activity per Isozyme Band

Isozyme	Gametophyte 1n		Gametophyte 4n	
	R_f	Percent total activity	Percent total activity	R_f
a	0.003	1.5		
b	0.013	2.5		
c	0.048	1.3	0.3	0.077
e	0.064	1.9	0.3	0.093
f	0.106	5.1	0.6	0.120
g	0.170	24.8	9.2	0.202
h	0.234	6.3	8.5	0.235
i	0.272	4.5	13.7	0.317
j	0.341	2.7		
k			6.7	0.378
m	0.425	8.1	11.0	0.454
n	0.468	17.9	15.9	0.497
p	0.564	5.7	5.8	0.574
s	0.713	2.2	6.0	0.744
t	0.765	2.8	7.0	0.776
w	0.852	2.8	2.2	0.869
y	0.915	1.9	3.8	0.940
z	1.00	5.1	7.7	1.00

The peroxidase banding patterns for $2n$ and $4n$ sporophytes were almost indistinguishable (Fig. 2). There were fewer differences in band intensity between the two genotypes in this series than there were in the gametophyte series. The profile of s and t bands differed in the $2n$ and $4n$ tissue. In the $2n$ plants s and t bands were unequal in intensity, while in the $4n$ plants both bands appeared alike in intensity. A similar, but not as pronounced, relationship was noted between m and n bands. In the $2n$ tissue the isozyme represented by band n was dominant, but in the $4n$ tissue the isozymes represented by bands m and n were present in equal concentrations. Two minor bands, l and u, were observed in the tetraploids and were absent from the diploids. Both bands were clearly separated from their nearest neighboring bands and had distinctive R_f values. Because these bands were not detected in any of the gels containing $2n$ material, we conclude that they represent peroxidase isozymes unique to $4n$ plants, or variants of existing isozymes.

There was less quantitative variation between isozymes of the two sporophyte genotypes than had been observed in the gametophytes (Table III). Peroxidase activity of most bands was the same for $2n$ and $4n$ sporophytes. A few isozymes (e, m, n, q, and t) had higher activities in the $2n$ than in the

Table III. Peroxidase Activity per Isozyme Band

Isozyme	R_f	Sporophyte 2n Percent total activity	Sporophyte 4n Percent total activity	R_f
a	0.035	0.4	1.0	0.026
b			0.9	0.041
d	0.052	1.7	0.7	0.062
e	0.080	3.2	2.3	0.103
f	0.115	1.1	1.5	0.129
g	0.179	7.6	8.3	0.196
h	0.247	14.8	17.8	0.248
l			3.6	0.407
m	0.431	9.6	6.3	0.443
n	0.483	11.3	5.8	0.500
o	0.540	2.6	2.4	0.532
p	0.569	4.4	3.4	0.562
q	0.644	1.9	1.1	0.619
r	0.678	0.7	2.1	0.655
s	0.724	3.0	7.9	0.727
t	0.770	13.3	8.8	0.773
u			1.8	0.835
v	0.857	0.8	1.3	0.855
w	0.873	0.6	1.0	0.876
x	0.885	1.0	1.1	0.887
y	0.925	1.6	2.8	0.931
z	1.00	15.5	14.8	1.00

4n plants, while isozymes h, r, and y were more active in the 4n than in the 2n plants.

A comparison of banding patterns between 4n gametophytes and 4n sporophytes provided the opportunity to examine peroxidase isozyme activity in plants of the same genotype but different phenotype. Many differences were observed both in the nature and in the activity of the isozymes present (Fig. 3). Gametophytes and sporophytes share 12 isozymes representing 78% and 80%, respectively, of the total peroxidase activity. In bands h and z more enzyme activity was found in the sporophyte than in the gametophyte, but in bands m and n higher levels of activity were found in the gametophyte. In the other shared bands, approximately the same levels of activity were present in both plants. The remaining 22% activity in the gametophyte was found in two major bands (i and k) and one minor band (c). The sporophyte contains ten unique bands, each of which has a very low level of peroxidase activity. Together they constitute 20% of the total enzyme activity.

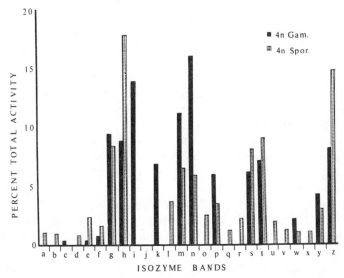

Fig. 3. Comparison of peroxidase activity in individual bands separated from 4n gametophytes and 4n sporophytes.

DISCUSSION

One of the major results of this study was the finding that the total spectrum of isolated, soluble proteins as well as the qualitative and quantitative distribution of peroxidase isozymes differed in gametophytes and sporophytes. The usual alternation of a 1n gametophyte generation with a 2n sporophyte generation is thus accompanied by marked alterations in proteins. Since these two plants are not only genotypically but also phenotypically distinct, it is not too surprising to find that distinctive patterns of proteins and enzymes are associated with each type of plant. Many studies have shown that changing protein patterns are correlated with morphological and developmental changes in a variety of organisms (Barber and Steward, 1968). It is interesting to note that total protein and peroxidase isozyme patterns did not vary to any great extent within either phenotypic group, gametophytes or sporophytes. Our data suggest that duplication of the genome by somatic polyploidy does not result in any significant changes in proteins or isozymes within each phenotype. However, duplication of the genome by the usual process of fertilization results in protein and isozyme differences.

In some respects, the situation is comparable to that recently reported for fungi (Wang and Raper, 1969, 1970; Ross *et al.*, 1973), where monokaryons (haploids) and dikaryons (diploids) also displayed different protein and enzyme patterns. In the case of fungi, the haploids and diploids are phenotypically alike and incompatibility factors are known to be involved in the reproductive process, Wang and Raper (1970) postulate a role for the incompatibility genes in controlling the transition from one morphogenetic state to the other. Despite the lack of information concerning genetic control of reproduction in the ferns, there is the striking similarity that in both fungi and ferns a significant change in protein pattern accompanies the transition from gametophyte or monokaryon to the meiosis-competent sporophyte or heterokaryon. Whether or not the changes are causal and primary determinants of the alternation of generations remains to be established.

Total peroxidase activity in the fern plants appears to be gene dosage related. The amount of enzyme produced in gametophytes is almost directly proportional to the number of genes. Figures are lacking for sporophytes, but based on our observations that cell size increases with increases in the genome we can expect enzyme activity per cell to be higher in $4n$ than in $2n$ plants. The regulation of peroxidase activity in ferns is, therefore, similar to the regulation of tryptophan synthetase in yeast (Ciferri *et al.*, 1969), collagen synthesis in rat fibroblast-like cells (Priest and Priest, 1969), phosphoglucose kinase activity in Chinese hamster cells (Westerveld *et al.*, 1972), and the regulation of other enzymes (Carlson, 1972; Lucchesi and Rawls, 1973) where a linear relationship exists between amount of protein and number of coding genes.

Carlson (1972) has studied the activity of a number of enzymes in primary and secondary trisomics of *Datura stramonium* and demonstrated that the presence of an additional structural gene causes a marked dosage effect. He theorizes that transcription is probably the rate-limiting process determining levels of enzyme activity. An increase in the number of structural genes increases transcription rates of those genes and increased activity follows. He proposes that the rate of transcription is constant for each structural gene and independent of the number of identical structural genes in the genome. The simplicity of this thesis is attractive and in the present investigation could explain the increase in peroxidase activity. The variation in peroxidase isozymes between $1n$ and $4n$ gametophytes and between $2n$ and $4n$ sporophytes seems to require a degree of regulation not provided by control of transcription rates. If all of the genes coding for specific isozymes were transcribed in the usual fashion, one would expect the isozymes to be the same in $1n$ and $4n$ gametophytes and in $2n$ and $4n$ sporophytes. In addition, the activity of each isozyme would probably increase as gene dosage increased. Since this situation does not exist and since isozyme activity varies with gene duplication, it appears necessary to consider some form of regulatory mechanism, in addition to transcription rate, controlling isozyme activity in the polyploid ferns.

In $4n$ gametophytes and $4n$ sporophytes, peroxidase isozymes show marked differences in banding patterns and activity. These plants possess identical genomes, yet the nature of their gene products is significantly different. We conclude that whatever the regulatory mechanisms influencing iso-zyme activity may be, they must be modulated by the extra-genomic environment.

Some consideration already has been given to regulatory mechanisms influencing gene expression (Tomkins *et al.*, 1969; Ohno, 1970) and hypothetical models have been proposed (Britten and Davidson, 1969). However, more experience with a wider variety of organisms and more quantitative data are needed before gene regulation in eukaryotes with duplicated genes is understood.

REFERENCES

Barber, J. T., and Steward, F. C. (1968). The proteins of *Tulipa* and their relation to morphogenesis. *Develop. Biol.* **17**:326.

Becak, W., and Pueyo, M. T. (1971). Gene regulation in the polyploid amphibian *Odontophrynus americanus*. *Exptl. Cell Res.* **63**:448.

Britten, R. J., and Davidson, E. H. (1969). Gene regulation for higher cells: A theory. *Science* **165**:349.

Caponetti, J., Harvey, W. H., and DeMaggio, A. E. (1972). Changes in soluble proteins of cinnamon fern leaves during development. *Canad. J. Bot.* **50**:1479.

Carlson, P. S. (1972). Locating genetic loci with aneuploids. *Mol. Gen. Genet.* **114**:273.

Ciferri, O., Sora, S., and Tiboni, O. (1969). Effect of gene dosage on tryptophan synthetase activity in *Saccharomyces cerevisiae*. *Genetics* **61**:567.

DeMaggio, A. E. (1961). Morphogenetic studies on the fern *Todea barbara* (L.) Moore. I. Life history. *Phytomorphology* **11**:46.

DeMaggio, A. E., and Stetler, D. A. (1971). Polyploidy and gene dosage effects on chloroplasts of fern gametophytes. *Exptl. Cell Res.* **67**:287.

DeMaggio, A. E., Wetmore, R. H., Hannaford, J. E., Stetler, D. A., and Raghavan, V. (1971). Ferns as a model system for studying polyploidy and gene dosage effects. *Bioscience* **21**:313.

Goldthwaite, J., and Bogorad, L. (1971). One-step method for the isolation and determination of leaf ribulose-1,5-diphosphate carboxylase. *Anal. Biochem.* **41**:57.

Hannaford, J. E., and DeMaggio, A. E. (1970). Cell cycle kinetics and cytophotometric analysis in polyploid fern prothalli. *Am. J. Bot.* **57**:741.

Lucchesi, J. C., and Rawls, J. M., Jr. (1973). Regulation of gene function: A comparison of enzyme activity levels in relation to gene dosage in diploids and triploids of *Drosophila melanogaster*. *Biochem. Genet.* **9**:41.

Luck, H. (1963). *Methods of Enzymatic Analysis*, Academic Press, New York, p. 895.

Ohno, S. (1970). *Evolution by Gene Duplication*, Springer, New York.

Priest, R., and Priest, J. H. (1969). Diploid and tetraploid clonal cells in culture: Gene ploidy and synthesis of collagen. *Biochem. Genet.* **3**:371.

Ross, I., Martini, E. M., and Thoman, M. (1973). Changes in isozyme patterns between monokaryons and dikaryons of a bipolar *Coprinus*. *J. Bacteriol.* **114**:1083.

Seevers, P. M., Daly, J. M., and Catedral, F. F. (1971). The role of peroxidase isozymes in resistance to wheat stem rust disease. *Plant Physiol.* **48**:353.

Tomkins, G. M., Gelehrter, T. D., Granner, D., Martin, D., Jr., Samuels, H. H., and Thompson, E. B. (1969). Control of specific gene expression in higher organisms. *Science* **166**:1474.

Wang, C., and Raper, J. (1969). Protein specificity and sexual morphogenesis in *Schizophyllum commune*. *J. Bacteriol.* **99**:291.

Wang, C., and Raper, J. (1970). Isozyme patterns and sexual morphogenesis in *Schizophyllum*. *Proc. Natl. Acad. Sci.* **66**:882.

Westerveld, A., Visser, R. P. L. S., Freeke, M. A., and Bootsma, D. (1972). Evidence for
    linkage of 3-phosphoglycerate kinase, hypoxanthine-guanine-phosphoribosyl trans-
    ferase, and glucose-6-phosphate dehydrogenase loci in Chinese hamster cells studied by
    using a relationship between gene multiplicity and enzyme activity. *Biochem. Genet.*
    7:33.

18

Reprinted from Hereditas **70**:69–73 (1972)

Oxygen requirement of root meristems in diploid and autotetraploid rye

OVE HALL

Swedish Seed Association, Svalöv, Sweden

The oxygen requirement of the root meristems of diploid and tetraploid rye seedlings has been determined. It was found that the polyploid seedlings require 40–50 per cent higher oxygen concentration at the root surface than the diploid seedlings for normal (maximum) respiration in the temperature interval 15–30° C. The critical oxygen concentration, when respiring air atmosphere, occurred at about 22° C in the diploid meristem and at about 15° C in the tetraploid meristem. A hypothesis is presented according to which the radial oxygen diffusion into the root meristems is one of the factors determining the geographical distribution of polyploid plant species.

In the course of studies at this institute on the biochemical properties and environmental requirements of polyploid plants an investigation was carried out on the oxygen demand of the root meristem of diploid and autotetraploid rye. The results of this investigation are reported in the present paper.

Hyperbolic relationship between meristem respiration and oxygen tension has been demonstrated by several workers (WANNER 1945; BERRY and NORRIS 1949; LUXMOORE et al. 1970 and others). The curves show that a rapid decrease in the rate of oxygen consumption occurs when a certain critical oxygen pressure is reached. Below this pressure apparently only a certain velocity of respiration can be supported. At pressures higher than the critical pressure the respiratory rate is constant regardless of the oxygen tension and the critical pressure thus represents the oxygen tension just supporting the maximum rate of respiration. It is supposed that diffusion is the primary limiting factor in rate of oxygen consumption by the meristems at oxygen tensions below the critical pressure.

The equation used by FENN (1927) and GERARD (1927) in investigations on nerve metabolism has been applied to studies on the oxygen consumption and diffusion in meristem and other parts of the root by WANNER (1945), BERRY and NORRIS (1949), LEMON (1962), and LEMON and WIEGAND (1962). Applied on the diffusion within the meristem this equation states:

$$C_i = C_R - \frac{q}{4 \times D_i}(R^2 - r_i^2) \qquad (1)$$

where C_i (g cm^{-3}) is the oxygen concentration inside the meristem at a radial distance r_i (cm) from the meristem axis, q (g cm^{-3} sec^{-1}) is the rate of oxygen consumption by the meristem tissue, C_R (g cm^{-3}) is the oxygen concentration at the meristem surface (where $r_i = R$) and D_i (cm^2 sec^{-1}) is the diffusion coefficient inside the meristem.

When $C_i = 0$ at $r_i = 0$ the equation is reduced to:

$$C_R = \frac{q \times R^2}{4 \times D_i} \qquad (2)$$

This value of C_R is the critical value of the oxygen concentration of the meristem respiration and is indicated C'_R. If C_R equals or exceeds C'_R there will be sufficient oxygen for maximum respiration and the reverse would indicate the existance of a core of anaerobic cells in the center of the meristem.

The oxygen concentration in the environment

of the meristem required for normal (maximum) respiration of all the cells of the meristem can thus be calculated according to equation (2) if the maximum rate of respiration of the meristem is known as well as the radius of the meristem and the diffusion coefficient.

The radius (R) and the maximum respiration rate (q) of the diploid and tetraploid rye root meristems have been experimentally determined. From these data the critical oxygen tension of the two meristems have been calculated according to equation (2) and compared to each other assuming the same value of the diffusion coefficients of the two meristems.

Material and methods

The material consisted of the spring rye variety Sv 0201 and the corresponding tetraploid variety Fourex. Seeds after being washed with Desivon and soaked in sodium hypochlorite for 15 minutes were grown on filter paper under glass in a growth cabinet at $19 \pm 2°$ C without light. The apical 5 mm of three-day old roots were excised and used in the determination of the respiration rate. No difference in growth rate between the diploid and the tetraploid group could be observed.

Maximum respiration rates were determined by the Warburg manometric method. The volumes of the flask were about 13 cm³. The main chamber contained the respiring material, 50 pieces of meristem in 2.5 cm³ of distilled water and the center well contained a filter paper wick dipped in 0.5 cm³ of 10 per cent potassium hydroxide to absorb carbon dioxide. A shaking rate of 100 strokes per minute was employed and readings of oxygen consumption were made every 15 minute for 90 minutes. The respiration flasks were flushed for 15 minutes with pure oxygen and shaken for 15 minutes for equilibration before the readings were started.

For the determination of the meristem radii a Visopan microscope equipped with a measuring device was used.

Experiments and results

The maximum respiration rates of the meristems were determined at three temperatures, 15, 22,

and 30° C. The results of these determinations together with the values of the radii of the meristems and the product $q \times R^2$ for each meristem and temperature are collected in Table 1.

The values of the radii are mean values of 50 separate determinations and the values of the maximum respiration rates are mean values of at least 10 separate determinations. The standard error of the values of the radii is ± 0.003 cm and of the maximum respiration rates $\pm 0.05 \times 10^{-7}$ g cm⁻³ sec⁻¹. The volume of the meristems were determined as the volume of a cylinder having a radius of 0.023 cm for the diploid and 0.027 cm for the tetraploid and a height of 0.5 cm.

After 3 days of growing the radius of the tetraploid was 16 per cent longer than the radius of the diploid meristem. This difference was found to persist for the 6 days determinations of the radii were carried out but the absolute values of the radii of the diploid as well as of the tetraploid were slightly decreasing with age.

The maximum respiration rates of the two meristems increased with temperature, Q_{10} being 1.7 for the diploid and 2.0 for the tetraploid when calculated over the whole temperature range. The temperature coefficient of the tetraploid meristem is 2.0 in the range $15-22°$ C and 1.9 in the range $22-30°$ C. The corresponding values for the diploid is 2.1 and 1.3. These results suggest a difference between the diploid and tetraploid meristem in respect to the relationship between temperature and oxygen consumption.

According to equation (2) the critical oxygen pressure for maximum meristem respiration is directly proportional to the product $q \times R^2$ and inversely proportional to the diffusion coefficient. Thus, assuming the same diffusion coefficient of the two types of meristems the relations between the critical oxygen tensions of the meristems at the different temperatures can be calculated from the values of $q \times R^2$. At 15° C this relation is $\frac{248}{169} = 1.46$, at 22° C $\frac{350}{248} = 1.41$, and at 30° C $\frac{459}{302} = 1.51$. On the assumption that oxygen diffuses with the same velocity in the diploid and tetraploid meristem these figures indicate that *the tetraploid rye root meristem in the temperature range 15–30° C requires 40–50 per cent higher oxygen tension at the root surface than the diploid meristem for normal respiration.*

Approximative values of the critical oxygen

Table 1. The radii (R) and the maximum respiration rates (q) at 15, 22 and 30° C of 2x and 4x rye root meristem and the product $q \times R^2$ for each meristem and temperature

Meri- stem	Radius, R cm	Max. respiration rate, $q \times 10^{-7}$ g cm^{-3} sec^{-1}			$q \times R^2 \times 10^{-12}$		
		15°	22°	30°	15°	22°	30°
2x	0.023	3.2	4.7	5.7	169	248	302
4x	0.027	3.4	4.8	6.3	248	350	459

Table 2. Critical oxygen concentration of 2x and 4x rye root meristem at 15, 22 and 30° C and values of the diffusion coefficients used
Values in brackets are oxygen percentage in equilibrium with the solution bathing the root sections.

Meri- stem	Diffusion coefficients $\times 10^{-6}$ cm^2 sec^{-1} (BERRY and NORRIS 1949)			Critical O_2-conc. $C'_k \times 10^{-6}$ g cm^{-3}		
	15°	22°	30°	15°	22°	30°
2x	7.077	8.028	9.843	6.0(13)	7.7(18)	7.7(23)
4v	7.077	8.028	9.843	8.8(19)	11.0(27)	11.7(35)

tensions of the rye meristems have been calculated by using the diffusion coefficients obtained by BERRY and NORRIS (1949) from studies of onion root respiration. These values are shown in Table 2 together with the values of the diffusion coefficients which have been used. As is clearly indicated, the values of the critical oxygen tension are dependent on temperature. The higher temperature the higher also the values of the critical oxygen tension. An approximative limit for maximum respiration in air lies for the diploid rye root meristem at 22° C and for the tetraploid meristem at 15° C. It must be pointed out, however, that these values apply only to three-day old roots since the radii of the meristems diminish with age of the roots.

Discussion

It has often been observed that doubling of the number of chromosomes causes an increase of organ size of the plant. The effect is, however, different dependent on species and organs.

SCHWANITZ (1949) determined the leaf thickness on 11 plant species and found that the increase of leaf thickness varied between 3 and 25 per cent depending on species. For the root meristem the present author has found (unpublished results) that the tetraploids of red clover and fodder kale have about 16 per cent longer radius than the corresponding diploids and that tetraploid barley has 24 per cent longer radius than the diploid. The allopolyploid rape has generally 40 per cent longer meristem radius than turnip rape, which is one of the parental species, when the comparison is carried out on the same sieve fraction of the seeds.

The autotetraploid rye variety Fourex was found in this investigation to have 16 per cent longer radius than the corresponding diploid Sv 0201. This value apparently agrees with similar values obtained from other studies.

Maximum respiration rates have been determined for maize and rice root meristem by LUX-MOORE et al. (1970) and for onion root meristem by BERRY and NORRIS (1949). At 25° C the maximum rate of respiration of maize root meristem was 1.9×10^{-7} g O_2 cm^{-3} sec^{-1} and maximum rate of respiration of rice root meristem 3.6×10^{-7} g O_2 cm^{-3} sec^{-1}. At 20° C onion root meristem had a maximum respiration rate of 2.3×10^{-7} g O_2 cm^{-3} sec^{-1} and at 30° C a maximum respiration rate of 4.5×10^{-7} g O_2 cm^{-3} sec^{-1}. The values of the maximum respiration rate of rye root meristem obtained in this study are somewhat higher than those rapported by LUXMOORE et al. and BERRY and NORRIS from respiration studies on maize, rice and onion.

LEMON and WIEGAND (1962) have calculated the critical oxygen concentration C'_R for maximum onion root meristem respiration using data from BERRY and NORRIS (1949). At 15° C this value was 7.2×10^{-6} g cm^{-3}, at 20° C 9.1×10^{-6} g cm^{-3}, and at 30° C 16.2×10^{-6} g cm^{-3}. These values are evidently very similar to the corresponding values for the rye root meristem obtained in the present work. An approximative limit for maximum respiration in air lies for the onion root meristem at 20° C since 9.1×10^{-6} g of oxygen per cm³ of water corresponds to a partial pressure of about 21 %.

The composition of soil atmosphere is very divergent depending on soil structure, water content of the soil, microbiotic activity and root respiration of higher plants. Under normal conditions soil atmosphere consists of about 20 per cent of oxygen but under bad conditions this value can decrease to some few per cent.

A comparison between the values of the critical oxygen tension obtained in these experiments and the oxygen concentration in soil shows that the root respiration of diploid rye is normal up to about 22° C under favourable conditions. The root respiration of the tetraploid rye, however, ceases to be normal already at 15° C even under good soil conditions. *It can thus be concluded that tetraploid rye is less tolerant towards high soil temperatures than diploid rye.* This is caused primarily by the larger volume of the tetraploid meristem as the respiration rates of the diploid and tetraploid meristem were found to be very similar.

A low tolerance towards higher soil temperature may be common to many newly arisen plant polyploids and this may constitute a hindrance for the developing of polyploids in warmer climates and thus be one of the factors of the increased frequency of polyploidy with increase in latitude or altitude, which has been observed by various investigators (cf. LÖVE and LÖVE 1948;

TISCHLER 1950; HASKELL 1952; BORGMANN 1964; MORTON 1966). Low oxygen tension in the soil may in a similar way act as a barrier for the spreading of polyploids.

The values of the critical oxygen pressures of the meristem respiration obtained in these studies are approximative values. The experiments have been carried out in vitro on isolated meristems, which means that the results have been obtained under unphysiological conditions. Further, diffusion coefficients obtained from experiments with onion root meristem have been used which clearly is not quite correct. When applying the Gerard-Fenn equation in calculating the critical pressures it is assumed that the root meristems are perfect cylinders which is not true as they have the form of a cone at the distal end. These facts decrease the reliability of the absolute values of the critical oxygen pressure which have been obtained in this work. Since all the values are subject to the same errors, the relations between the absolute values are, however, more reliable than the absolute values of the critical pressures themselves. Thus, on the assumption that the diffusion coefficients of the diploid and tetraploid root meristem have the same value, the conclusion that the critical oxygen pressure of the autotetraploid variety is higher at all temperatures than that of the diploid variety must be reliable. For the same reason it can be concluded that the critical pressure of the two meristems as well as the difference between them will increase with temperature.

Literature cited

BERRY, L. J. and NORRIS JR., W. E. 1949. Studies of onion root respiration. II. The effect of temperature on the apparent diffusion coefficient in different segments of the root tip. — *Biochim. Biophys. Acta 3*: 607—614.

BORGMANN, E. 1964. Anteil der Polyploiden in der Flora des Bismarcksgebirges von Ostneuguinea. — *Z. Bot. 52*: 118—172.

FENN, W. O. 1927. The oxygen consumption of frog nerve during stimulation. — *J. Gen. Physiol. 10*: 767—779.

GERARD, R. W. 1927. Studies on nerve metabolism. II. Respiration in oxygen and nitrogen. — *Am. J. Physiol. 82*: 381—404.

HASKELL, G. 1952. Polyploidy, ecology and the British flora. — *J. Ecol. 40*: 265—282.

LEMON, E. R. 1962. Soil aeration and plant root relations. I. Theory. — *Agron. J. 54*: 167—170.

LEMON, E. R. and WIEGAND, C. L. 1962. Soil aeration and plant root relations. II. Root respiration. — *Ibid. 54*: 171—175.

Löve, Á. and Löve, D. 1948. Chromosome numbers of northern plant species. — *Icel. Univ. Inst. Appl. Sci. Dept. Agric. Rep. B. 3:* 1—131.

Luxmoore, R. J., Stolzy, L. H., and Letey, J. 1970. Oxygen diffusion in the soil-plant system. II. Respiration rate, permeability, and porosity of consecutive excised segments of maize and rice roots. — *Agron. J. 62:* 322—324.

Morton, J. K. 1966. The role of polyploidy in the evolution of a tropical flora. — *In Chromosomes Today (Eds.* C. D. Darlington and K. R. Lewis*), Oliver and Boyd, Edinburgh, Vol. 1:* 73—76.

Schwanitz, F. 1949. Untersuchungen an polyploiden Pflanzen. IV. Zum Wasserhaushalt diploider und polyploider Pflanzen. — *Züchter 19:* 221—232.

Tischler, G. 1950. Die Chromosomenzahlen der Gefässpflanzen Mitteleuropas. — *Junk, Den Haag,* 263 p.

Wanner, H. 1945. Sauerstoffdiffusion als begrenzender Faktor der Atmung von Pflanzenwurzeln. — *Vierteljahrsschrift Naturf. Ges. Zürich 90:* 98.

Communications from the Swedish Seed Association No. 396.

Part VI

CYTOGENETICS AND CONTROL OF CHROMOSOME PAIRING

Editors' Comments
on Papers 19 Through 27

The papers of Part VI are concerned with mechanisms of chromosome pairing, chiasmata distribution, and fertility of polyploids. These papers have lead directly to some of the more important ideas on chromosome pairing models in polyploids and diploids discussed in the Epilogue of this book.

The first three articles of this series (Papers 19, 20, and 21) show that a single chromosome can carry a gene (or genes) that dramatically affects chromosome pairing. Two of the papers from Sears's laboratory (Papers 19 and 20) arrive at the same conclusion, as did Riley working independently in England. These articles have done much to change ideas on mechanisms of chromosome pairing, and they have demonstrated how it would be possible to transfer valuable genes from wild relatives to bread wheat, as Riley points out in Paper 21.

Paper 22 continues the work on the effects of the $5B^L$ gene(s) in wheat discovered by authors of the previous three papers. Feldman shows that logically the premeiotic chromosomes must occupy fixed regions in the nucleus and that different doses of the gene or genes on the long arm of chromosome $5B$ can bring about changes in pairing relationships. The author suggests that the $5B^L$ gene(s) regulates the premeiotic association of both homologous and homoeologous chromosomes.

The effects of colchicine on the meiotic process are shown by Driscoll and colleagues in Paper 23. It is demonstrated here and in a later paper by Driscoll and Darvey that the alkaloid does not affect synapsis or chiasma formation. The effect is on the premeiotic cells, and the authors make the clear distinction that premeiotic associations and synapsis are two different and distinct phenomena.

Paper 24 is important because it associates a specific genotype with pairing behavior in an autopolyploid. Avivi suggests here that the greater tendency for bivalent formation in one of the two kinds of autoploids is due to association in premeiotic cells of homologous chromosomes in sets of two. This idea is a logical extension of the concepts expressed in Papers 22 and 23.

The article by John and Henderson (Paper 25) should be considered as an exemplar for cytological analyses. It is extremely well done and contains the kind of quantitative data that are needed for a rigorous analysis of polyploidy. It is also of great importance because it sets forth a model for synapsis in autotetraploids that follows a brief suggestion of Hughes-Schrader (1943). This model can be used effectively to predict chromosomal configurations in autotetraploids with a maximum of two chiasmata per bivalent, equal-sized chromosomes, and a high and partly random distribution of chiasmata.

All too often data about chiasma frequency per cell and per bivalent and variations among bivalents are not included in studies of diploids and polyploids. Paper 26 clearly shows the importance of such data in estimating the expected frequencies of quadrivalents, trivalents, bivalents, and univalents in autotetraploids. The effect of the numbers of chiasmata on the frequencies of such configurations and the relationships of the configurations to fertility are unequivocally demonstrated in this article. Even today, articles are still being published that attribute sterility to quadrivalents per se despite this and later works by Rees clearly demonstrating that unbalanced gametes caused by unequal disjunction of univalents and trivalents are the major causes of chromosomal imbalance.

The importance of Levan's article (Paper 27) is that it shows the potential for error if we make extrapolations from metaphase I configurations to pachytene. In the case of autotetraploid *Allium porrum* analyzed by Levan, quadrivalents were present rarely at metaphase I and probably in less than 1 percent of the cells. However, at least one and up to eight quadrivalents were observed at pachytene. This discrepancy is due to extreme localization of the normal two chiasmata in opposite arms of the pachytene quadrivalent. The lesson to be learned from this analysis is that the pachytene stage is important and fundamental for cytogenetic analysis, and critical studies should not be solely based on late prophase I and metaphase I stages.

REFERENCE

Hughes-Schrader, S., 1943, Meiosis Without Chiasmata in Diploid and Tetraploid Spermatocytes of the Mantid *Callimantis antillorum* Saussure, *J. Morphol.* **73:**111–141.

19

Reprinted from *Wheat Inf. Serv.* **5**:6 (1957)

Asynaptic effect of chromosome V

M. Okamoto

Curtis Hall, University of Missouri, Columbia, Missouri, U.S.A.

It is known that chromosome III of the Chinese Spring variety has a marked effect on chromosome pairing, and that chromosome II of the same variety has a similar but less pronounced effect (Sears 1944). The other chromosomes of Chinese Spring have not been suspected of any effect on chromosome pairing.

In the F_1 between plants which were monosomic for a telocentric chromosome V and AADD plants (amphidiploid *T. aegilopoides* × *Ae. squarrosa*), 34-chromosome plants which did not carry the telocentric chromosome showed unexpectedly much better pairing than 35-chromosome plants, as is shown in the following table of data from typical plants :

	No. of PMC's	Average number of							
		univs.	bivalents		tri-valents	quadri-valents	5–valents	6–valents	7–valents
			closed	open					
34-chr. plant	200	8.24	5.07	3.445	.735	.685	.19	.22	.02
35-chr. plant	200	23.82	1.025	4.115	.320	.005	0	0	0

If chromosome V belongs to the A genome (Larson 1953), the 34-chromosome plants would be expected to give more univalents and fewer bivalents than 35-chromosome plants. But the above table shows that the average number of univalents is much less and that of bivalents is much more in 34-chromosome plants than in 35- chromosome plants.

If chromosome V belongs to the B genome, 34-chromosome plants would be expected to give fewer univalents than 35-chromosome plants, and the number of bivalents would remain the same. This very slight increase in the expected frequency of univalents comes nowhere near explaining the observed increase of 15 per cell.

A possible explanation for the above facts is that the telocentric chromosome V and hence chromosome V of Chinese Spring carries a gene or genes for asynapsis. Then it follows that chromosome pairing is good in 34-chromosome plants due to the absence of the asynaptic effect of chromosome V, while 35-chromosome plants show poor chromosome pairing due to the asynaptic effect of the telocentric chromosome V. This asynaptic effect of chromosome V has only been observed in this pentaploid hybrid, where two sets of each of the A and D genomes and one set of the B genome are present.

[*Editors' Note:* No references were included in the original publication.]

INTERGENOMIC CHROMOSOME RELATIONSHIPS IN HEXAPLOID WHEAT

E. R. Sears and M. Okamoto

The 21 chromosomes of *Triticum aestivum* can be placed in 7 homoeologous groups of 3 according to the ability of tetrasomes to compensate for particular nullisomes. Within these groups all the 42 possible tetrasomic-nullisomic combinations are nearer normal than the nullisomics themselves. Between the groups nearly 50 different combinations have been tested, without any of them showing compensation. The high degree of compensation exhibited in most of the combinations within groups indicates that almost every chromosome is very similar to its two homoeologues in gene content. The homoeologous chromosomes show little tendency to pair, however, for the bivalent frequency in haploids normally averages only about one.

That the pairing in haploids usually involves homoeologous chromosomes has been shown by analysis of translocations recovered from haploids, where each translocation is presumably the result of haploid pairing and crossing over. Of the 13 translocations analysed, 9 involved homoeologous chromosomes: VI–XIX (4 occurrences), II–XX (2), XI–XXI (2), and XII–XVI. The other 4 were: II–VIII, IV–XII, VI–XVIII, and IX–XI.

The virtual failure of homoeologous chromosomes to pair appears to be due to suppression of pairing by chromosome V. When chromosome V is missing, hybrids of hexaploid wheat (genomic formula AABBDD) with amphidiploid *T. aegilopoides* × *Aegilops squarrosa* (AADD) show many microsporocytes with the equivalent of the expected 14 bivalents (average over 12), but only rarely have as many as 13 bivalents (average about 8) when V is present. Hybrids of AABBDD with AA, which normally have only 2 to 7 pairs, have the equivalent of 5 to 13 pairs (average about 10) when chromosome V is missing. It seems likely that the acquisition of a mutation on chromosome V which decreases pairing intensity is the means by which hexaploid wheat has achieved diploid pairing. As demanded by this hypothesis, pairing is definitely less intense in hexaploid than in diploid wheat.

GENETIC CONTROL OF THE CYTOLOGICALLY DIPLOID BEHAVIOUR OF HEXAPLOID WHEAT

By Dr. RALPH RILEY and VICTOR CHAPMAN

Plant Breeding Institute, Cambridge

COMMON wheat, *Triticum vulgare*, is a hexaploid species with 42 chromosomes in which the genomes of diploid wheat and of the diploid species *Aegilops speltoides* and *Aegilops squarrosa* are combined together[1-5]. There is considerable similarity between the chromosomes of these three diploids. Indeed, in hybrids between diploid wheat and *Aeg. squarrosa* and between diploid wheat and *Aeg. speltoides*, complete pairing at meiosis occasionally results in the formation of seven bivalents, although the mean bivalent frequency is somewhat less, being between three and four per cell[5,6]. Moreover, the attraction which results in bivalent formation in the hybrids is of such strength as to cause multivalent formation in tetraploids induced from them. However, each chromosome normally conjugates with its complete homologue at meiosis in *T. vulgare*, and only bivalents are formed. Thus, the pairing attractions between the equivalent, homoeologous, chromosomes of different genomes either no longer exists or is no longer expressed. Polyploid wheat has therefore evolved to behave cytologically as a diploid. Further, the shift has been of such efficiency that there is very little chromosome pairing even in 21-chromosome haploids of *T. vulgare*, in which intergenome pairing is not restricted by the affinity of complete homologues. The cytological distinctiveness of the equivalent chromosomes of different genomes is all the more striking in view of the close genetic relationships which have enabled Sears', by nullisomic-tetrasomic compensation, to recognize the seven homoeologous groups into which the complement of wheat can be arranged.

Recent results with various aneuploids have revealed something of the genetic control of the diploid behaviour of *T. vulgare*. The evidence centres on the derivatives of a 41-chromosome monosomic line, designated *HH*, which arose at Cambridge in the variety Holdfast. The formation of twenty bivalents and one univalent (Fig. 6) in *HH* monosomics is like the behaviour of all other monosomics of common wheat[8]. Meiosis is normal in the 42-chromosome segregants in the monosomic line, and in the 42-chromosome hybrids from crosses of *HH* monosomics to euploid plants of Holdfast. Therefore, the chromosomes of *HH* monosomics are structurally unaltered compared with those of Holdfast. One arm of the chromosome monosomic in the *HH* line is rather more than twice the length of the other, but the chromosome has not yet been positively identified relative to the numbering based on the variety Chinese Spring.

Monosomics of *HH* have been used in crosses with 44-chromosome plants which had the full complement of wheat chromosomes plus a single pair of rye chromosomes. From these crosses five 20-chromosome, nulli-haploid, plants have been obtained, which, by the nature of their origin, must have had the haploid complement of Holdfast minus the chromosome which is monosomic in the *HH* line. The nulli-haploids arose in (a) the F_1 of the cross *HH* monosomic × rye chromosome III disomic addition, and in (b) the F_2 of crosses of *HH* monosomic × rye II, or rye III, disomic additions[9]. All the F_1's which produced nulli-haploids in F_2 had twenty bivalents composed of wheat chromosomes and the *HH* chromosome and a rye chromosome as univalents. In each situation, therefore, the 20-chromosomes in the embryo sacs which functioned parthenogenetically must have been those which had segregated from bivalents—that is, all the wheat chromosomes except *HH*.

The meiotic pairing of the *HH* nulli-haploids has been compared with the pairing in 21-chromosome euhaploids of Holdfast. The meiotic behaviour of the euhaploid plant listed in Table 1 was similar to that of twelve other 21-chromosome plants of Holdfast examined over a period of four years. The number of bivalents in these plants never exceeded four, with a mean frequency of between 1·3 and 1·7 per cell, and trivalents were rare (Fig. 1). There was usually only one chiasma per bivalent.

By contrast, in the *HH* nulli-haploid enumerated in Table 1, which is typical of five such plants examined in two successive years, the mean bivalent frequency exceeded the maximum observed in euhaploids. Trivalents were quite frequent with as many as five in some cells (Fig. 3), although there was never more than one per cell in the euhaploids. Very many more chromosomes per cell, a mean of 11·0 compared with a mean of 2·9, were involved in chiasma-associations and a much higher frequency of closed bivalents and of 'pan-handle' trivalents resulted from more intimate pairing. As many as nineteen of the twenty chromosomes of the nulli-haploid have been observed in various associations simultaneously, although never more than nine were involved in associations in the same cell in euhaploids.

The most reasonable hypothesis to account for these observations was that the deficient chromosome carried a gene, or genes, which restricted

Table 1. MEIOTIC PAIRING IN EUHAPLOID AND *HH* NULLI-HAPLOID OF *T. vulgare*

Type	Chromosome No.	Cells	Mean pairing			Proportion of bivalents as rings	Conjugated chromosomes per cell	
			Bivalents	Trivalents	Quadrivalents		Mean	Range
Euhaploid	21	100	1·38 ± 0·09	0·07	—	0·00	2·86 ± 0·23	0-9
Nulli-haploid	20	75	4·16 ± 0·12	0·96 ± 0·03	0·02	0·23	10·95 ± 0·33	4-19

intergenomic, homoeologous, pairing ; the alternative being that association between randomly distributed duplicate segments was normally inhibited. On either view, there being no other independent mechanism, in the absence of the pairing restriction normally undetected affinities were expressed. Crucial evidence that the pairing was homoeologous rather than random was afforded by the high frequency of trivalents and the infrequency of quadrivalents.

If homoeologous pairing could take place in HH nulli-haploids, it was argued that 40-chromosome nullisomics deficient for the HH pair should deviate from the strictly bivalent-forming regime of *T. vulgare*. Consequently, three HH nullisomics have been examined. All had large and frequent multivalents, and associated univalents, at meiosis. No method has yet been found of consistently preparing well-spread pollen mother cells of this material, but the complexity of the behaviour is readily apparent (Fig. 4). More than half the cells have at least one multivalent, usually a quadrivalent, and many have several. Associations of three, four, five and six are common, but no higher multivalents have been observed. However, the greater the number and size of multivalents the less is the chance of disentangling the snarl, so there is danger of observational bias. No doubt, also, there are homoeologous, or allosyndetic, bivalents which would be undetected unless markedly heteromorphic. The magnitude of the meiotic disturbance is therefore hard to assess ; nevertheless, it is clear that, in the absence of the HH chromosome, *T. vulgare* ceases to be a stable bivalent former, a classical example of the autosyndetic pairing allohexaploid, and behaves as an intermediate, autoallopolyploid.

The diploidizing mechanism is effective in the hemizygous state in monosomics and in euhaploids. Furthermore, in a haploid plant which had twenty normal wheat chromosomes plus the iso-chromosome formed from the long arm of the HH chromosome (Fig. 5), meiotic pairing was the same as in euhaploids. Moreover, there was no multivalent formation in a 41-chromosome monosomic which had twenty normal pairs of Holdfast chromosomes and in which the HH chromosome was represented only by a single telocentric of the long arm. The long arm alone is thus effective in prohibiting homoeologous association ; but no evidence is available on the effects of the short arm alone. It may be that both arms have genes with equally effective control over the pairing behaviour, but this seems unlikely. The control of bivalent formation is probably restricted to one arm, and may indeed be effected by quite a localized region.

The implications of these results are both theoretical and practical. First, there has been considerable discussion on the methods by which many polyploids have attained their cytologically diploid character, with its obvious advantages in fertility and genetic stability. Several authors have favoured the theory that this situation has been achieved by the selective accumulation of many small changes of chromosome structure, which lead to the divergence of homoeologues. However, the selection of the localized product of a single mutational step, such as seems likely to have happened in wheat, is very much simpler to envisage, and would clearly be a more rapid and efficient process. To compare the frequencies of the genetical, or cytological, determination of diploid meiotic behaviour, other allopolyploid species should be examined for evidence of a mechanism similar to that in wheat.

Secondly, knowledge of the pairing restricting mechanism may be useful in wheat improvement.

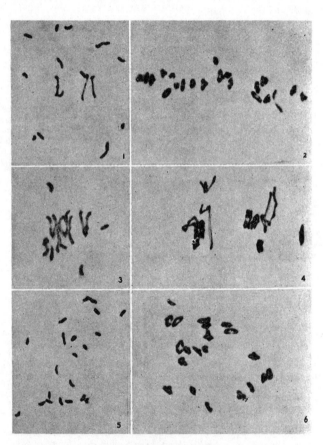

First metaphase of meiosis in Feulgen- and orcein-stained squashes of pollen mother cells of various derivatives of *T. vulgare*

(1) 21-Chromosome euhaploid with 4 bivalents (one widely stretched) and 13 univalents. (2) Euploid with 21 bivalents. (3) 20-Chromosome HH nulli-haploid with one 'pan-handle' and 4 chain trivalents, one bivalent and 3 univalents. (4) 40-chromosome HH nullisomic with one ring of six, 2 chains of four and 13 bivalents. (5) 21-Chromosome haploid in which the HH chromosome is an isochromosome of the long arm ; one stretched bivalent, 18 normal univalents and the isochromosome paired interbrachially with two chiasmata. (6) 41-Chromosome HH monosomic with 20 bivalents and one univalent. (×770)

The utilization of related diploid species, for example, in the genus *Aegilops*, in wheat breeding is hampered because the chromosomes of the diploids do not pair with those of wheat, in intergeneric hybrids. To overcome this obstacle, Sears[10] was forced to use X-rays to induce the translocation of a disease-resistance gene of *Aeg. umbellulata* on to a wheat chromosome, and others[11,12] have been compelled to explore alien chromosome addition and substitution lines. However, this intergeneric allosyndetic pairing may well take place in the absence of the *HH* chromosome, just as it does between chromosomes of the different genomes ; alien genes could then be introduced into wheat chromosomes by normal recombination. Critical hybrids of this constitution can be extracted from crosses of *HH* monosomics by the appropriate diploid.

Finally, nullisomic plants of this constitution may be used either within single varieties or in intervarietal hybrids in 'intergenome exchange' breeding. Thus a useful gene which showed a dose effect could be obtained duplicated or triplicated, represented on each chromosome of the homoeologous group. The chromosome structure within a variety could be re-patterned by breeding from *HH* nullisomics. Further, if an *HH* nullisomic were also an intervarietal hybrid, the release of variation, and the range of segregation in later generations, would be very much greater than from a euploid hybrid. Thus, the *HH* nullisomic may afford access to otherwise unavailable genetic variation.

However, the exploitation of *HH* nullisomics depends upon their fertility. There is no natural self-fertility, but a reasonable seed set has been obtained by pollinating *HH* nullisomics with Holdfast euploids, although the reciprocal cross has been unsuccessful. There were several large multivalents at meiosis in the one 41-chromosome derivative, so far examined, of an *HH* nullisomic pollinated by a euploid. The *HH* chromosome regained from the euploid pollen was always a univalent, and homoeologous pairing must have been prohibited by its presence. The multivalents were thus indicative of translocation heterozygosity, the outcome of homoeologous pairing in the nullisomic parent. Homozygotes for new structural conditions can be derived from such plants, although further back-crossing may first be necessary to reduce the extent of structural alteration.

Knowledge of the situation described already begins to make plain some of the problems of wheat cytogenetics and perhaps of polyploidy in general. Further investigation cannot fail to extend this advantage and may well contribute to the cytogenetic manipulation of wheat for practical breeding purposes.

The interest of Dr. G. D. H. Bell in the development of this work is gratefully acknowledged.

[1] Kihara, H., *Bot. Mag., Tokyo*, **32**, 17 (1919).
[2] Sax, K., *Genetics*, **7**, 513 (1922).
[3] McFadden, E. S., and Sears, E. R., *J. Hered.*, **37**, 81, 107 (1946).
[4] Sarkar, P., and Stebbins, G. L., *Amer. J. Bot.*, **43**, 297 (1956).
[5] Riley, R., Unrau, J., and Chapman, V., *J. Hered.* (in the press).
[6] Sears, E. R., *Res. Bul. Mo. Agric. Exp. Sta.*, **337** (1941).
[7] Sears, E. R., *Res. Bul. Mo. Agric. Exp. Sta.*, **572** (1954).
[8] Sears, E. R., *Amer. Nat.*, **87**, 245 (1953).
[9] Riley, R., and Chapman, V., *Heredity*, **12**, 301 (1958).
[10] Sears, E. R., Brookhaven Symp. in Biology. **9**, 1 (1956).
[11] Jenkins, B. C., Proc. Int. Genet. Symp., 1956, 295 (1953).
[12] Hyde, B. B., *Amer. J. Bot.*, **40**, 174 (1955).

22

Reprinted from *Natl. Acad. Sci. (USA) Proc.* **55**:1447–1453 (1966)

THE EFFECT OF CHROMOSOMES 5B, 5D, AND 5A ON CHROMOSOMAL PAIRING IN TRITICUM AESTIVUM*

By Moshe Feldman

CURTIS HALL, UNIVERSITY OF MISSOURI, COLUMBIA

Communicated by E. R. Sears, April 14, 1966

The common wheat *Triticum aestivum* is an allohexaploid ($2n = 6\times = 42$) consisting of three different genomes, A, B, and D. Genome A was derived from the diploid wheat, *T. monococcum*,[1, 2] genome B from *T. speltoides* (= *Ae. speltoides*), or from a close relative of this species,[3, 4] and genome D from *T. tauschii* (= *Ae. squarrosa*).[5–7] (The nomenclature used is after Morris and Sears.[8]) In hybrids between these diploid ancestors, considerable chromosome pairing (in some cells even complete pairing) is observed regularly,[9–11] indicating that the homoeologous chromosomes of the diploids are still very closely related.

In spite of this close relationship, *T. aestivum* behaves like a typical allopolyploid and forms only bivalents at meiosis. No association takes place between the homoeologous chromosomes. Even in haploids there is very little pairing between homoeologues.[12]

The failure of the homoeologous chromosomes to pair is due to suppression of pairing by chromosome 5B.[12–14] In plants nullisomic for chromosome 5B, multivalent associations are formed regularly which involve homoeologous chromosomes of all the three genomes.[15] Similarly, haploids deficient for chromosome 5B exhibit almost the same amount of pairing as is expected from the pairing observed in hybrids between the three diploid ancestors, *T. monococcum*, *T. speltoides*, and *T. tauschii*.[11]

Although genes for synapsis are known on other chromosomes, e.g., 3B (III) and 2A (II),[16] none of these differentially affect pairing of homoeologous chromosomes. This activity is restricted to the long arm of chromosome 5B (5B^L).[17]

From the fact that no alteration in chiasma frequency was found in nullisomic 5B, Riley[12] concluded that 5B interferes with the processes which lead to synapsis, decreasing the attraction to such a level that only pairing between fully homologous chromosomes can take place.

The results of Okamoto[18] and Sears and Okamoto[18] indicate that under certain circumstances chromosome 5B can also reduce pairing between homologous chromosomes. The expected 14 bivalents (or their equivalent) were found in *T. aestivum* (AABBDD) × *monococcum-tauschii* (AADD) when 5B was absent, but only about 10 bivalents when 5B was present.

Study of chromosomal pairing in plants having different doses of the long arm of chromosome 5B and of the homoeologous chromosomes 5A and 5D has shown that the gene or genes concerned are operative at a different time than previously supposed.

Materials and Methods.—Lines of *T. aestivum* var. *Chinese Spring* monosomic for isochromosomes of the long arm of chromosomes 5A, 5B, and 5D were kindly provided by E. R. Sears. From these mono-isosomic lines, plants with four doses of the long arm of each chromosome (di-isosomics) were obtained through selfing, and these in turn gave rise to plants with six doses (tri-isosomics).

The material was grown under normal greenhouse conditions except as otherwise specified. Spikes were fixed in Carnoy's solution and the anthers squashed in aceto-carmine.

In addition, cytological preparations were made available by E. R. Sears from plants which were, respectively, nullisomic 5D, nullisomic 5D trisomic 5A, and nulli-somic 5D tetrasomic 5A.

Results.—Chromosomal pairing at first meiotic metaphase in pollen-mother cells was normal in all the mono- and di-isosomic combinations and in tri-isosomic $5A^L$ and $5D^L$ (Table 1). Twenty bivalents, mostly rings, were formed almost regu-larly, and the average number of chiasmata per cell was 40 or more except in di-iso $5B^L$, where it was 37.7. The various arms of the isochromosomes paired to make a bivalent or two univalents in the di-isosomics; and in tri-isosomic $5A^L$ and $5D^L$ they appeared as a trivalent (sometimes ring), a bivalent, and a univalent, or three univalents. A univalent isochromosome usually appears as a small ring due to pairing between its two arms.

In tri-isosomic $5B^L$, in contrast, the chromosomal pairing was much reduced (Table 1, Fig. 1). Asynapsis was found consistently in the four tri-isosomic $5B^L$

TABLE 1

MEAN AND RANGE OF CHROMOSOMAL PAIRING AT FIRST METAPHASE OF *T. aestivum* VAR. *Chinese Spring* WITH DIFFERENT DOSES OF THE LONG ARM OF CHROMOSOMES 5B AND 5D

Chromosomal constitution	2n	No. cells	Uni-valents	Bivalents			Tri-valents	No. Xta per cell
				Rod	Ring	Total		
Disomic	42	30	0.06 (0–2)	2.10 (0–4)	18.87 (17–21)	20.97 (20–21)	0.00	43.40 (40–47)
Tri-isosomic $5D^L$	43	30	1.67 (0–3)	1.94 (0–6)	18.23 (14–21)	20.17 (20–21)	0.33 (0–1)	42.20 (39–46)
Mono-isosomic $5B^L$	41	30	1.13 (1–3)	2.97 (0–7)	16.96 (13–20)	19.93 (19–20)	0.00	40.30 (34–43)
Di-isosomic $5B^L$	42	30	1.14 (0–6)	4.30 (1–8)	16.13 (13–20)	20.43 (18–21)	0.00	37.70 (33–41)
Tri-isosomic $5B^L$ Plant 1	43	50	14.20 (2–39)	7.34 (1–13)	6.34 (0–14)	13.68 (2–19)	0.48 (0–1)	21.36 (2–34)
Plant 2	43	50	16.74 (1–39)	7.74 (2–12)	4.88 (0–16)	12.62 (2–21)	0.34 (0–1)	18.80 (2–36)

plants examined. (Detailed counts were made in only two plants.) The number of univalents varied greatly, from two or four in a very few cells to 20 or more in many cells. The bivalents were mostly rods with one, usually terminal, chiasma. The average number of chiasmata per cell was 20 (from 2 to 36).

In many cells the three isochromosomes paired and formed a chain or ring tri-valent, or a bivalent and univalent (Table 1, Fig. 1). In a few cells they appeared as three ring univalents showing interarm pairing. Because the isochromosomes were indistinguishable from other univalents when they failed to pair intrachromosomally, it was impossible to determine their exact frequency of pairing with each other. However, there was no clear evidence that pairing of the 5B arms was affected any differently than the pairing of other chromosomes.

Although it is difficult to analyze earlier stages of meiosis in *T. aestivum*, a few observations were made at prophase. In tri-isosomic $5D^L$, only double strands were seen at late pachytene and mostly bivalents at diplotene and diakinesis. (The iso-chromosomes appeared sometimes as univalents.) But in late pachytene of tri-

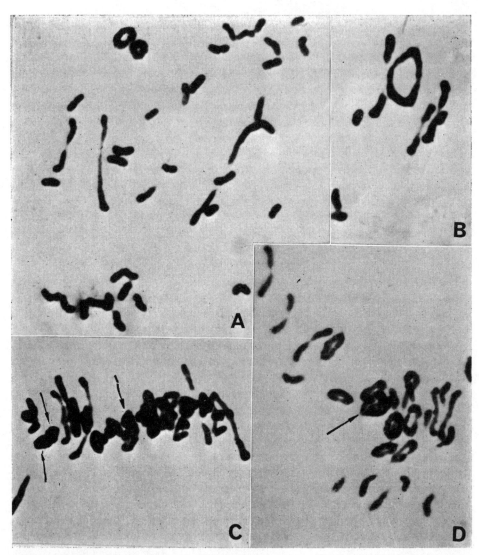

FIG. 1.—Meiotic first metaphase in *Triticum aestivum* var. *Chinese Spring* having six doses of the long arm of chromosome 5B (tri-isosomic 5B^L). (*A*) A cell showing much asynapsis. The three isochromosomes are paired as a chain trivalent at right. The central rod bivalent is heteromorphic, presumably involving homoeologous chromosomes. (*B*) A ring trivalent of the three isochromosomes of 5B^L. (*C*) A quadrivalent at right presumably representing pairing between homoeologous chromosomes. The three isochromosomes appear as ring univalents (indicated by arrows). (*D*) Interlocking of three ring bivalents (indicated by arrow).

isosomic 5B^L single strands as well as double strands were seen, and at diplotene and diakinesis many univalents were counted. Clearly, there was partial failure of synapsis in tri-iso 5B^L.

In each of the three tri-isosomic 5B^L plants examined, multivalent associations indicating pairing between homoeologous chromosomes were observed at first metaphase (Fig. 1*C*). The multivalents, mostly tri- and tetra- and very rarely pentavalents, were formed in very low frequency, one per 100 or more cells. Heteromorphic rod bivalents were also observed (Fig. 1*A*).

Frequent interlocking of ring and rod bivalents, of two rings, and even of three rings was seen in tri-isosomic 5B^L (Fig. 1D). This was in spite of the large reduction in the number of bivalents (Table 1).

Multivalent associations and frequent interlocking, as well as asynapsis, were also found in di-isosomic 5B^L plants which had been placed several days before meiosis at high (30°C) and low (15°C) temperatures. These extreme temperatures did not increase the amount of asynapsis in tri-isosomic 5B^L.

The one available plant nullisomic for chromosome 5D exhibited a degree of asynapsis. Two, four, and rarely more univalents were counted in many cells. In addition, rare multivalent associations (trivalents and quadrivalents) and frequent interlocking bivalents were observed.

Plants nullisomic for chromosome 5D and trisomic for chromosome 5A showed a similar picture in first metaphase; many cells exhibited interlocking bivalents and a small amount of asynapsis. However, a plant nullisomic for 5D and tetrasomic for 5A showed nearly regular pairing, with rare univalents and almost no interlocking bivalents. It seems that with respect to pairing, four doses of 5A compensate for the nullisomic condition of 5D.

Discussion.—It is generally assumed that the chromosomes are distributed at random in the premeiotic nucleus. However, some doubt is cast on this assumption by the demonstration by Kasha and Burnham[19] that pairing in barley is initiated regularly in the terminal parts of the chromosomes. With pairing commencing at the chromosome ends, bivalents formed from randomly distributed homologues should frequently be interlocked, but this is not the case.

Difficulty is also encountered in understanding the action of chromosome 5B in wheat if the premeiotic chromosomes are randomly distributed. If two doses of the 5B gene simply reduce the attraction forces between homoeologues (and also homologues, to a lesser extent) as Riley[12] suggested, then by chance alone two homoeologues may be so much closer together than either is to its homologue that even the reduced attraction forces will be strong enough to cause them to pair. Such homoeologous pairing is almost never observed in disomic *T. aestivum*.

Although extra dosage of the long arm of chromosome 5B reduced the pairing of fully homologous chromosomes as would be expected on the basis of Riley's hypothesis, it also induced pairing between homoeologous chromosomes and caused a high frequency of interlocking bivalents. Therefore, the action of the 5B gene is not simply to reduce chromosomal attraction at meiotic prophase.

The apparently paradoxical effect of extra dosage of 5B^L can be explained in the following way: in plants nullisomic for chromosome 5B the chromosomes are not randomly scattered in the premeiotic nucleus, but homoeologues as well as homologues lie near each other and may perhaps even associate throughout most of their length. The meiotic pairing follows as a second step and brings the closely oriented (or loosely paired) homologous segments to a full and more intimate contact.

It may be assumed that the 5B gene suppresses the premeiotic association and thus tends to cause random distribution of chromosomes in the premeiotic nucleus. Two doses of this gene, while scarcely affecting the homologous chromosomes, keep the homoeologues apart. This leads to exclusively homologous pairing. The homoeologues do show secondary association,[20] indicating that two doses of 5B do not completely prevent them from approximating each other.

285

Six doses of the 5B gene (in tri-isosomic 5B^L) are assumed to suppress premeiotic association completely. This leads to a random distribution of chromosomes in which the homologues may presumably be separated by a distance of as much as several microns. The meiotic attraction forces which usually cause pairing of very closely oriented homologues are not sufficient to bring about pairing of such distantly separated homologues, and as a result many chromosomes fail to pair. Since homoeologous chromosomes may by chance lie close to each other, homoeologous pairing can take place. Although multivalent associations were rare in first metaphase of tri-isosomic 5B^L because of the general reduction of pairing, heteromorphic rod bivalents occurred which were presumably also the result of homoeologous pairing. Finally, the wide occurrence of interlocking bivalents, in spite of the reduction in the total number of bivalents, shows clearly that some of the pairing here was between widely separated chromosomes.

Four doses of 5B^L affected pairing only slightly in optimal temperature. This effect was much increased in high and low temperatures, with the result that asynapsis, homoeologous pairing, and interlocking bivalents were quite common. Because higher temperature did not increase the effect of six doses of 5B^L, it is assumed that the maximum possible effect, complete suppression of premeiotic association, was brought about by six doses at normal temperature.

Although one or both of the homoeologues of chromosome 5B might also be expected to have a gene affecting synapsis, six doses of the long arms of 5A and 5D did not provide evidence for the existence of such genes. However, from the fact that nullisomic 5D is partially asynaptic and yet exhibits homoeologous pairing and interlocking bivalents, it seems clear that this chromosome does carry a gene conditioning premeiotic association. Its effect is inverse to that of the 5B gene.

Since four doses of 5A compensate for the absence of 5D, it seems likely that chromosome 5A also has a gene which promotes premeiotic association. However, since three doses of 5A do not compensate for nullisomic 5D, the 5A gene must have a weaker effect. If it were as strong as the 5D gene, nulli-5D tri-5A would be equivalent to mono-5D and would therefore have normal synapsis.

Plants mono-isosomic for the long arms of chromosomes 5D and 5A have regular pairing. Thus, the synapsis genes of these chromosomes are located on their long arms, like that of 5B.[17] Since the long arms of 5A, 5B, and 5D are homoeologous,[21] all three genes are probably derived from one ancestral locus.

Every diploid species in the wheat group may be assumed to have a gene similar to the one on chromosome 5D which promotes premeiotic association. The genes in the different species evidently differ in potency, for it was found[21] that the genomes of *T. speltoides* and *T. tripsacoides* (*Ae. mutica*) are much more effective than those of other diploids in reducing the effect of chromosome 5B. The first tetraploid wheat, believed to have been a doubled hybrid of *T. monococcum* and *T. speltoides*, presumably had homoeologous pairing and resultant meiotic irregularity until a mutation occurred on chromosome 5B to suppress the homoeologous pairing.[4] The simplest assumption is that the gene on 5B which promoted premeiotic association mutated to a form with the opposite effect.

The conclusion that premeiotic association serves as an essential step in regulating meiotic pairing invites reconsideration of certain assumptions concerning chromosome behavior. One of these is the existence of long-range forces. Since it has been assumed that homologues are randomly distributed throughout the nucleus, the existence of a mechanism capable of bringing homologous segments together over long distances has been widely accepted. However, the fact that in *T. aestivum* with six doses of 5B^L the presumably randomly distributed chromosomes fail to pair suggests that the meiotic forces do not operate effectively at long range.

Smith[22] thought that synapsis begins at the preceding anaphase and terminates at late pachytene. However, premeiotic association and meiotic pairing should be considered as two distinct phenomena which are conditioned by different factors. This is clearly apparent from the fact that the 5B gene suppresses only the premeiotic association.

It may be that association of homologues (and homoeologues in nulli-5B) takes place in all the somatic cells throughout the life of the plant. However, this association may be difficult to observe, for it may be more pronounced at interphase than during division stages. Close association during interphase would bring homologous genes into proximity at the time when they are most active, and this might be advantageous. During mitosis the chromosomes move and divide, and as a consequence may become more scattered. Yet somatic association was found in dividing cells in tissue culture of *Haplopappus gracilis*,[23] and it has been shown that homologous human chromosomes tend to lie closer to each other on the metaphase plate than expected by chance.[24]

It has been suggested[25] that the chromosomes occupy definite positions in the interphase nucleus. It is possible that they are attached by their centromeres or telomeres in a specific order to the nuclear membrane[26, 27] with homologues lying close to each other. Support for this assumption comes from reports[28−30] that the chromosomes take an active part in the formation of the nuclear membrane in mitotic telophase.

Summary.—In normal dosage, the long arm of chromosome 5B (5B^L) prevents the pairing of homoeologues. In four doses (di-isosomic 5B^L) it slightly reduces chiasma frequency between homologues. In six doses (tri-isosomic 5B^L) it causes considerable asynapsis of homologues but permits some pairing between homoeologues and induces a high frequency of interlocking bivalents. The explanation is advanced that 5B^L regulates the premeiotic association of homologous and homoeologous chromosomes. Two doses of 5B^L suppress partially this premeiotic association, especially that of the homoeologues. Six doses of 5B^L suppress all the premeiotic association, causing random distribution of the chromosomes in the premeiotic nucleus. The meiotic attraction forces are not strong enough to cause pairing of the distantly separated chromosomes, and thus partial asynapsis results. Under these conditions, homoeologous pairing takes place when two homoeologues happen to be closer to each other than to their respective homologues. Pairing between separated chromosomes results in a high frequency of interlocking bivalents.

Chromosomes 5D and 5A have an effect on premeiotic association opposite to that of their homoeologue 5B. It is assumed that the 5B gene was derived as an antimorphic mutation from an association-promoting gene like those carried by 5D and 5A.

The author is grateful to Professor E. R. Sears for his encouragement and advice during the course of this work and for his invaluable contribution in the interpretation of the data and the preparation of the manuscript. Thanks are also due to Drs. L. M. Steinitz-Sears and T. Mello-Sampayo for their critical reading of the manuscript.

* This work was supported by a grant from the National Science Foundation to Dr. E. R. Sears. Journal series no. 3085 of the Missouri Agricultural Experiment Station. Approved by the Director.

[1] Kihara, H., *Botan. Mag. (Tokyo)*, **32**, 17 (1919).

[2] Sax, K., *Genetics*, **7**, 513 (1922).

[3] Sarkar, P., and G. L. Stebbins, *Am. J. Botany*, **43**, 297 (1956).

[4] Riley, R., J. Unrau, and V. Chapman, *J. Heredity*, **49**, 91 (1958).

[5] McFadden, E. S., and E. R. Sears, *Rec. Genet. Soc. Am.*, **13**, 26 (1944).

[6] McFadden, E. S., and E. R. Sears, *J. Heredity*, **37**, 81 (1946).

[7] Kihara, H., *Agri. Hort. (Tokyo)*, **19**, 889 (1944).

[8] Morris, R., and E. R. Sears, in *Wheat and Wheat Improvement*, ed. K. S. Quisenberry and L. P. Reitz, in press.

[9] Sears, E. R., *Res. Bull. Missouri Agri. Exptl. Sta.*, 337 (1941).

[10] Mochizuki, A., and M. Okamoto, *Chromosome Inform. Serv.*, **2**, 12 (1961).

[11] Kimber, G., and R. Riley, *Can. J. Genet. Cytol.*, **5**, 83 (1963).

[12] Riley, R., *Heredity*, **15**, 407 (1960).

[13] Sears, E. R., and M. Okamoto, *Proc. Intern. Congr. Genet., 10th*, **2**, 258 (1958).

[14] Riley, R., and V. Chapman, *Nature*, **182**, 713 (1958).

[15] Riley, R., and C. Kempanna, *Heredity*, **18**, 287 (1963).

[16] Sears, E. R., *Res. Bull. Missouri Agri. Exptl. Sta.*, 572 (1954).

[17] Riley, R., V. Chapman, and G. Kimber, *Nature*, **186**, 259 (1960).

[18] Okamoto, M., *Wheat Inform. Serv.*, **5**, 6 (1957).

[19] Kasha, K. J., and C. R. Burnham, *Can. J. Genet. Cytol.*, **7**, 620 (1965).

[20] Kempanna, C., and R. Riley, *Heredity*, **19**, 289 (1964).

[21] Riley, R., and V. Chapman, *Nature*, **203**, 156 (1964).

[22] Smith, S. G., *Can. J. Res.*, **20**, 221 (1942).

[23] Mitra, J., and F. C. Steward, *Am. J. Botany*, **48**, 358 (1961).

[24] Schneiderman, L. J., and C. A. B. Smith, *Nature*, **195**, 1229 (1962).

[25] Mirsky, A. E., and S. Osawa, in *The Cell*, ed. J. Brachet and A. E. Mirsky (New York: Academic Press, 1961), vol. 2, p. 695.

[26] Sved, J. A., *Genetics*, **53**, 747 (1966).

[27] Vanderlyn, L., *Botan. Rev.*, **14**, 270 (1948).

[28] Barer, R., S. Joseph, and G. A. Meek, *Exptl. Cell Res.*, **18**, 179 (1959).

[29] Porter, K. R., and R. D. Machado, *J. Biophys. Biochem. Cytol.*, **7**, 167 (1960).

[30] Ito, S., *J. Biophys. Biochem. Cytol.*, **6**, 433 (1960).

EFFECT OF COLCHICINE ON MEIOSIS OF HEXAPLOID WHEAT

C. J. Driscoll, N. L. Darvey, and H. N. Barber

HEXAPLOID wheat, *Triticum aestivum*, behaves cytologically as a diploid organism in that only bivalents are formed at meiosis. This diploid behaviour is principally the result of the activity of a gene or genes in the long arm of chromosome 5*B* (refs. 1–4). In the absence of this gene or genes multivalents are formed involving homoeologous chromosomes[5].

Feldman[6] has demonstrated that an increased dosage of this gene or genes results in asynapsis, interlocking of bivalents and some multivalent formation. The hypothesis put forward to account for these various behaviours was that in the absence of 5*B* both homologues and homoeologues are associated premeiotically; with the normal two doses of 5*B*, however, this association is restricted to homologues; with six doses of the long arm of 5*B*, or four doses at extreme temperatures, there is no premeiotic association, all chromosomes being randomly arranged before meiosis.

The phenomena of asynapsis and interlocking of bivalents were also observed by Barber[7] in *Fritillaria meleagris* treated with colchicine before meiosis. Because of the similarities observed in these dissimilar experiments colchicine was applied to hexaploid wheat shortly before the onset of meiosis. Plants of the variety 'Chinese Spring' grown in a glasshouse at approximately 20° C were used. A single application of approximately 0·25 ml. of either a 1 per cent or a 2·5 per cent solution of colchicine was injected into tillers above the tip of the immature spike by means of a hypodermic syringe.

Spikes were subsequently collected for meiotic study some days later, fixed in Carnoy's 6 : 3 : 1 and anthers were stained in acetocarmine.

From a spike fixed 7 days after treatment with colchicine anthers containing pollen mother cells (PMCs) with twice the normal number of chromosomes at metaphase I were obtained from three different florets. These dodecaploid cells apparently arose by means of C mitosis as described by Levan[8]. Thus the last premeiotic mitosis apparently took place about 7 days before metaphase I of meiosis in these cases. The uniformity of PMCs with eighty-four chromosomes in this material indicates that the last premeiotic mitosis is fairly well synchronized in all cells of the same anther.

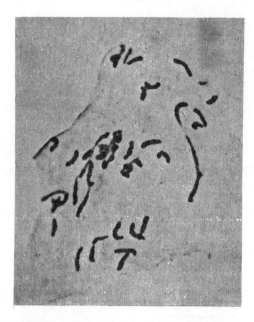

Fig. 1. Metaphase I of meiosis of a 6× cell after treatment with colchicine, showing marked asynapsis and one trivalent.

In eight different spikes fixed 3–6 days after colchicine treatment anthers possessing PMCs with the normal forty-two chromosomes at metaphase I were observed. Thus in these cases colchicine had been applied after completion of the last premeiotic mitosis. In these cells, asynapsis and multivalent formation were observed. Multivalent formation was quite infrequent, only an occasional trivalent or quadrivalent being observed (Fig. 1). The mean pairing observed in these cells, as shown in Table 1, is slightly lower than that observed by Feldman[6] in plants with six doses of the gene or genes in the long arm of chromosome 5B.

Table 1. MEAN AND RANGE OF CHROMOSOMAL PAIRING AT METAPHASE I OF MEIOSIS IN PLANTS TREATED WITH COLCHICINE

Type of cell	No. of cells	Univalents	Bivalents	Trivalents	Quadri-valents
6×	50	21·28 (12–36)	10·32 (3–15)	—	0·02 (0–1)
12×	50	8·32 (2–26)	37·38 (29–40)	0·04 (0–1)	0·20 (0–2)

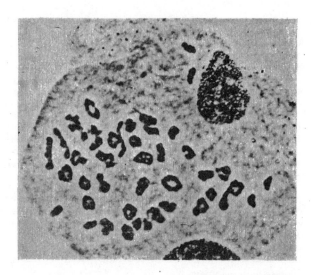

Fig. 2. Metaphase I of meiosis of a 12× cell after treatment with colchicine, showing thirty-eight bivalents and eight univalents. Six of the univalents are obviously in three groups of two morphologically similar chromosomes.

From these observations it can be concluded that colchicine applied before meiosis inhibits premeiotic association of homologues.

The dodecaploid cells chiefly possessed bivalents (Fig. 2), with only an occasional multivalent as shown in Table 1. Multivalents find a ready explanation in these cells, for homologues are present in four doses. The scarcity of multivalents may be explained by the presence of colchicine. It is known that colchicine persisted in this material, for double dodecaploid cells undergoing C mitosis have also been observed in these anthers. Colchicine presumably suppresses quadrivalent formation by preventing the homologues associating into fours. Thus multivalent formation is limited to cases where more than two homologues happen to be in close proximity.

These cells, however, exhibit fairly regular bivalent formation. This is because the members of each bivalent were sister chromatids before C mitosis. Thus these automatically became premeiotically associated in formations referred to as "pairs of skis" by Nebel and Ruttle[9]. Synapsis and chiasma formation then ensued.

Bivalents were formed by these chromosomes which were associated because of failure of anaphase movement, and so it can be concluded that colchicine does not inhibit synapsis or chiasma formation. Also it can be concluded that colchicine cannot disrupt chromosome association once it has taken place but can only inhibit such association from being initiated. This clearly shows that premeiotic association and synapsis are two distinct phenomena.

Colchicine is unable to disrupt chromosome association, and so the absence of such association in the hexaploid cells in this study indicates that the chromosomes of these cells were not premeiotically associated when colchicine was applied. This, however, does not necessarily mean that chromosomes are only associated immediately before meiosis, for there may be an association–disassociation cycle operating. This would involve normal association of homologues with disruption to some degree when the chromosomes go through the processes of successive mitoses. Such behaviour has been suggested by Feldman, Mello-Sampayo and Sears[10] following their detection of a vestige of chromosome association in mitotic cells of hexaploid wheat.

Received September 11, 1967.

[1] Sears, E. R., and Okamoto, M., *Proc. Tenth Intern. Cong. Genet.*, **2**, 258 (1958).

[2] Riley, R., and Chapman, V., *Nature*, **182**, 713 (1958).

[3] Riley, R., *Heredity*, **15**, 407 (1960).

[4] Riley, R., Chapman, V., and Kimber, G., *Nature*, **186**, 259 (1960).

[5] Riley, R., and Kempanna, C., *Heredity*, **18**, 287 (1963).

[6] Feldman, M., *Proc. US Nat. Acad. Sci.*, **55**, 1447 (1966).

[7] Barber, H. N., *J. Genet.*, **43**, 359 (1942).

[8] Levan, A., *Hereditas*, **24**, 471 (1938).

[9] Nebel, B. R., and Ruttle, M. L., *J. Heredity*, **29**, 3 (1938).

[10] Feldman, M., Mello-Sampayo, T., and Sears, E. R., *Proc. US Nat. Acad. Sci.*, **56**, 1192 (1966).

24

Reprinted from *Can. J. Genet. Cytol.* **18**:357-364 (1976)

THE EFFECT OF GENES CONTROLLING DIFFERENT DEGREES OF HOMOEOLOGOUS PAIRING ON QUADRIVALENT FREQUENCY IN INDUCED AUTOTETRAPLOID LINES OF *TRITICUM LONGISSIMUM*

LYDIA AVIVI

Department of Plant Genetics, The Weizmann Institute of Science, Rehovot, Israel

Different genotypes of *Triticum longissimum* are known to either promote or suppress chromosome pairing in crosses with polyploid wheats. Lines that promote homoeologous pairing are here designated as intermediate pairing lines, while those which have no such effect or suppress pairing are known as low pairing lines. To determine a possible effect of these genotypes on homologous pairing, tetraploidy was induced in both lines and chromosomal pairing was studied at first metaphase of meiosis. While the two induced autotetraploids did not differ in chiasma frequency or in the number of paired chromosomal arms, they differed significantly in multivalent frequency; the intermediate-pairing autotetraploid exhibited the same multivalent frequency as that expected on the basis of two telomeric initiation sites, while the low pairing autotetraploid exhibited a significantly lower frequency. It is assumed that in the autotetraploid the low pairing genotype does not affect meiotic pairing *per se*, but modifies the pattern of homologous association in a similar manner to that known in polyploids and caused by diploidization genes. It is speculated that the tendency for bivalent pairing in the low pairing autotetraploid is due to spatial separation of the four homologous chromosomes in somatic and premeiotic cells into two groups of two.

On sait que certains génotypes de *Triticum longissimum*, dans les croisements avec des blés polypoïdes, semblent favoriser ou bien nuire à l'appariement des chromosomes. Les lignées favorisant l'appariement homéologue sont désignées ici comme lignées d'appariement intermédiaire, alors que celles chez lesquelles on n'observe aucun de ces effets ou qui réduisent l'appariement sont désignées comme lignées d'appariement faible. Afin de connaître l'effet possible de ces génotypes sur l'appariement homologue on a provoqué la tétraploïdie chez les deux lignées et l'appariement chromosomal a été étudié à MI. Alors que les deux autotétraploïdes résultants n'ont pas manifesté de différence de fréquence de chiasma ou de nombre de bras chromosomiques accouplés, on a observé une différence significative dans la fréquence des multivalents. L'autotétraploïde résultant de l'appariement intermédiaire a fait preuve d'une fréquence des multivalents en accord avec l'hypothèse des deux sites d'initiation télomérique, alors que celui résultant d'un appariement faible a révélé une fréquence des multivalents significativement plus basse. On suppose que chez l'autotétraploïde, le génotype à faible fréquence d'appariement n'affecte pas en soi l'appariement méiotique mais modifie la forme d'association homologue de manière similaire à celle connue chez les polyploïdes et causée par les gênes de diploïdisation. On formule la conjecture que la tendance, chez l'autotétraploïde issu de l'appariement faible, à l'appariement bivalent est occasionnée par la séparation physique des quatre chromosomes homologues en leur groupes de deux dans les cellules somatiques et préméiotiques.

[Traduit par le journal]

Introduction

Cytological diploidization of hexaploid wheat is caused by the action of genes such as that located on the long arm of chromosome 5B (Okamoto, 1957; Sears and Okamoto, 1958; Riley, 1958; Riley and Chapman, 1958), that on the β arm of chromosome 3D (Mello-Sampayo, 1968, 1971a; Upadhya and Swaminathan, 1967), and that on the β arm of chromosome 3A (Driscoll, 1972). These diploidizing genes suppress synapsis of homoeologous chromosomes so that in their presence only full homologues pair. The same genes also interfere with pairing of homoeologous chromosomes in hybrids between common wheat and related species.

293

Hexaploid wheat also contains genes that promote pairing of homoeologous chromosomes in interspecific hybrids. These genes are located on the long and short arms of both chromosomes 5A and 5D and on the short arm of chromosome 5B (Feldman, 1966, 1968; Feldman and Mello-Sampayo, 1967; Riley *et al.*, 1966), as well as on the α arm of chromosomes 3D and 3A (Driscoll, 1972; Mello-Sampayo and Canas, 1973). The net effect on homoelogous pairing in hexaploid wheat stems from a very delicate balance between suppressors and promoters.

The diploid species of the wheat group are also known to contain genes that affect pairing of homoeologous chromosomes. *Triticum speltoides, T. tripsacoides, T. longissimum* and *T. caudatum* contain genes that counteract the effect of the suppressors and bring about homoeologous pairing in hybrids with polyploid wheats (Riley, 1960, 1963; Riley *et al.*, 1961; Mello-Sampayo, 1971b; Kihara and Lilienfeld, 1935; Upadhya, 1966). Riley, (1960, 1963) assumed that these genes inhibit the suppressive effect of the $5B^L$ gene, i.e., the diploidizing gene on the long arm of chromosome 5B. Alternatively, they can be considered as promoters. Feldman and Mello-Sampayo (1967) showed that the *speltoides* genotype, actively promotes pairing of homoeologous chromosomes. The promoters are presumably homoeoalleles to the promoters of the polyploid wheats. The amount of homoeologous pairing achieved in different interspecific hybrids depends, therefore, on the kind of balance between the suppressors and the promoters that exists in these hybrids. Genetic intraspecific variation in the ability of different genotypes to balance the suppressors in hybrids with common wheat has been found in *T. speltoides* (Dvorak, 1972; Kimber and Athwal, 1972), *T. tripsacoides* (Dover and Riley, 1972), and *T. Longissimum* (Mello-Sampayo, 1971bg Kimber and Sallee, cited in Kimber, 1973). Accordingly, the respective genotypes of these diploids have been designated as high-, intermediate-, or low-homoeologous pairing promoters. There is at present no evidence to indicate whether a genotype that induces low-homoeologous pairing contains relatively weak promoters, or whether its action is caused by strong suppressors.

That various diploid species may also contain genes that suppress homoeologous pairing has been suggested by Okamoto and Inomato (1974) and Waines (1975). While producing a series of alien addition lines in common wheat using *T. longissimum* as the donor parent, Feldman (unpublished data) indeed found that one of the lines behaves cytologically like tetrasomic 5B, i.e., it exhibits partial asynapsis and interlocking bivalents at high (30°C) temperatures. This may indicate the presence of a suppressor on the diploid level that is homoeoallelic to the $5B^L$ gene.

If the diploid species contain suppressors that are homoeoalleles to those of the polyploid taxa, then it is reasonable to assume that the diploidizing mechanisms did not arise on the polyploid level but originated in diploids. It is thus of interest to see whether suppressors from the two levels have similar effects. One peculiarity of the $5B^L$ gene is that it can also affect homologous pairing under certain conditions (Feldman, 1966). One can ask whether the suppressors present in the diploids share this ability.

Larsen and Kimber (1973) found no differences between the pairing patterns of induced autotetraploids of *T. speltoides* containing genes controlling either high or low homoeologous pairing. They concluded that these genes only affect the pairing between homoeologues and not that between homologues. However, since *T. speltoides* is a cross-pollinator and as such is highly heterozygous (Zohary and Imber, 1963), and since no plants have been found so far that are homozygous for the low-pairing effect, it is probable that the induced autotetraploids of *T. speltoides* studied by Larsen and Kimber (1973) were heterozygous and not sufficiently extreme in their genetic constitution to exert a differential effect on homologous pairing. A self-pollinator diploid is much more likely to be homozygous at the critical loci and, therefore, more suitable material for a study of homologous pairing effects.

Triticum longissimum is a self-pollinator and, moreover, contains contrasting genotypes which, when crossed with common wheat, either promote or suppress homoeologous pairing (Mello-Sampayo, 1971b). The present study was initiated in *T. longissimum* with the aim of comparing pairing patterns of homologous chromosomes in induced autotetraploids derived from genotypes with a known intermediate and with a known low homoeologous pairing effect.

Materials and Methods

Two lines of *T. longissimum* were chosen for this comparative study. The first, designated as 7214, was obtained from T. Mello-Sampayo, who found that the genotype of this particular line induces intermediate pairing of homoeologous chromosomes in hybrids with common wheat (Table I). This line carries a promoter of homoeologous pairing. The seond line of *T. longissimum*, designated as 7011, was collected by M. Feldman south of Beersheva, in the Negev of Israel. This line contains a suppressor of homoeologous pairing; its genotype induces low homoeologous pairing in hybrids with common wheat (Table I). Moreover, in one of the addition lines in which 7011 of *T. Longissimum* was used as donor and common wheat as recipient, an effect similar to that of the suppressor on $5B^L$ has been found by Feldman (unpublished).

Forty-eight hours after germination, seedlings of the two lines were placed in a $3 \times 10^{-4}M$ solution of colchicine for four h. After treatment the seedlings were planted in the greenhouse, where the temperature ranged from 22 to 25°C.

Samples of young ears were fixed in Carnoy's solution, and the anthers were squashed in 2% aceto-carmine. To embrace the variation due to fluctuations in environmental conditions, replicate samples were fixed on different days. Samples of untreated diploid material were also prepared for analysis.

Results and Discussion

The pairing behavior of the two lines 7214 and 7011 was quite similar on the diploid level (Table II). In both cases pairing was complete with minor differences in the mean number of ring- or rod-bivalents, i.e., the number of paired chromosomal arms did not differ significantly at the 0.01 level of probability ($t_{67} = 2.28 < 2.66$). However, the two lines differed in chiasma frequency (Table II); the intermediate pairing line exhibited a significantly higher chiasma frequency than the low pairing line at the 0.01 probability level ($t_{67} = 4.16 > 2.66$). This difference results from a higher frequency of interstitial chiasmata in the intermediate pairing line (Table II).

One explanation for the difference in interstitial chiasma frequencies of the two lines is to assume that terminalization is more rapid and/or more efficient in the low pairing line. Another is that, in the low pairing line, pairing is initiated only at the telomeres, while the higher pairing line contains interstitial pairing initiation sites as well as telomeric ones. If this is the case, the induced autotetraploid of this line should exhibit a higher frequency of multivalents than that expected on the basis of two initiation sites at the telomeres. But since the quadrivalent frequency here is equal to the expected (Table II), the number of pairing initiation sites should be regarded as two. However, since the diploidizing genes that affect homoeologous pairing do so by controlling the chromosomal spatial relationships in somatic and premeiotic cells (Feldman, 1966, 1968; Feldman and Avivi, 1973), one can speculate that the homologues are much more intimately associated in premeiotic cells of the higher than in those of the low pairing line. When premeiotic association of homologues is intimate, the telomeres can start to pair immediately at the beginning of zygotene and the process of pairing can proceed quickly and involve larger segments; as a result, interstitial chiasmata are formed at higher frequencies.

The two colchicine induced tetraploids exhibited complete pairing of homologous chromosomes; no detectable signs of asynapsis were found in either the intermediate or the low pairing line (Table II). The chromosomes were usually associated as quadrivalents and as bivalents; very low frequencies of univalents and trivalents were observed. There is no significant difference, at the 0.05 level, in the mean number of paired chromosomal arms ($t_{151} = 0.63 < 1.98$), nor in the chiasma frequency (Table II). The fact that the degree of pairing in the autotetraploids of the intermediate- and low-pairing lines is similar indicates that the low pairing genotype does not suppress pairing of homologues under the set of experimental conditions. A similar conclusion was reached by Kimber and Athwal (1972), who found that the low-pairing forms of *T. speltoides* do not interfere with homologous chromosomal pairing: in an amphiploid derived by colchicine treatment from a common wheat × *T. speltoides* hybrid exhibiting low-pairing, regular bivalent formation was observed.

The two autotetraploid lines differ significantly, however, in frequency of multivalents, mostly quadrivalents (Table II). The intermediate pairing autotetraploid exhibited a significantly higher frequency of quadrivalents than that of the low pairing autotetraploid ($t_{151} = 4.91 > 2.61$). When all four homologues have an equal chance to pair with each other, and when pairing is initiated only or mainly at the telomeres, i.e., at two sites, as is probably the case in the Gramineae (Tabata, 1963; Kasha and Burnham, 1965), the expectation of quadrivalent frequency is two-thirds (Sved, 1966). The frequency of quadrivalent formation in the intermediate pairing autotetraploid is similar to the expected value, i.e., 4.53 per cell or 64.8% of the chromosome complement (Table II). This good fit indicates that for each chromosome all four homologues have an equal chance to pair with each other. In contrast to this, the quadrivalent frequency in the low-pairing autotetraploid is significantly lower than the expected, thus indicating a clear tendency towards bivalent pairing. Clearly, the low pairing genotype modifies the homologous pairing pattern and induces bivalent pairing, i.e., cytological diploidization, even in the autopolyploid condition.

From the above, it seems that an easy way to screen diploid species for suppressors is to study multivalent frequency in induced autotetraploids; multivalent frequency which is lower than two-thirds will indicate the activity of a diploidizing gene.

It is expected that the diploidizing system of polyploid wheats will reduce quadrivalent frequency in $8x$ plants of durum wheat or $12x$ plants of common wheat in a similar manner. To my knowledge, no cytological observations have been made in induced autotetraploids of common wheat ($2n = 12x = 84$). In $8x$ durum wheat, Morrison and Rajhathy (1960a) found that on the average, 63% of the chromosomes were involved in quadrivalent formation, namely 8.8 multivalents per cell, which is lower than the expected (9.3). In $4x$ autotetraploid *T. monococcum* (Morrison and Rajhathy, 1960b), on the other hand, the observed frequency of quadrivalents was higher; 73% of the chromosomes formed multivalents. Coucoli (1970) found that the observed values of quadrivalent frequencies in tetrasomic lines of *T. aestivum* var. Chinese Spring were generally lower than those expected and concluded that there is a preferential tendency towards bivalent formation in these tetrasomic lines.

TABLE I

Mean chromosomal pairing per cell at the first meiotic metaphase in F_1 hybrids between *Triticum aestivum* var. Chinese Spring and the intermediate- and low-pairing lines of *Triticum Longissimum*

Genotype of *T. longissimum*	Univalents	Bivalents	Trivalents	Quadrivalents	Xmata	Remarks
Intermediate pairing (7214)	17.46	4.69	0.32	0.05	5.84	*
Low pairing (7011)	24.56	1.69	0.02	–	1.75	**

* Means taken from data of Mello-Sampayo (1971).
** Means taken from data of M. Feldman (unpublished).

TABLE II

Mean chromosomal pairing per cell at first meiotic metaphase in diploid and autotetraploid plants derived from intermediate- (7214) and low- (7011) pairing lines of Triticum longissimum

	No. of cells	Univalents	Bivalents			Multivalents		Xmata			No. of paired chromosome arms	% of chromosomes in multivalent association
			Rod	Ring	Total	III	IV	Terminal	Interstitial	Total		
Diploid level (2n = 2x = 14)												
7214	36	–	0.28	6.72	7.00	–	–	13.19	1.28	14.47 ±0.13	27.44 ±0.20	–
7011	33	0.06	0.58	6.39	6.97	–	–	13.12	0.43	13.55 ±0.18	26.72 ±0.34	–
Tetraploid level (2n = 4x = 28)												
7214	100	0.01	0.90	4.02	4.92	0.01	4.53 ±0.11	23.94	1.94	25.93	51.36 ±0.26	64.82
7011	53	0.09	1.45	5.68	7.13	0.02	3.40 ±0.20	25.19	0.47	25.66	51.21 ±0.41	48.79

297

The similarity between the effect of the low-pairing genotype of *T. longissimum* on bivalent formation in the induced autotetraploid and that of diploidizing genes, such as the 5B^L gene in polyploid wheats, is easily seen. The effect of the polyploid gene is probably so much more pronounced and results in strict bivalent formation because of the presence of homoeologues here, which are more readily affected by this genetic system than the homologues (Feldman, 1966).

Genetic systems that induce cytological diploidization similar to that in polyploid wheat have also been found in other allopolyploids, such as *Avena* (Rajhathy, 1971), *Lolium* (Jauhar, 1975), *Gossypium* (Kimber, 1961; Endrizzi, 1962), and *Delphinium* (Legro, 1961). Moreover, genetic systems that determine strict or almost strict bivalent pairing in autotetraploids are also known. Bivalent pairing in the assumed autotetraploid *Avena barbata* is achieved through the activity of a single gene (Ladizinsky, 1973). A bivalent forming genetic system, closely akin to that of polyploid wheat, was found in autotetraploids of the Brassiceae (Harberd, 1972). In *Phleum pratense* ($2n = 6x = 42$) there is a genetic restraint on pairing, as a result of which only bivalents are formed (Nordenskiold, 1957). Strict bivalent pairing was found also in the autotetraploid *Lotus corniculatus* (Dawson, 1941). *Lotus corniculatus* appears to be an example of a naturally occurring tetraploid showing tetrasomic inheritance, but having a low frequency of quadrivalents.

Riley and Law (1965) differentiate between the diploidizing systems of allopolyploids and those of autopolyploids. There can be no doubt that the genetic and evolutionary consequences of strict bivalent pairing, with random choice of partners in autopolyploids, are quite different from those arising out of the strict pairing of full homologues in allopolyploids. However, the mechanism of cytological diploidization in the two types of polyploids is probably the same, and differences are quantitative rather than qualitative.

There are several ways of explaining how cytological diploidization is achieved. To account for a quadrivalent frequency lower than two-thirds, Sved (1966) assumed disturbances in pairing initiation sites and the commencement of pairing from only one end, or by a failure in chiasma formation. However, in the case presented here the two autotetraploid lines did not differ in the number of paired chromosomal arms or in chiasma frequency. The present results can be explained by Feldman's (1966) spatial hypothesis. According to this hypothesis the spatial distribution of homologous chromosomes in somatic and premeiotic cells determines their meiotic behavior.

Normally, chromosomes are arranged in the nucleus in a polarized manner, i.e., their centromeres are attached to the nuclear membrane at the polar site and their telomers to the nuclear membrane at the opposite side of the nucleus (Avivi and Feldman, 1973). Homologous chromosomes that are destined to pair are arranged in intimate association with each other (Feldman *et al.*, 1966; Feldman and Avivi, 1973). It is suggested that in the high pairing autotetraploid, all four homologues of a chromosomal type are perhaps intimately associated already in premeiotic cells and have therefore equal chances to pair with each other, while homologues in the low pairing autotetraploid tend to be arranged in groups of two. In other words, it is suggested that the two genotypes differ from each other in their premeiotic patterns of chromosomal distribution.

Acknowledgments

This study was supported in part by a grant from the Stiftung Volkswagenwerk, Az 11 2073. Thanks are due to M. Feldman for his encouragement and help during the course of this work and to A. Horovitz for assistance in the preparation of the manuscript.

References

Avivi, L. and Feldman, M. 1973. Mechanism of non-random chromosome placement in common wheat. Proc. 4th Int. Wheat Genet. Symp., Columbia, Missouri, pp. 627-633.

Coucoli, H. D. 1970. The frequency of multivalents formation and their orientation in the tetrasomes of *Triticum aestivum* var. Chinese Spring. A thesis submitted to the Department of Botany, University of Thessaloniki, Greece.

Dawson, C. D. R. 1941. Tetrasomic inheritance in *Lotus corniculatus* L. J. Genet. **42**: 49-72.

Dover, G. A. and Riley, R. 1972. Variation at two loci affecting homoeologous meiotic chromosome pairing in *Triticum aestivum* × *Aegilops mutica* hybrids. Nature (London), New Biol. **235**: 61-62.

Driscoll, C. J. 1972. Genetic suppression of homoeologous chromosome pairing in hexaploid wheat. Can. J. Genet. Cytol. **14**: 39-42.

Dvorak, J. 1972. Genetic variability in *Aegilops speltoides* affecting homoeologous pairing in wheat. Can. J. Genet. Cytol. **14**: 371-380.

Endrizzi, J. E. 1962. The diploid-like cytological behavior of tetraploid cotton. Evolution, **18**: 325-329.

Feldman, M. 1966. The effect of chromosomes 5B, 5D and 5A on chromosomal pairing in *Triticum aestivum*. Proc. Natl. Acad. Sci. U.S.A. **55**: 1447-1453.

Feldman, M. 1968. Regulation of somatic association and meiotic pairing in common wheat. Proc. 3rd Int. Wheat Genet. Symp., Canberra, pp. 169-178.

Feldman, M. and Avivi, L. 1973. The pattern of chromosomal arrangement in nuclei of common wheat and its genetic control. Proc. 4th Int. Wheat Genet. Symp., Columbia, Missouri, pp. 675-684.

Feldman, M. and Mello-Sampayo, T. 1967. Suppression of homoeologous pairing in hybrids of polyploid wheats × *Triticum speltoides*. Can. J. Genet. Cytol. **9**: 307-313.

Feldman, M., Mello-Sampayo, T. and Sears, E. R. 1966. Somatic association in *Triticum aestivum*. Proc. Natl. Acad. Sci. U.S.A. **56**: 1192-1199.

Harberd, D. J. 1972. Bivalent-forming natural autotetraploids in the *Brassiceae*. Heredity, **29**: 394.

Jauhar, P. P. 1975. Genetic control of diploid-like meiosis in hexaploid tall fescue. Nature (London), **254**: 595-597.

Kasha, K. J. and Burnham, C. R. 1965. The location of interchange breakpoints in barley: II. Chromosome pairing and intercross method. Can. J. Genet. Cytol. **7**: 620-632.

Kihara, H. and Lilienfeld, F. 1935. Genomanalyse bei *Triticum* und *Aegilops*. VI. Weitere Untersuchungen an *Aegilops* × *Triticum*-und *Aegilops* × *Aegilops* — Bastarden. Cytologia, **6**: 195-216.

Kimber, G. 1961. The basis of diploid-like meiotic behavior of polyploid cotton. Nature (London), **191**: 98-100.

Kimber, G. 1973. The relationships of the s-genome diploids to polyploid wheats. Proc. 4th Int. Wheat Genet. Symp., Columbia, Missouri, pp. 81-85.

Kimber, G. and Athwal, R. S. 1972. A reassessment of the course of evolution of wheat. Proc. Natl. Acad. Sci., U.S.A. **69**: 912-915.

Ladizinsky, G. 1973. Genetic control of bivalent pairing in the *Avena strigosa* polyploid complex. Chromosoma, **42**: 105-110.

Larsen, J. and Kimber, G. 1973. The effect of the genotype of *Triticum speltoides* on the pairing of homologous chromosomes. Can. J. Genet. Cytol. **15**: 233-236.

Legro, R. A. H. 1961. Species hybrids in *Delphinium*. Euphytica, **10**: 1-23.

Mello-Sampayo, T. 1968. Homoeologous chromosome pairing in pentaploid hybrids of wheat. Proc. 3rd Int. Wheat Genet. Symp., Canberra, pp. 179-184.

Mello-Sampayo, T. 1971a. Genetic regulation of meiotic chromosome pairing by the chromosome 3D of *Triticum aestivum*. Nature (London), New Biol. **230**: 22-23.

Mello-Sampayo, T. 1971b. Promotion of homoeologous pairing in hybrids of *Triticum aestivum* × *Aegilops longissima*. Genet. Iber. **23**: 1-9.

Mello-Sampayo, T. and Canas, A. P. 1973. Suppressors of meiotic chromosome pairing in common wheatm Proc. 4th Int. Wheat Genet. Symp., Columbia, Missouri, pp. 709-713.

Morrison, J. W. and Rajhathy, T. 1960a. Frequency of quadrivalents in autotetraploid plants. Nature (London), **187**: 528-530.

Morrison, J. W. and Rajhathy, T. 1960b. Chromosome behavior in autotetraploid cereals and grasses. Chromosoma, **11**: 297-307.

Nordenskiold, H. 1957. Segregation ratios in progenies of hybrids between natural and synthesized *Phleum pratense*. Hereditas, **43**: 525-540.

Okamoto, M. 1957. Asynaptic effect of chromosome V. Wheat infor. Serv. **5**: 6.

Okamoto, M. and Inomato, N. 1974. Possibility of 5B-like effect in diploid species. Wheat Infor. Serv. **38**: 15-16.

Rajhathy, T. 1971. The alloploid model in *Avena*. Stadler Genet. Symp., **3**: 71-87.

Riley, R. 1958. Chromosome pairing and haploids in wheat. Proc. 10th Int. Congr. Genet., Montreal, **2**: 234-235.

Riley, R.   1960.   The diploidization of polyploid wheat.   Heredity, 15: 407-429.

Riley, R.   1963.   The genetic regulation of meiotic behavior in wheat and its relatives.   Proc. 2nd Int. Wheat Genet. Symp., Lund, pp. 395-408.

Riley, R. and Chapman, V.   1958.   Genetic control of the cytologically diploid behavior of hexaploid wheat.   Nature (London), 182: 713-715.

Riley, R. and Law, C. N.   1965.   Genetic variation in chromosome pairing.   Adv. Genet. 13: 57-114.

Riley, R., Kimber, G. and Chapman, V.   1961.   The origin of genetic control of diploid-like behavior of polyploid wheat.   J. Hered. 52: 22-25.

Riley, R., Chapman, V., Young, R. M. and Biefield, A. M.   1966.   The control of meiotic chromosome pairing by the chromosomes of homoeologous group 5 of Triticum aestivum.   Nature (London), 212: 1475-1477.

Sears, E. R. and Okamoto, M.   1958.   Intergenomic chromosome relationships in hexaploid wheat.   Proc. 10th Inter. Congr. Genet., Montreal, 2: 258-259.

Sved, J. A.   1966.   Telomere attachment of chromosomes.   Some genetical and cytological consequences.   Genetics, 53: 747-756.

Tabata, M.   1963.   Chromosome pairing in intercrosses between stocks of interchanges involving the same two chromosomes in maize. II. Pachytene configurations in relation to breakage positions.   Cytologia, 28: 278-292.

Upadhya, M. D.   1966.   Altered potency of chromosome 5B in wheat — caudata hybrids.   Wheat Infor. Serv. 22: 7-9.

Upadhya, M. D. and Swaminathan, M. S.   1967.   Mechanism regulating chromosome pairing in Triticum.   Biol. Zentralbl., Suppl. 86: 239-255.

Waines, J. G.   1975.   A model for the origin of diploidizing and mechanisms in polyploid species.   Am. Nat. In press.

Zohary, D. and Imber, D.   1963.   Genetic dimorphism in fruit-types in Aegilops speltoides.   Heredity, 18: 223-231.

25

Reprinted from pages 111–118 and 129–147 of *Chromosoma* **13**:111–147 (1962)

ASYNAPSIS AND POLYPLOIDY IN SCHISTOCERCA PARANENSIS

By

B. JOHN and S. A. HENDERSON

With 61 Figures in the Text

(Received March 15, 1962)

A. Introduction

Acridoid chromosomes are unique in a number of respects. First, they are sufficiently large in size, few in number and of such a size range as to permit an unambiguous and consistent classification of the members of the complement. Secondly, diplotene in male meiosis offers an unparalleled opportunity for the study of chiasma frequency and distribution. The genus *Schistocerca* combines both these advantages with an additional marker: the positions of the centromeres at diplotene are revealed by precocious pro-centric condensation. In this paper we are concerned with showing how these several chromosome markers can be used to clarify two main controversial issues, namely the problem of chiasma localisation and that of multivalent formation in autotetraploid cells.

B. Materials and Methods

The South American locusts used in this study were all supplied by courtesy of the Anti-Locust Research Centre, London. They originate from Nicaragua and have been provisionally identified by Sir BORIS P. UVAROV as *S. paranensis*.

The testes of male imagines were removed by vivisection under insect saline and fixed in 1:3 acetic-alcohol. Squash preparations were made in acetic orcein and aceto-carmine.

C. Observations

I. *The Normal Complement*

The male karyotype of *S. paranensis*, like that of its sister species *S. gregaria*, consists of 23 acrocentric chromosomes. It agrees with *S. gregaria* also in the number of size groups present (3 L, 5 M and 3 S pairs) and in the relative sizes of the members of each group (Table 1). But it differs in one clear respect. The S-chromosomes in *S. paranensis* are considerably smaller than those of *S. gregaria* (Figs. 1—4, Table 1). Chromosome size is sometimes a difficult character on which to base comparisons, since it can be influenced by a number of variables. But this difference between the small chromosomes of *S. gregaria* and

S. paranensis is consistent both at mitosis and at meiosis. Indeed, the two species can usually be distinguished on this character alone.

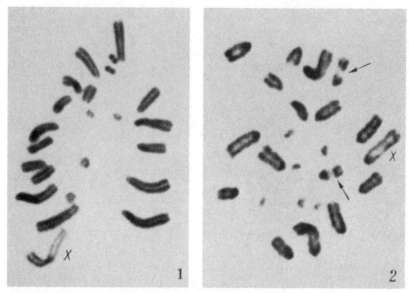

Figs. 1 and 2 (both 1750 ×). 1: Male mitotic complement (2n = 23), *X* negatively hetero-pycnotic. 2: Mitotic metaphase showing secondary constrictions in the M_6 chromosomes

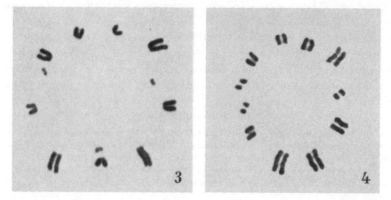

Figs. 3 and 4. Second meiotic metaphase (both 1000 ×). 3: *Schistocerca paranensis*. 4: *Schistocerca gregaria*. Note difference in size of the three small chromosomes in the two species

There is a further difference in their chromosome phenotype which involves nucleolar organisation. In the mitotic complement of *S. paranensis* the M 6 chromosomes sometimes exhibit submedian secondary constrictions (Fig. 2). At first prophase of meiosis, aceto-carmine preparations reveal a nucleolus attached at this locus. This persists until diplotene when it becomes detached. Thereafter it either gradually diminishes in size or else breaks up into a number of smaller nucleoli. At zygotene-pachytene it is possible to show that the nucleolus is in fact attached to a prominent chromomere, which we have called the nucleolar chromomere (Fig. 5). Indeed, this chromomere can be recognised even in acetic-orcein preparations where, of course, the nucleolus does not stain (Fig. 6). Phase-contrast examination of such preparations, however, reveals the presence of a nucleolus both at these early stages and until diplotene (Fig. 11).

At early prophase of the first meiotic division only one nucleolus is usually found; one presumes that nucleolar fusion is a common occurrence during the premeiotic interphase. Sometimes, though rarely, the nucleolus is associated with the X-chromosome (Fig. 14). At diplotene the locus of attachment frequently appears as a non-staining chromosome segment (Fig. 9). These achromatic gaps presumably represent localised regions which remain unspiralised, though their inability to stain may be enhanced by the accumulation of the products of their own activity.

The nucleolar gaps are usually heteromorphic in expression in the sense that they are found on only one side of a bivalent. Such heteromorphism may indeed be evident as early as pachytene. Homomorphism is rare (Fig. 16). In some heteromorphic cases two disproportionate nucleoli may in fact be formed. Here the larger nucleolus is associated with the achromatic segment. Comparable heteromorphism of non-staining segments can also be found at first or even second anaphase, though this phenomenon is not regular in occurrence (Fig. 20). We have

Table 1. Comparison of chromosome length in *S. paranensis* and *S. greguria*

| Species | Mean mitotic length in micra with standard deviations | | | | | | | | | | | |
	L 1	L 2	L 3	M 4	M 5	M 6	M 7	M 8	S 9	S 10	S 11	X
S. paranensis (mean of 10 nuclei)	8.5 ±0.33	7.5 ±0.26	6.5 ±0.26	5.7 ±0.18	5.1 ±0.18	4.7 ±0.17	4.3 ±0.18	3.8 ±0.13	1.7 ±0.05	1.5 ±0.04	1.3 ±0.03	7.8 ±0.41
S. greguria[1] (mean of 16 nuclei)	7.4 ±0.1	6.7 ±0.2	5.8 ±0.1	4.9 ±0.1	4.3 ±0.1	3.9 ±0.1	3.6 ±0.1	3.3 ±0.1	2.4 ±0.1	2.1 ±0.1	1.9 ±0.1	6.7 ±0.2

[1] Data from JOHN and NAYLOR (1961).

also seen such heteromorphism in a number of other orthopterans, for example *Tetrix* (Henderson 1961, Fig. 14 and 15), *Pyrgomorpha* (Lewis and John 1960, Fig. 13) and *Chorthippus*. Ohno *et al.* (1961)

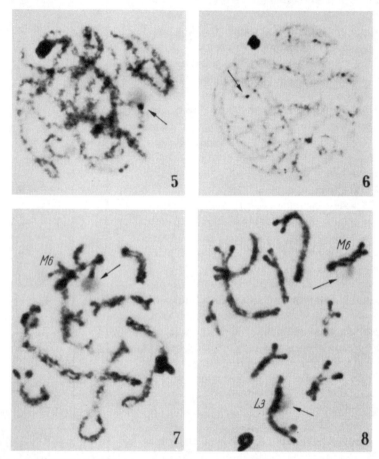

Figs. 5—8 (all 1250 ×). 5: Zygotene in *S. paranensis* with nucleolus attached to prominent chromomere (aceto-carmine). 6: Early zygotene in *S. paranensis* showing prominent nucleolar chromomere (acetic orcein). 7: Pachytene in *S. paranensis* with single nucleolus attached to M_6 (aceto-carmine). 8: Pachytene in *S. gregaria* with two nucleoli, attached to L_3 and M_6 respectively (aceto-carmine)

have recently described a more consistent heteromorphism for nucleolar organisation in a pair of autosomes of *Cavia cobaya*. This is expressed in somatic and germinal cells of both sexes. A similar consistency had been reported earlier in *Disporum sessile* (Kayano 1960) where two different pairs of autosomes were involved.

S. gregaria, like *S. paranensis*, also organises nucleolar material, and with similar consequences. As in *S. paranensis* the M 6 bivalent produces one of the nucleoli, while an additional nucleolus is organised by the L_3 bivalent (Figs. 8, 10 and 12). This clarifies our earlier report (JOHN and NAYLOR 1961, p. 189) of achromatic gaps in the long and medium

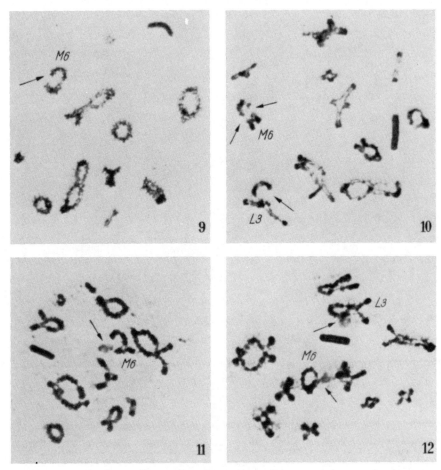

Figs. 9—12 (all 1000 ×). 9: Diplotene in *S. paranensis* showing heteromorphism for an achromatic gap in the M_6 (acetic orcein). 10: Diplotene in *S. gregaria* showing heteromorphism for achromatic gaps in L_3 and M_6 (acetic orcein). 11: Diplotene in *S. paranensis* with nucleolus attached at achromatic gap in M_6 (acetic orcein, phase contrast). 12: Diplotene in *S. gregaria* with nucleoli attached at achromatic gaps in L_3 and M_6 (acetic orcein, phase contrast)

chromosomes of this species. We have re-examined these preparations and confirm that these consistently involve the L_3 and M_6.

These observations on nucleolar development in *Schistocerca spp.* are instructive in three respects:

1. It has been claimed (DARLINGTON 1947; JOHN and LEWIS 1959) that in male orthopterans nucleolar formation and heteropycnosity

of the X chromosome are mutually exclusive states. Our observations
here show that this is clearly not the case. Indeed, we are not the first
to notice this concurrence (Corey 1940; Sinoto 1944). We have also
found a nucleolus and a positively heteropycnotic X chromosome to

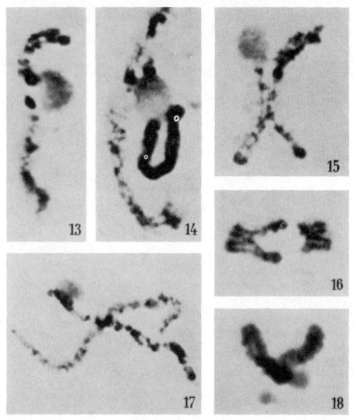

Figs. 13—18. 13: Pachytene in *S. paranensis*, M_6 chromosome with attached nucleolus
(aceto-carmine, 3000 ×). 14: Pachytene in *S. paranensis* showing non-specific association of
nucleolus with the ends of the X-chromosome (aceto-carmine, 3500 ×). 15: Early, diplotene
in M_6 chromosome of *S. paranensis* (acetocarmine, 3500 ×). 16: M_6 chromosome of *S.
paranensis* homomorphic for the achromatic gaps (acetic orcein, 2250 ×). 17: Pachytene
in *Tylotropidius gracilipes* showing nucleolus attached to terminal region of the precocious
bivalent (aceto-carmine, 2000 ×). 18: Diplotene in *T. gracilipes* with fragmenting nucleolus
still attached to the precocious bivalent (aceto-carmine, 2000 ×)

be present simultaneously in the male of *Tylotropidius gracilipes*.
Furthermore, in this species the nucleolus is organised by a positively
heteropycnotic precocious bivalent (Figs. 17 and 18). In *Blaberus discoidalis* (John and Lewis 1959) the reduced heteropycnosity of the
X chromosome is presumably not causally related to the nucleolus
which is present in the male of this species.

A further orthopteran in which a nucleolus has been described is *Mantis religiosa* (CALLAN and JACOBS 1957), though in this species the sex chromosomes lose their positive heteropycnosity by zygotene. Finally, DARLINGTON (1936) has described the presence of a small nucleolus organised by the L_3 chromosomes in *Chorthippus brunneus* (= *C. bicolor*). He concluded that the organisers were located near the ends of the chromosomes, but from our experience they are submedian in location.

2. These findings in *S. paranensis* and *S. gregaria* parallel those of BEERMAN (1960). He reported a difference in the number and distribution of nucleolar organisers in the sibling species *Chironomus tentans* and *C. pallividivatus*. *C. tentans* has two nucleoli, one on chromosome 2 and one on chromosome 3. The other sibling species possesses only one nucleolus, and this is found on chromosome 2. Moreover, salivary gland chromosome analysis in F_1 hybrids reveals that the site of attachment of nucleoli on the second chromosome differs in these two sibling species.

On the basis of the banding patterns in the hybrids BEERMAN concluded that these differences were not due to structural rearrangement of the chromosomes. In hybrids the nucleolar organisers behave as Mendelian loci.

That nucleolar organisers can be lost by mutation is clear from the findings of FISHBERG and WALLACE (1960) in *Xenopus laevis*. But here it is not possible to decide unambiguously whether the mutation is chromosomal or genic in character. Nucleolar organisers must also be capable of increasing in number since in man, for example, five pairs of chromosomes are satellited (FERGUSON-SMITH and HANDMAKER 1961; OHNO et al. 1961) and all of these appear to be involved in nucleolar production. We cannot say whether loss or gain is involved in the case of these two species of *Schistocerca*. Similarly it is impossible to decide whether the organisers are at exactly the same loci on the M_6 chromosome in both species, though they certainly occupy comparable positions.

3. The presence of such sites of nucleolar attachment might be expected to influence chiasma formation in their proximity. Indeed, it has been shown that the presence of a nucleolus in *Fritillaria* (DARLINGTON 1935) and *Eremurus* (UPCOTT 1935) can interfere with chiasma distribution. Significantly, the organisers in the *Schistocerca spp.* occur at positions where chiasma formation takes place infrequently. This siting is, however, consequential, not causal. The distribution of chiasmata in the nucleolar organising chromosomes is not modified, but is in fact similar to that of other members of the complement in the same size range (HENDERSON, in preparation).

These differences we have mentioned between the two species of *Schistocerca*, namely, the difference in the size of the S-chromosomes and the difference in the number and distribution of nucleolar organisers, are the only ones we can detect. They do not differ significantly in mean or variance of chiasma frequency, or in chiasma distribution (Table 2; Fig. 19). And this, despite the fact that *S. paranensis* is confined to the new world, while *S. gregaria* is present only in the old.

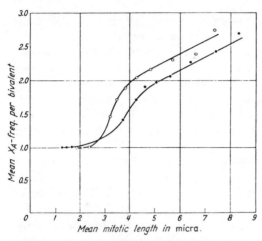

Fig. 19. Relationship between the mean mitotic chromosome length and mean chiasma frequency per bivalent in *Schistocerca gregaria* (——o——) and *S. paranensis* (——•——)

However, chromosome morphology is not always a good criterion of chromosome homology. In the grasshopper genus *Eyprepocnemis*, for instance, two of the so-called subspecies which have been recognised, *E. plorans ornatipes* and *E. p. meridionalis*, possess complements which are indistinguishable numerically and morphologically, though they do differ significantly in chiasma frequency. Experimentally produced hybrids between these two sub-species reveal that the complements have undergone considerable structural alteration (John and Lewis, in preparation).

Evidently, the outward appearance of chromosomes may conceal their own change. Attempts to hybridise the two species of *Schistocerca* have unfortunately failed.

Table 2. *Comparison between the chiasma frequencies of S. paranensis and S. gregaria.* 25 first metaphases were scored in each of 12 individuals from the two species, making a total of 300 observations in each case

Species	Chiasma frequency per bivalent $\left(\dfrac{\text{Mean}}{\text{Variance}}\right)$											Cell total $\left(\dfrac{\text{Mean}}{\text{Variance}}\right)$
	L 1	L 2	L 3	M 4	M 5	M 6	M 7	M 8	S 9	S 10	S 11	
S. para-nensis	2.67	2.40	2.25	2.05	1.96	1.89	1.70	1.40	1.00	1.00	1.00	19.33
	0.02	0.02	0.03	0.01	0.01	0.01	0.02	0.05	0.00	0.00	0.00	0.64
S. gregaria	2.71	2.36	2.29	2.13	2.01	1.87	1.69	1.43	1.00	1.00	1.00	19.43
	0.03	0.03	0.03	0.02	0.03	0.03	0.04	0.02	0.00	0.00	0.00	1.02

Evidently, the outward appearance of chromosomes may conceal their own change. Attempts to hybridise the two species of *Schistocerca* have unfortunately failed.

[*Editors' Note:* Material has been omitted at this point.]

b) The Polyploid Cells

Polyploid cells were present in several of the specimens examined but their frequency varied in different individuals. Two types of polyploid cells were found: those containing and those which did not contain

Table 4. *Analysis of 25 tetraploid cells in S. paranensis, summarising the frequencies and distributions of the different chromosome associations throughout the 3 size classes of the complement*

Cell no.	IV				III + I				II				2 I			
	L	M	S	Total	L	M	S	Total	L	M	S	Total	L	M	S	Total
1	3	5	1	9	—	—	—	—	—	—	4	4	—	—	—	—
2	3	5	1	9	—	—	—	—	—	—	4	4	—	—	—	—
3	3	3	1	7	—	—	—	—	—	4	4	8	—	—	—	—
4	3	3	—	6	—	—	—	—	—	4	6	10	—	—	—	—
5	3	2	—	5	—	—	—	—	—	6	6	12	—	—	—	—
6	3	3	—	6	—	1	—	1	—	2	6	8	—	—	—	—
7	3	4	1	8	—	—	—	—	—	2	4	6	—	—	—	—
8	3	3	2	8	—	—	1	1	—	4	—	4	—	—	—	—
9	3	5	—	8	—	—	—	—	—	—	6	6	—	—	—	—
10	1	3	2	6	—	—	—	—	4	4	2	10	—	—	—	—
11	3	4	1	8	—	—	—	—	—	2	4	6	—	—	—	—
12	3	5	—	8	—	—	—	—	—	—	5	5	—	—	1	1
13	3	4	—	7	—	—	—	—	—	2	6	8	—	—	—	—
14	3	4	—	7	—	—	—	—	—	2	6	8	—	—	—	—
15	3	5	—	8	—	—	—	—	—	—	6	6	—	—	—	—
16	3	5	1	9	—	—	—	—	—	—	3	3	—	—	1	1
17	3	2	—	5	—	—	—	—	—	6	6	12	—	—	—	—
18	1	3	1	5	—	—	1	1	2	4	4	10	—	—	—	—
19	3	4	—	7	—	—	—	—	—	2	5	7	—	—	1	1
20	3	5	1	9	—	—	—	—	—	—	4	4	—	—	—	—
21	3	3	1	7	—	—	1	1	—	4	2	6	—	—	—	—
22	3	3	—	6	—	—	—	—	—	4	6	10	—	—	—	—
23	3	3	—	6	—	—	—	—	—	4	6	10	—	—	—	—
24	2	4	—	6	—	—	—	—	2	2	6	10	—	—	—	—
25	3	2	—	5	—	—	1	1	—	6	4	10	—	—	—	—
Total	70	92	13	175	—	1	4	5	8	64	115	187	—	—	3	3
Mean	2.8	3.7	0.5	7.0	—			0.2	0.3	2.6	4.6	7.5	—	—	0.1	0.1

multivalents. Like their counterparts in *Pyrgomorpha* (Lewis and John 1960), the latter, which were more common, originate in multinucleate cells produced following a failure of cytoplasmic division. Multivalent containing cells were all tetraploid. Though less common they were much more instructive. Twenty-five such cells were analysed in detail with regard to the formation and distribution of multivalents throughout the complement (Tables 4, 5). This analysis revealed the following points of interest.

1. In all cases the two X chromosomes do not form a chiasmate association (Figs. 42—47).

2. All size classes can form multivalents.

3. The L-chromosomes almost invariably form multivalents (93.3%), the M-chromosomes do so very commonly (73.6%) but the S-chromosomes rarely form them (17.3%).

4. Occasionally trivalents and univalents were formed, but in the medium and small classes only (Table 5, Fig. 45).

Table 5. *Chiasma frequency analysis in 16 tetraploid cells of S. paranensis*

Total chiasmata per chromosome group			Cell total	
L (3)	M (5)	S (3)		
19	18	6	43	
20	21	7	48	
18	20	6	44	
16	16	5	37	
16	18	7	41	
16	19	6	41	
17	24	6	47	
20	20	8	48	
24	21	7	52	
17	19	5	42	
14	16	6	36	
18	22	6	46	
16	20	9	45	
24	21	6	51	
19	22	6	47	
14	20	9	43	
Cell mean 18.00	19.81	6.56	44.44	4x Mean
Mean per chromosome pair				
4x 3.00	1.98	1.09		
2x 2.44	1.80	1.00	19.33	2x Mean

5. Univalent production in the small chromosomes, unaccompanied by trivalents, also occurred in three cells (Fig. 43).

6. Chain quadrivalents were never observed in the long chromosomes. They were sometimes present in the medium class (Figs. 42 and 45) while almost all quadrivalents involving the small chromosomes were of this type (Figs. 45 and 46).

7. Many of the large quadrivalents had a complex structure, with a large number of chiasmata involved (Figs. 42—47).

8. Because of this last feature it was not possible to score chiasma frequency accurately in the multivalents of all 25 cells. But 16 cells were analysed in this way (Table 5) and in these the total chiasma frequency per cell varied from 36 to 49, with a mean of 44.44. The corresponding mean in diploid cells is 19.33. The average number of

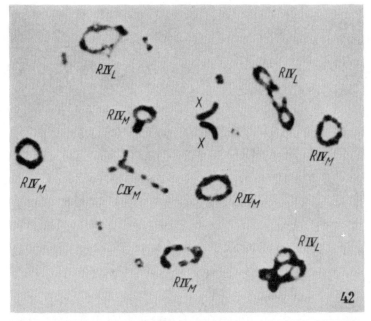

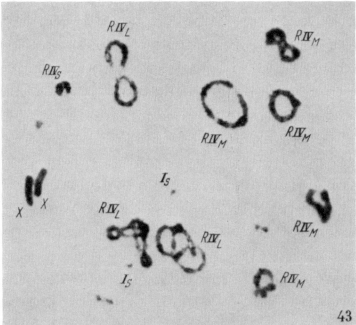

Figs. 42 and 43. Tetraploid cells from *S. paranensis* (both 1000 ×). 42: 8 *RIV* + 1 *CIV* + 6 *II* + 2 *X*, all the S-chromosomes have formed bivalents in this cell. 43: 9 *RIV* + 3 *II* + 2 *I* + 2 *X*, two of the S-chromosomes are unpaired, there are 3 S-bivalents and one S-quadrivalent

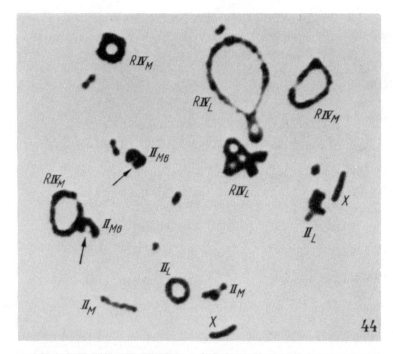

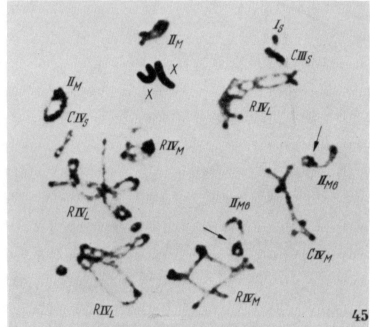

Figs. 44 and 45 (Legend p. 133)

chiasmata present in the tetraploid cells is thus usually well in excess of twice the average diploid number.

9. Not only the number of quadrivalents, but also the number of chiasmata they contain, and hence their complexity, is a reflection of the diploid chiasma frequencies of the bivalents concerned.

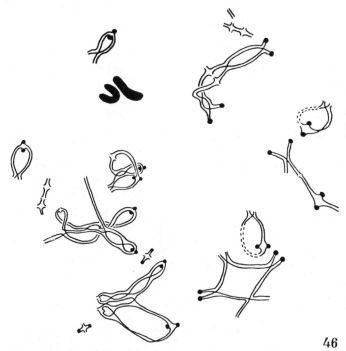

46

Fig. 46. Tetraploid cell from *S. paranensis* drawn to show chromatid structure — 5 *RIV* + 2 *CIV* + (*CIII* + *I*) + 6*II* + 2 *X* (compare Fig. 45: 1000 ×)

These observations on *S. paranensis* are sufficiently instructive to merit a more general consideration of the properties of multivalent formation. There are six relevant points to deal with:

1. WHITE (1934) has studied tetraploid spermatocytes in *S. gregaria*, though unfortunately he only obtained two such cells. In these a total of 6 quadrivalents were found, which were confined exclusively to the larger chromosomes. Since *S. gregaria* and *S. paranensis* do not differ in either their mean chiasma frequency or its variance, and since the chromosomes have a similar size range, one might expect comparable behaviour to obtain in tetraploid cells of both species. To date we have

Figs. 44 and 45. Tetraploid cells from *S. paranensis* (both 1000 ×). 44: 5 *RIV* + 12 *II* + 2 *X*, all the S-chromosomes have formed bivalents, the two M_6-bivalents both show a nucleolar gap (arrow). 45: 5 *RIV* + 2 *CIV* + (*CIII* + *I*) + 6 *II* + 2 *X*, one of the chain quadrivalents is composed of S-chromosomes as is the (*CIV* + *I*) and the 2 M_6-bivalents are again heteromorphic for pronounced nucleolar gaps (arrows) found only one complete and fully analysable tetraploid cell in *S. gregaria*. This had 6 quadrivalents, one of which was a chain of four S-chromosomes (Fig. 57). The cell as a whole was similar to the tetraploid cells found in *S. paranensis*, both with regard to the high total chiasma frequency (53) and with regard to the number of chiasmata present in the complex L-multivalents. On of these, involving the L_1 chromosome, possessed a total of 10 chiasmata (Figs. 56 and 58). We would conclude, therefore that the quadrivalent and chiasma frequency in *S. gregaria* would not appear to be as low as WHITE'S use of microtomy would lead us to believe.

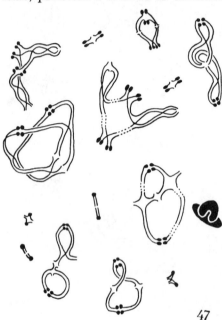

2. All the evidence we have obtained from tetraploid orthopteran cells supports the contention that, within the limits imposed by the genotype, quadrivalent formation is predominantly a function of relative chromosome length and of chiasma frequency.

In *Pyrgomorpha kraussi* (LEWIS and JOHN 1960) there are 19 chromosomes in the diploid male complement, one of which is an X chromosome. This species has a mean diploid chiasma frequency of 9.8 per cell. The L-chromosomes often form rings, but these are uncommon in M's and unknown in S's. Single chiasmata are either interstitial or distal in position. In tetraploid cells, multivalents are rarely formed and those that are possess few chiasmata. For this reason chains are more common than rings, though most are full quadrivalents: trivalents are rare. Never more than five quadrivalents have been found in any cell at one time and usually only one or two are present. These invariably involve the larger autosomes (Table 6). The low chiasma and multivalent frequency of the tetraploid cells in this species are reflections of the low diploid chiasma frequency. However, here again (Table 7), the mean chiasma frequency per tetraploid cell (22.3) is in excess of twice the mean diploid value (9.8 per cell).

We have found isolated polyploid cells in other orthopteran species (Table 8), for example *Chorthippus brunneus* (Fig. 59) and *Locusta*

Fig. 47. Tetraploid cell from *S. paranensis* drawn to show chromatid structure — 8 *RIV* + 6 *II* + 2*X* (1000 ×)

Table 6

Analysis of 10 tetraploid cells in Pyrgomorpha kraussi, summarising the frequencies and distributions of the different chromosome associations throughout the 3 size groups

Cell no.	IV				III + I				II				2I			
	L	M	S	Total	L	M	S	Total	L	M	S	Total	L	M	S	Total
1	2	3	—	5	—	—	—	—	—	4	4	8	—	—	—	—
2	1	3	—	4	—	—	—	—	2	4	4	10	—	—	—	—
3	—	1	—	1	—	—	—	—	4	7	4	15	—	1	—	1
4	1	1	—	2	—	—	—	—	2	8	4	14	—	—	—	—
5	1	1	—	2	—	—	—	—	2	8	4	14	—	—	—	—
6	—	—	—	—	—	—	—	—	4	10	4	18	—	—	—	—
7	1	—	—	1	—	—	—	—	2	10	4	16	—	—	—	—
8	1	—	—	1	—	—	—	—	2	9	4	15	—	1	—	1
9	1	—	—	1	—	—	—	—	2	10	4	16	—	—	—	—
10	2	—	—	2	—	—	—	—	—	10	4	14	—	—	—	—
Total	10	9	—	19	—	—	—	—	20	80	40	140	—	2	—	2
Mean	1.0	0.9	—	1.9	—	—	—	—	2.0	8.0	4.0	14.0	—	0.2	—	0.2

migratoria (Figs. 51 and 52). These two cases further support our above findings. Thus in *Chorthippus* (2n = 17 ♂) the two longest pairs of chromosomes only were involved in quadrivalent formation. One of these possessed 6, the other 10 chiasmata, raising the chiasma frequency of the tetraploid cell as a whole to 35. The mean diploid chiasma frequency in this species is 13.05 per cell (LEWIS and JOHN, in preparation) and the long chromosomes usually possess 2 or 3 chiasmata.

Finally in *Locusta* (2n = 23 ♂), where a chiasma frequency between that of *Pyrgomorpha* and *Schistocerca* obtains (mean diploid value of 14.3 per cell), the only cell so far found had 3 quadrivalents, all of which involved the long chromosomes and which possessed either 4 or 5 chiasmata (Figs. 51 and 52). In diploid cells such chromosomes have only one or two chiasmata. The

Table 7

Chiasma frequency analysis of the 10 tetraploid cells from Pyrgomorpha kraussi listed in Table 6

Cell no.	Total Xta. per chromosome group			Cell total	
	L (2)	M (5)	S (2)		
1	10	16	4	30	
2	7	15	4	26	
3	7	10	4	21	
4	8	11	4	23	
5	6	12	4	22	
6	5	10	4	19	
7	7	10	4	21	
8	6	9	4	19	
9	7	10	4	21	
10	6	11	4	21	
Cell Mean	6.9	11.4	4.0	22.3	4x Mean
Mean per chromosome pair					
$4x$	1.73	1.14	1.00		
$2x$[1]	1.42	1.02	1.00	9.8	2x Mean

[1] Scored in 25 cells taken from the individual in which all 10 tetraploid cells were obtained.

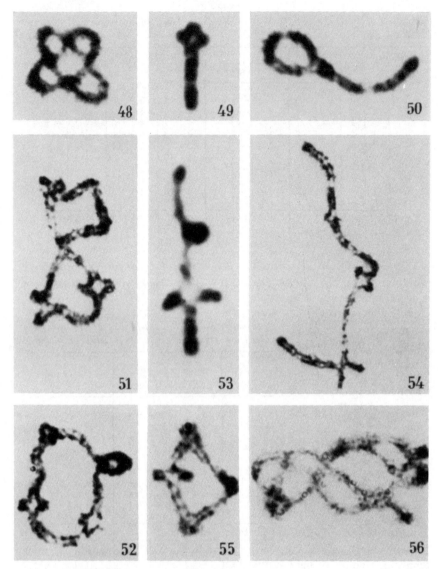

Figs. 48—56. Isolated multivalents from tetraploid cells of *S. paranensis* (48—50 2000 ×);
L. migratoria (51—52, 2500 ×); *P. kraussi* (53—54, 1500 ×) and *S. gregaria* (55—56, 2000 ×).
48: Ring quadrivalent with 8 chiasmata. 49: Chain quadrivalent. 50: "Frying pan"
quadrivalent. 51 and 52: Ring quadrivalents with 5 chiasmata. 53: Trivalent. 54: Chain
quadrivalent. 55: Ring quadrivalent with 4 chiasmata. 56: Complex ring quadrivalent
with 10 chiasmata. (See also Fig. 58)

total chiasma frequency of this tetraploid cell was 35, again more
than twice the diploid value.

Essentially the same correlations can be drawn from the studies of
Rothfels (1950) on *Neopodismopsis abdominalis*, White (1954) on
Tettigonia viridissima, and Klingstedt (1939) on *Chrysochraon dispar*.
The latter author tabulates information which had, to that date, been

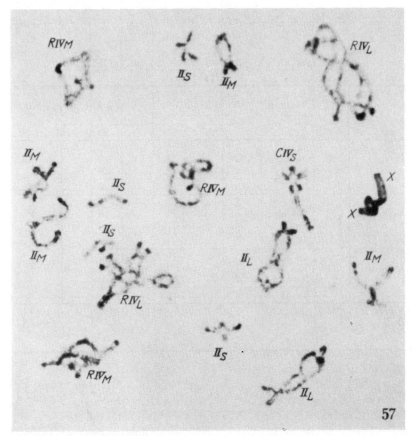

Fig. 57. Tetraploid cell from *S. gregaria* possessing 5 *RIV* + 1 *CIV* + 10 *II* + 2 *X*. In
this cell two of the bivalents involve L-chromosomes and the single chain quadrivalent is
composed of S-chromosomes (1000 ×)

obtained on the occurrence of multivalents in tetraploid cells in species
with chromosomes of different types (His Table 2, p. 199). This table
clearly demonstrates a correlation between relative length, chiasma
frequency and the occurrence of multivalents.

3. Callan (1949) has made a study of tetraploid cells at first meiotic
metaphase in the dermapteran *Forficula auricularia*. In this species
the eleven pairs of metacentric autosomes normally possess a single
chiasma in one arm only, though ring bivalents having a single chiasma

Table 8. *Frequency and distribution of chromosome associations within the complements of 6 different species of insect*

Species	2x	Autosomal complement		Observed numbers of				Total cells analysed
		Type	haploid number	IV	III + I	II	2I	
Schistocerca paranensis	23	Long	3	70 (93.3%)	—	8	—	25
		Medium	5	92 (73.6%)	1	64	—	
		Short	3	13 (17.3%)	4	115	3	
		Total	11	175 (63.6%)	5	187	3	
Schistocerca gregaria	23	Long	3	2 (66.6%)	—	2	—	1
		Medium	5	3 (60.0%)	—	4	—	
		Short	3	1 (33.3%)	—	4	—	
		Total	11	6 (54.5%)	—	10	—	
Pyrgomorpha kraussi	19	Long	2	10 (50.0%)	—	20	—	10
		Medium	5	9 (18.0%)	—	80	2	
		Short	2	— (0.0%)	—	40	—	
		Total	9	19 (21.1%)	—	140	2	
Locusta migratoria	23	Long	3	3 (100.0%)	—	—	—	1
		Medium	5	— (0.0%)	—	10	—	
		Short	3	— (0.0%)	—	6	—	
		Total	11	3 (27.2%)	—	16	—	
Chorthippus brunneus	17	Long	3	2 (66.6%)	—	2	—	1
		Medium	4	— (0.0%)	—	8	—	
		Short	1	— (0.0%)	—	2	—	
		Total	8	2 (25.0%)	—	12	—	
Forficula auricularia	24	All small	11	37 (24.0%)	12	206	4	14

in each arm are occasionally found. This restriction of chiasmata he attributes to chiasma interference operating across the centromere. In 14 tetraploid cells of this species (Table 8) CALLAN found that from 0—6 quadrivalents were present per cell, with a mean value of 2.6. Trivalents and univalents, as in our cases, were far less common (range 0—3; mean value 0.85 per cell). Only rarely were independent univalents produced (range 0—2 pairs per cell; mean value 0.28 per cell). Most of the quadrivalents were chains. The mean chiasma frequency per nucleus in these tetraploid cells was 24.71 which, as CALLAN points out, is significantly greater than twice the comparable diploid mean value of 11.1 chiasmata per nucleus. As might be expected, the variance of these tetraploid cells, like those we have studied, was greater than that of the diploid. He attributes this disproportionality between tetraploid and diploid chiasma frequency values to a breakdown of chiasma interference across the centromere.

In all the cases we have reported here, the average chiasma frequency in the tetraploid cells was also higher than twice the average diploid value. This excess can invariably be accounted for in terms of the particularly high chiasma frequencies of some of the longer quadrivalents.

Since at any one site only two homologues can pair, one might have expected pairing to be less efficient in the multivalent, where it is of necessity interrupted at one or more points. Far from reducing the efficiency of the process of chiasma formation, however, such an interruption obviously increases the chiasma potential of a system of homologues. And the reason for this is clear: where pairing partners change there must be a marked drop in chiasma interference (see Fig. 58). There is a notable difference between our cases and the one studied by CALLAN. In the earwig, chiasma interference is operative across the centromere i.e. it is inter-arm. But where, as in most of our examples, all the chromosomes are acrocentric, the interference which breaks down in tetraploid cells is of necessity intra-arm. The

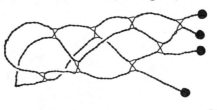

58

Fig. 58. Wire model showing chromatid structure of the complex ring quadrivalent of *S. gregaria* shown in Figs. 56 and 57. For clarity some chromatids have been displaced. (2000 ×)

same is true for the long metacentrics of *Chorthippus*, where there is no pronounced chiasma interference across the centromere.

4. It is necessary to distinguish between two kinds of tetraploid meiotic cells capable of forming multivalents. First, those which originate spontaneously, either in a newly established tetraploid or as occasional cells within an otherwise diploid reproductive tissue. Secondly, those present in established tetraploids.

In the former, meiotic behaviour necessarily reflects the potentialities inherent in the initial diploid. But this may not be true of the latter. Here meiotic behaviour may have been adjusted by natural or artificial selection. Furthermore it is not always clear with what diploid form the comparison should in fact be made. With these qualifications in mind, the available data on established tetraploids suggests three possible relationships with related diploids. In these cases the tetraploid chiasma frequency is:

i) Less than twice the diploid. Such a situation has been claimed to obtain in *Tulipa* (UPCOTT 1939).

ii) Exactly double that of the diploid. This has been shown to hold in some subspecies of the *Dactylis glomerata* complex by McCOLLUM (1958). He has also found the same relationship in colchicine-induced tetraploids of this species complex.

iii) Higher than twice the diploid value. This occurs, for example, in *Agrostis canina* (Jones 1956), which exists in both diploid (*A. c. canina*, 2n = 14) and tetraploid (*A. c. montana*, 2n = 28) subspecies.

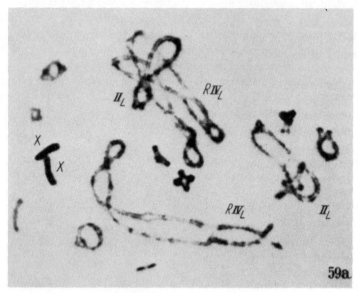

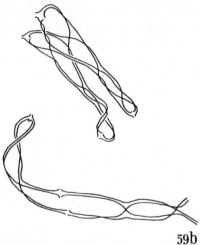

59b

Fig. 59a. Tetraploid cell from *Chorthippus brunneus*, with 2 *RIV* + 12 *II* + 2 *X*. Of the 12 long chromosomes present in this cell, 4 form 2 bivalents and the remainder are associated as 2 quadrivalents. One of these has 6 chiasmata, the other 10 chiasmata. Fig. 59b. The two quadrivalents drawn to show chromatid structure. (Both 1000 ×)

Though selection would appear to have produced a reduction in chiasma frequency in polyploid tulips, it does not seem to have had any marked effect in *Dactylis glomerata*. This may also be true for *Agrostis canina montana* since, as we have demonstrated, an increase in chiasma frequency is usually found in spontaneous polyploidy.

This heterogeneity of established polyploids thus stands in marked contrast to the greater homogeneity of those newly originated. For this reason we would argue that meaningful generalisations concerning the initial relationship of a tetraploid to its diploid progenitor may best be obtained from the latter.

5. MORRISON and RAJHATHY (1960 a, b) have compared quadrivalent frequencies in several unrelated autotetraploid plant species, some of which differed considerably in chromosome number and size. In one case, *Asparagus*, they also made a comparison between multivalent formation in the long and short chromosomes of the same complement. On the basis of these comparisons

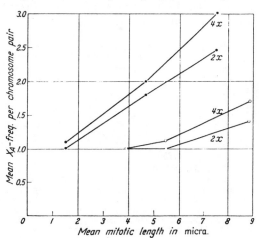

Fig. 60. Relationship between the mean mitotic chromosome length and mean chiasma frequency per chromosome pair in diploid and tetraploid cells of *S. paranensis* (——•——) and *P. kraussi* (——○——)

they imply that approximately two-thirds of a tetraploid complement will usually form multivalents, irrespective of chromosome size (within or between species), chiasma frequency or genotype.

As a generalisation there can be no doubt that this is incorrect. We have shown above, from our own and previous data, that there is an indisputable correlation between multivalent frequency and both (i) chiasma frequency between species, and (ii) relative length and chiasma frequency within the complement of a species.

Assuming that their figures are accurate and reliable, an assumption which may be incorrect (see pg. 143), one can reconcile the discrepancies between the species studied on the one hand by MORRISON and RAJHATHY, and on the other by other workers in this field in terms of a simple model (Fig. 61). Consider a metacentric chromosome where terminal initiation of pairing at both ends is more or less obligatory, i.e. there are two terminal pairing blocks. Such behaviour may normally characterise all those chromosomes within a complement which usually form two terminal chiasmata per bivalent and are about the same

321

relative size. In tetraploid cells where these conditions are satisfied, pairing can be defined, with statistical rigour. In homologues of the type A^1B^1—A^4B^4, let us assume that pairing is initiated between A^1A^2. This automatically means that A^3 can now only pair with A^4. Of the remaining B ends, any one may pair with any of the three others (e.g. B^1 with B^2, B^3 or B^4). But once it does so, pairing between the two remaining ends is again obligatory. The consequences of these various patterns of pairing are not uniform. Thus in this example, pairing between B^1 and B^2 would result in the production of two bivalents. But

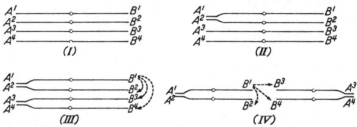

Fig. 61. Simplified model indicating how obligatory pairing for all chromosome ends in a system of four homologues in a tetraploid cell will give rise to only two types of chromosome association — quadrivalents and bivalents. The former will be formed twice as commonly as the latter in such a system. i) The four homologues, A^1B^1—A^4B^4. ii) Pairing between any two ends, e.g. A^1A^1, automatically makes pairing for the other two (A^3A^4) obligatory. iii) With the four A ends paired, any one B end, e.g. B^1 may pair with any of the other three ends $(B^1$—$B^4)$. iv) The same as iii), opened out to clarify the pairing relationships. If B^1 pairs with B^2, then B^3 can only pair with B^4, and two bivalents will result. However, if B^1 pairs with either B^3 or B^4, a quadrivalent will be formed

pairing of B^1 with either B^3 or B^4 will produce a quadrivalent. This means that provided all ends pair and have an equal chance of doing so, quadrivalents should form twice as commonly as bivalents. And as a result, the total number of quadrivalents present will equal the total number of bivalents. For simplicity of description we have dealt with pairing in a stepwise manner. Exactly the same end result would be achieved if pairing took place simultaneously at more than one site. The model can also clearly be extended to include acrocentric chromosomes. Furthermore, this argument holds irrespective of the number of chromosomes in a complement and irrespective of their absolute size.

How this operates within a complement is demonstrated by our data on *S. paranensis*. Small chromosomes, which hardly ever form two chiasmata in the diploid state, rarely form multivalents. Long chromosomes have a mean chiasma frequency well in excess of two and, from the nature of the multivalents formed (Fig. 58), apparently possess several points where pairing may be initiated (pairing blocks). These chromosomes almost invariably form quadrivalents (93.3%). Just over two thirds of the M-chromosomes (73.6%) form quadrivalents and significantly their mean chiasma frequency in the diploid is only a little below two. Furthermore, when two chiasmata are present in such bivalents they are located at the ends of the chromosomes.

One must therefore presume that if their quadrivalent scores were reliable, many of the plants with which MORRISON and RAJHATHY (1960 a, b) were dealing may have fulfilled the conditions required by our model. The most obvious exception they present is that of *Lilium* which they have estimated as possessing three to five chiasmata per bivalent. The mean numbers of bivalents and quadrivalents per cell were 5.0 and 9.4 respectively. This is a significant departure from the equal frequencies expected. In six of the cereal and grass species which they studied (1960 a) the same number of chromosomes was present ($2n = 4x = 28$). These were all metacentric and similar in relative size both within and between species. Their chiasma frequencies were also similar. Under these circumstances markedly different quadrivalent frequencies would not be expected.

It is regrettable that MORRISON and RAJHATHY were unable to score chiasma frequencies accurately in the plants they studied. But clearly they do not regard this as important in qualifying the relationship between chiasma and quadrivalent frequencies, since they believe that: "... *to be of any value ... a chiasma count would have to be made at the actual time of union or crossing over.*" Those workers who first equated cytological chiasmata so successfully with genetical crossovers were of course similarly handicapped! But this did not stop them making valuable conclusions which still apply. It is agreed that in many plants, particularly those possessing small chromosomes, chiasma recognition at diplotene is virtually impossible. But this is certainly not true for all organisms. In the Acridoid species we have dealt with in this paper, terminalization during diplotene and diakinesis is so slight that chiasma frequency scores made at first metaphase do not differ from those made at diplotene. Indeed, at early diplotene, one can score with confidence not only the number of chiasmata per bivalent but also the positions of these chiasmata along the chromosome and even, in most cases, the chromatids involved at each exchange. Observations made in species which meet these requirements must surely be of more value if generalisations are to be made, than those from species where the accurate detection of quadrivalents must be viewed with reserve (cf. Figs. 2 and 3, MORRISON and RAJHATHY 1960 b), let alone the assessment of chiasma frequency.

Because those species with small chromosomes which they studied were held to form quadrivalents with a high frequency, they implied that size is unimportant, not only between, but also within species since: "*the shortest chromosomes are long enough to undergo the reactions necessary to produce quadrivalents*". In this they have missed the point. While it is true that the absolute size of chromosomes in different species is not quite so important for chiasma or multivalent frequency, what is important within the complement of a given species is the relative size. Thus, within complements in which there is a distinct size range, there is unequivocal evidence for a direct correlation between chiasma frequency and chromosome length (cf. Fig. 19). And as we have shown above, this is also exhibited in the distribution of quadrivalents throughout the complement (Fig. 60). We defy even Morrison and Rajhathy to demonstrate how two-thirds of the complement are involved in quadrivalent formation in *Pyrgomorpha*, for example, where only one or two quadrivalents are usually formed in tetraploid cells (Table 5).

In their papers, Morrison and Rajhathy make a number of other rather naïve statements, two of which should not be allowed to pass without comment. One of these doubts the reality of chiasma localisation which they suggest may be *"an artefact arising from our* (i.e. their?) *inability to trace chiasmata in cytological preparations"*. As we have demonstrated in this paper, chiasma localisation is not *"difficult to detect by cytological methods"* in suitable material. And when present it will unquestionably be an extremely important factor in quadrivalent formation. This is particularly true when it is based on an incompleteness of pairing. Even where pairing is complete, however, it can still play a key role. An excellent example of this is provided by *Allium porrum*, where in all the metacentric chromosomes chiasmata are localised proximally: one usually forms on either side of the centromere (Levan 1940). During pachytene quadrivalents were found in most cells but at first metaphase quadrivalents are rarely found (10 IV's in 380 cells analysed). Clearly, as a result of chiasma localisation, homologues which have succeeded in pairing fail to maintain their association (Levan 1940, Fig. 1, p. 436). Where quadrivalents do persist localisation is invariably maintained, resulting in the development of a novel type of quadrivalent (Levan 1940, Fig. 2f—h).

Morrison and Rajhathy also doubt the role of the genotype in determining quadrivalent frequency. That chiasma frequency is under genotypic control is now beyond question (Rees 1961). Roseweir and Rees (1961, 1962) have also shown that there is heritable variation in the frequency with which multivalents form among autotetraploids produced from F_2 families in rye. Genes control the chiasma frequency per cell and hence, in polyploids, the multivalent frequency per cell. Relative size then determines chiasma and multivalent distribution throughout the complement (Fig. 60).

6. Durrant has, on a theoretical basis, derived the association frequencies expected in tetraploids on the assumption of random chiasma formation between the four homologues (Tables 2—4, Durrant

1960). Our data show a significant departure from his expectations. Quadrivalents and pairs of bivalents are much more common than DURRANT'S predictions require. Conversely, trivalents and univalents, or associations of bivalents and two univalents, are extremely rare (Tables 4, 5). The same paucity of these latter two configuration types in tetraploid cells has also been commented on by CALLAN (1949), MORRISON and RAJHATHY (1960a), McCOLLUM (1958) and ROSEWEIR and REES (1962). These discrepancies can be accounted for in terms of the simplified model we have outlined in section 5 (see pg. 142). If pairing, though random, is usually obligatory for all four homologues then the presence of one or two univalents will be rare.

Summary

1. *Schistocerca paranensis* is a South American locust which possesses the same number of chromosomes as the related, though allopatric, species, *S. gregaria* ($2n = 22$ AA $+$ X). The haploid complements also agree in the relative sizes of their members (3 L, 5 M and 3 S), in their mean chiasma frequency and in chiasma distribution throughout each size group. They differ, however, in two principal ways:

a) All three S-chromosomes of *S. paranensis* are consistently smaller than their counterparts in *S. gregaria*.

b) Both species organise nucleoli at first prophase of meiosis. But *S. paranensis* organises only one at a sub-median nucleolar chromomere on the M_6. *S. gregaria* has two such submedian organisers, one on the M_6, the other on the L_3. Sites of nucleolar production are often marked at diplotene by achromatic gaps.

2. One individual of *S. paranensis* proved to be partially asynaptic. In this individual, homologues which succeeded in forming bivalents showed a reduced chiasma frequency and a marked chiasma localisation. The pattern of localisation was characteristic. Single chiasmata were usually sited proximally; two chiasmata were proximal-distal in location. In this, *S. paranensis* differs from *S. gregaria* where, following partial asynapsis, single chiasmata are predominantly distally localised. The presence of univalents at first anaphase in *S. paranensis* led to the suppression of cleavage at telophase. Cleavage was also subsequently suppressed at second division in these restitution cells. This resulted in the formation of giant tetraploid spermatids.

3. An analysis was made of the number and distribution of multivalents throughout the complement of 25 tetraploid cells obtained from several individuals. The chiasma frequencies of 16 of these cells were also scored. This analysis revealed that:

i) Multivalent frequency in tetraploid cells was directly correlated with diploid chiasma frequency and relative chromosome size.

ii) The chiasma frequency of the multivalents, and hence their complexity, was directly related to the diploid chiasma frequency of the homologues involved in the multivalents.

These conclusions were supported by comparisons, both published and unpublished, with a number of other species.

Acknowledgments: Once again it is our pleasure to thank Mr. P. Hunter-Jones of the Anti-Locust Research Centre, London, for continued co-operation and supplying the material. We would also like to express our gratitude post-humously to the members of the D.R.O.I.C.M.A., Kara, all of whom tragically lost their lives in an air disaster and who supplied the specimens of *Duronia tricolor* and *Tylotropidius gracilipes*.

References

Beermann, W.: Der Nukleolus als lebenswichtiger Bestandteil des Zellkernes. Chromosoma (Berl.) 11, 263—296 (1960).

Callan, H. G.: Chiasma interference in diploid, tetraploid and interchange spermatocytes of the earwig, *Forficula auricularia*. J. Genet. 49, 209—213 (1949).

—, and P. Jacobs: The meiotic process in *Mantis religiosa*, L. males. J. Genet. 55, 200—217 (1957).

—, and L. Lloyd: Lampbrush chromosomes of crested newts *Triturus cristatus* (Laurenti). Phil. Trans. B 243, 135—219 (1960).

—, and H. Spurway: A study of meiosis in interracial hybrids of the newt, *Triturus cristatus*. J. Genet. 50, 235—249 (1951).

Corey, H. I.: Chromomere vesicles in orthopteran cells. J. Morph. 66, 299—317 (1940).

Darlington, C. D.: The internal mechanics of the chromosomes II Relational coiling and crossing over in *Fritillaria*. Proc. roy. Soc. B 87, 74—96 (1935). — Crossing over and its mechanical relationships in *Chorthippus* and *Stauroderus*. J. Genet. 33, 465—500 (1936). — Recent advances in Cytology. London: Churchill 1937. — Nucleic acid and the chromosomes. Symp. Soc. exp. Biol. 1, 252—269 (1947). — Evolution of genetic systems, 2nd edit. Edinburgh: Oliver & Boyd 1958.

—, and L. F. LaCour: The genetics of embryo-sac development. Ann. Bot., N.S. 5, 547—562 (1941).

—, and K. Mather: The elements of genetics. London: Allen & Unwin 1949.

Durrant, A.: Expected frequencies of chromosome associations in tetraploids with random chiasma formation. Genetics 45, 779—783 (1960).

Ferguson-Smith, M. A., and S. D. Handmaker: Observations on the satellited human chromosomes. Lancet 1961, 638—640.

Fishberg, M., and H. Wallace: A mutation which reduces nucleolar number in *Xenopus laevis*. The Cell Nucleus, pp. 30—33. London: Butterworth's 1960.

Henderson, S. A.: The Chromosomes of the British *Tetrigidae (Orthoptera)*. Chromosoma (Berl.) 12, 553—572 (1961).

John, B., and K. R. Lewis: Selection for interchange heterozygosity in an inbred culture of *Blaberus discoidalis*. Genetics 44, 251—267 (1959).

— — and S. A. Henderson: Chromosome abnormalities in a wild population of *Chorthippus brunneus*. Chromosoma (Berl.) 11, 1—20 (1960).

—, and B. Naylor: Anomalous chromosome behaviour in the germ line of *Schistocerca gregaria*. Heredity 16, 187—198 (1961).

Jones, K.: Species differentiation in *Agrostis*. I. Cytological relationships in *Agrostis canina*. L. J. Genet. 65, 370—376 (1956).

KAYANO, H.: Chiasma studies in structural hybrids. III. Reductional and equational separation in *Disporum sessile*. Cytologia (Tokyo) 25, 461—467 (1960).

KLINGSTEDT, H.: On some tetraploid spermatocytes in *Chrysochraon dispar (Orthoptera)*. Mem. Soc. pro Fauna et Flora Fenn. 12, 194—209 (1937).

LEVAN, A.: Meiosis of *Allium porrum*, a tetraploid species with chiasma localisation. Hereditas (Lund) 26, 454—462 (1940).

LEWIS, K. R., and B. JOHN: Breakdown and restoration of chromosome stability following inbreeding in a locust. Chromosoma (Berl.) 10, 589—618 (1959).

McCOLLUM, G. D.: Comparative studies of chromosome pairing in natural and induced tetraploid *Dactylis*. Chromosoma (Berl.) 9, 571—605 (1958).

MAEDA, T.: Chiasma studies in the silkworm, *Bombyx mori*. Jap. J. Genet. 15, 118—127 (1939).

MORRISON, J. W., and T. RAJHATHY: Chromosome behaviour in autotetraploid cereals and grasses. Chromosoma (Berl.) 11, 297—309 (1960a). — Frequency of quadrivalents in autotetraploid plants. Nature (Lond.) 187, 528—530 (1960b).

OHNO, S., J. M. TRUJILLO, W. D. KAPLAN and R. KINOSITA: Nucleolus-organisers in the causation of chromosomal anomalies in man. Lancet 1961, 123—126.

-- C. WEILER and C. STENIUS: A dormant nucleolus-organiser in the guinea pig, *Cavia cobaya*. Exp. Cell Res. 25, 498—503 (1961).

REES, H.: Genotypic control of chromosome behaviour in rye. I. Inbred lines. Heredity 9, 93—116 (1955). — Distribution of chiasmata in an "asynaptic" locust. Nature (Lond.) 180, 559 (1957). — Genotypic control of chromosome form and behaviour. Bot. Rev. 27, 288—318 (1961).

—, and B. THOMPSON: Localisation of chromosome breakage at meiosis. Heredity 9, 399—407 (1955).

ROSEWEIR, J., and H. REES: Fertility and chromosome pairing in autotetraploid rye. Heredity 16, 524 (1961). — Fertility and chromosome pairing in autotetraploid rye. Nature (Lond.) 195, 203—204 (1962).

ROTHFELS, K. H.: Chromosome complement, polyploidy and supernumeries in *Neopodiomopsis abdominalis (Acrididae)*. J. Morph. 87, 287—315 (1950).

SINOTO, Y.: Über die Struktur des X-Chromosomes und die Trabant-Chromosomen an *Homoeogryllus japonicus*. Cytologia (Tokyo) 13, 210—213 (1944).

TANAKA, Y.: Genetics of the silkworm *Bombyx mori*. Adv. Genet. 5, 239—317 (1953).

UPCOTT, M.: The origin and behaviour of chiasmata. XII. *Eremurus spectabilis*. Cytologia (Tokyo) 7, 118—130 (1936). — The genetical structure of *Tulipa* III. Meiosis in polyploids. J. Genet. 37, 303—339 (1939).

WHITE, M. J. D.: Tetraploid spermatocytes in a locust, *Schistocerca gregaria*. Cytologia (Tokyo) 5, 135—139 (1934). — An extreme form of chiasma localisation in a species of *Bryodema (Orthoptera; Acrididae)*. Evolution 8, 350—358 (1954a). — Animal cytology and evolution, 2nd Edit. Cambridge: Univ. Press 1954b.

26

FERTILITY AND CHROMOSOME PAIRING IN AUTOTETRAPLOID RYE

J. Roseweir and H. Rees

IT is generally accepted that the fertility of auto-tetraploids depends in part on the types and distributions of chromosome associations at meiosis. On this basis selection for heritable change in the distributions of the associations should be effective in improving fertility. There is, however, little experimental evidence to confirm this[1-3]. On the contrary the evidence of Morrison[4] and of Morrison and Rajhathy[5,6] indicates that selection would be ineffective. They find no evidence either for heritable variation in the pattern of chromosome association at meiosis or for its effect on fertility.

These results are surprising. There are at least two good reasons for expecting heritable variation in the chromosome association pattern: (1) Theoretically, the types and frequencies of the associations: quadrivalent, trivalent plus univalent, bivalent pair, bivalent plus two univalents and four univalents, must depend partly on chiasma frequency[7]. (2) Chiasma frequencies are genotypically controlled[8].

That both factors are relevant and operate to control the fertility of autotetraploids is in fact established by our own investigations on rye. The results which follow here are from induced auto-tetraploids of two F_2 families derived from crossing three diploid inbred rye lines in pairs. Firstly, the range of chiasma frequencies found in these and other F_2 families indicates segregation of genes controlling chiasma frequencies. Secondly, when quadrivalent and univalent frequencies are plotted against the chiasma frequencies of the F_2 plants, as in Fig. 1 A and B, it will be seen that the distributions of the various chromosome configurations are dependent on the chiasma frequencies. In each case the regressions are significant. The frequencies of the other associations show exactly the same dependence on chiasma frequencies. Clearly the adjustment of chromosome association frequency is a consequence of change in chiasma frequency which, as we have inferred earlier, is genotypically controlled. It will also be observed in Fig. 1 that whereas in both F_2 families the association frequencies change with chiasma frequencies, 6 × 13, over the same range of chiasma frequencies, has more quadrivalents and more univalents than

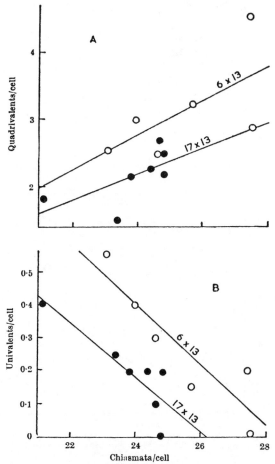

Fig. 1. Regression of association frequencies on chiasma frequencies in the two F_2 families. *A*, Quadrivalent distribution; *B*, univalent distribution calculated from the frequencies of the associations with univalents, III. I and II. 2I

17×13 ($P = 0\cdot001$). There is, therefore, variation in the mean number of quadrivalents and univalents in the two families that is partly independent of their chiasma frequencies. This also must be assumed to be heritable, and provides an additional source of variation on which selection could act in altering the pattern of chromosome associations.

When we come to consider the relation between chromosome behaviour and fertility it is surely to be expected that variation in univalent frequency alone must influence the viability of the gametes produced.

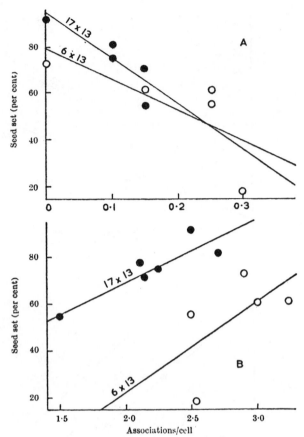

Fig. 2. Regression of fertility (percentage of seed set) on association frequencies in the two F_2 families. *A*, Trivalent frequencies;
B, quadrivalent frequencies

Also since many of the univalents are associated with
trivalent formation a negative correlation between
trivalent frequency and fertility should be expected.
In Fig. 2 *A* fertility, measured as per cent seed set, is
plotted against trivalent frequency. The regressions
are significant and a negative correlation is confirmed.
When, as in Fig. 2 *B*, the per cent seed set is plotted
against quadrivalent frequency the regressions are
positive.

From these results it is evident that the fertility
of autotetraploid rye is dependent on the pattern of
chromosome association at meiosis. Since the association pattern is under genotypic control it follows that
selection for high fertility in rye must be effective

and should be based on increasing the quadrivalent frequency and reducing the trivalent frequency. Both, as we have seen, would result from increasing the chiasma frequency. Finally, it is worth pointing out that the somewhat surprising relation between high quadrivalent frequency and high fertility found in rye may not hold for all other autotetraploids. In rye the chiasmata are distal and relatively few in number. This allows for efficient and regular separation of quadrivalent chromosomes at first anaphase of meiosis.

One of us (J. R.) acknowledges the award of an Agricultural Research Council postgraduate studentship.

[1] Gilles, A., and Randolph, L. F., *Amer. J. Bot.*, **38**, 12 (1951).
[2] Bremer, G., and Bremer-Reinders, D. E., *Euphytica*, **3**, 49 (1954).
[3] Hilpert, G., *Hereditas* (Lund), **43**, 318 (1957).
[4] Morrison, J. W., *Canad. J. Agric. Sci.*, **36**, 157 (1956).
[5] Morrison, J. W., and Rajhathy, T., *Chromosoma* (Berlin), **11**, 297 (1960).
[6] Morrison, J. W., and Rajhathy, T., *Nature*, **187**, 528 (1960).
[7] Durrant, A., *Genetics*, **45**, 779 (1960).
[8] Rees, H., *Bot. Rev.*, **27**, 288 (1961).

27

Copyright ©1940 by Hereditas, Institute of Genetics, Lund, Sweden
Reprinted from Hereditas **26**:454–462 (1940)

MEIOSIS OF ALLIUM PORRUM, A TETRA-PLOID SPECIES WITH CHIASMA LOCAL-ISATION

BY *ALBERT LEVAN*

CYTO-GENETIC LABORATORY, SVALÖF, SWEDEN

WHILE meiosis of diploid forms with localised chiasmata has been analysed on several occasions within various Liliaceous genera (*Fritillaria, Allium, Trillium*), this characteristic course of meiosis has not been studied so far in any polyploid form. Such a study affords an approach to problems of very great interest. DARLINGTON (1937) points to one special problem: »In polyploid forms with localised pairing and chiasmata we must expect fewer multivalents than in corresponding forms with complete pairing, since the effective length for pairing is shorter» (l. c. p. 129). This conclusion is drawn from a study of diploid *Fritillaria* species, where the pachytene pairing in forms with local-isation of chiasmata only reaches about 50 % of the total length of the chromosomes. In *Allium*, on the other hand, the pachytene pairing is morphologically complete in localised species such as *Allium fistulosum* (LEVAN, 1933), it is complete even in asynaptic species, where no single chiasma is formed. Thus, in an autopolyploid *Allium* with localised chiasmata a normal frequency of multivalents may be expected at earlier meiotic stages, and this frequency should decrease during the course of the prophase. At metaphase I only such multivalents as are associated by chiasmata should be left, even if the mutual orientation of some bivalents should indicate that earlier they had been joined to quadrivalents.

During my work on the cytology of *Allium* I came across an auto-tetraploid species, *Allium Porrum* L., which showed almost complete localisation of chiasmata at meiosis. I studied several forms of the species, both Swedish commercial varieties and material procured from botanical gardens. All these forms were tetraploids and had localised chiasmata. In one form I found a regular occurrence of two small somatic chromosome fragments, which sometimes paired at meiosis, in other respects no cytological differences were found between the different forms.

In *Allium Porrum* ordinary Navashin fixations of whole buds, preceded by a short dipping in Carnoy, gave better results than smears in osmic acid fixations. The pachytene stage in particular is very beautiful in Navashin fixed material, permitting a detailed study of whole quadrivalents at this stage.

The somatic chromosomes of *Allium Porrum* are of the ordinary *Allium* type, 28 chromosomes have a median centric constriction and 4 chromosomes have a subterminal constriction. These latter are the s_1 chromosomes and their satellite is quite small. The length of the longest chromosomes of *Allium Porrum* is about 10 μ.

I. THE COURSE OF MEIOSIS.

As already mentioned, the prophase stages show complete chromosome pairing. Small unpaired segments may, of course, be found even at mid-pachytene, but they do not occur more frequently towards the chromosome ends. Such unpaired segments are found also in species with random-distribution of chiasmata. The chromomeric structure is very clear at pachytene, and the chromomeres are sometimes rather broad, almost band-like. There is one nucleolus in each cell. At the nucleolus two deeply stained lumps, evidently the end portions of the s_1 bivalents, may be seen quite regularly. No differences in the pairing can be seen between the proximal and the distal parts of the chromosomes. The diplotene loops, however, seem to appear at first in the distal parts of the chromosomes, and this might be an indication of a less intimate pairing in these parts.

During pachytene quadrivalents are easily found in most cells (Fig. 1 *a—b*). The exchange of threads within the quadrivalents usually occurs in one place, but also quadrivalents with two exchanges have been met with. If there is one exchange it may be located in any region of the chromosomes, medially or terminally. It is difficult at this stage to analyse whole cells, but I have often observed more than one quadrivalent in the same cell. Sometimes all the chromosomes seem to be joined into 8 quadrivalents.

At diplotene the quadrivalents are still present in great number (Fig. 1 *c—d*). The spiralisation is prominent and the quadrivalents are held together by the torsion. During diakinesis, however, many of the earlier quadrivalents are seen to fall apart into pairs of bivalents (Fig. 1 *e—h*). Now the remarkably regular distribution of chiasmata is clearly seen: always 2 chiasmata per bivalent, one at each side of the centromere. Owing to the size of the chromosomes and their great number diakinesis is not a suitable stage for studying the quadrivalents.

At the first metaphase the chromosomes are arranged regularly into an equatorial plate. The characteristic structure of the bivalents is now clear. Owing to the chromosome contraction, the two chiasmata of each bivalent have been pulled together close to the centromere, so the chromosomes seem to touch each other only at one point. In

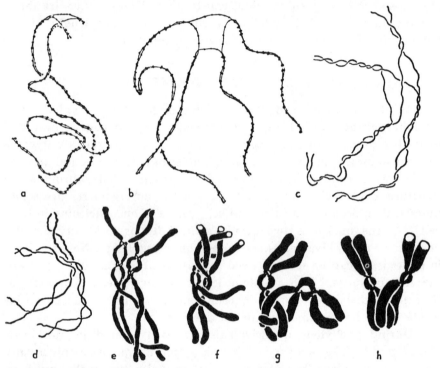

Fig. 1. *a—b*: quadrivalents at pachytene, *c, d*: at diplotene, *e—h*: quadrivalents at diakinesis, held together only by torsion. — × 3900.

polar view (Fig. 2 *a*) it is difficult to see the chiasmata but in side-view (Fig. 2 *b*) the chromatid arrangement is made quite clear. Most plates consist of only this kind of bivalents. The regularity is striking and gives a good illustration of DARLINGTON's term »semi-clonal» inheritance.

There may occur, however, certain exceptions to the scheme. The most common exception is the occurrence of bivalents with only one chiasma (Fig. 2 *c—d*). In one pollen sac the following frequency of such bivalents was counted:

Number of deviating bivalents:	0	1	2	3	4	5	Total	M/cell
Number of cases:	7	6	9	3	5	1	31	1,9

Compared with the conditions in *Allium fistulosum*, a diploid species with chiasma localisation, the frequency of deviating bivalents is rather high. *Allium fistulosum* had from 0,15 to 0,55 such bivalents per cell. In the next chapter it is suggested that the formation of pachytene

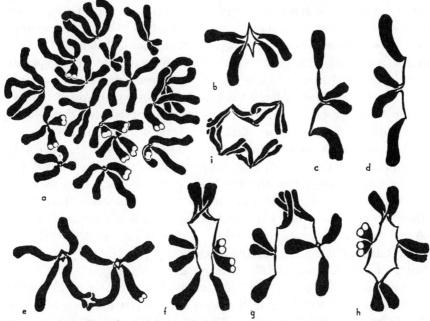

Fig. 2. *a*: metaphase I, a plate consisting of 16 cruciform bivalents seen from the pole, *b*: side-view of a cruciform bivalent, *c*, *d*: bivalents with one chiasma, *e*: a quadrivalent held together by a non-localised chiasma, *f*, *g*: localised chain quadrivalents, *h*: a localised ring quadrivalent, *i*: a ring quadrivalent at anaphase I. — *a*: × 1400, *b—i*: × 2800.

quadrivalents may predispose to the origin of pairs of bivalents with one chiasma.

In these bivalents with only one chiasma, usually the portion between the centromeres and the chiasma is extended and is somewhat narrower than the rest of the chromosomes. The chiasma is often more terminalised than in the cruciform bivalents within the same plate. The chromatid pairing evidently yields more easily under the strain from the centromeres in the bivalents with only one chiasma. In the cruciform bivalents, where four chromatids resist the strain of the centromeres, the chiasmata seem to be more stationary.

Quadrivalents are present so rarely at metaphase I that an intense study was needed merely to demonstrate their occurrence. This study was made difficult by the large size of the 64 long cross-arms present, which quite fill up the equatorial plates, so that the exact analysis of each element is often rendered impossible. It was detected, however, that now and then two cross-arms joined by a clear-cut chiasma were stretched out extremely far towards one pole. In the cases where the connections of these arms could be followed also inside the plate they were found to form each one chiasma with two other chromosomes. In fact there were present instances of chain quadrivalents of a type hitherto unrecorded (Fig. 2 *f*—*g*). The chiasmata of these quadrivalents were gathered as close to the four centromeres as possible. After an extensive search the corresponding ring-type of quadrivalents was also found (Fig. 2 *h*). These rings were observed only three times. I wish to point out, however, that the occurrence of quadrivalents is probably somewhat more frequent than these data indicate, since they can be demonstrated only in especially favourable cases. Even if this is taken into account the quadrivalents must be very rare at metaphase I, and certainly much rarer than in autotetraploids with random-distribution of chiasmata. I counted in one slide 6 quadrivalents in 250 analysed cells and in another slide 4 quadrivalents in 130 cells. In normal tetraploids of *Allium* the average frequency is 1—5 quadrivalents per cell, i. e. about 100 times greater frequency.

The appearance of the quadrivalents is characteristic. They differ from ordinary quadrivalents in the same qualities as localised bivalents differ from ordinary bivalents. Thus the chiasmata of the quadrivalents are very close to the centromeres, which brings about the formation of large cross-arms at each chiasma. The orientation on the spindle of the quadrivalents is typical, the 4 centromeres form a rectangle with 2 corners at each side of the equator. Neighbouring centromeres are orientated towards the same pole. Anaphase formations indicate that sometimes also zig-zag arrangement may occur, but such quadrivalents were not observed at metaphase. Anaphase of a ring quadrivalent is seen in Fig. 2 *i*.

It was often noticed that in the quadrivalents the portions between the centromeres and the chiasmata were longer and more extended than the corresponding portion of the cruciform bivalents. The chiasmata of the quadrivalents were evidently more terminalised. This is probably due to the same condition as that causing the greater terminalisation of the bivalents with one chiasma. It may be that also quadrivalents with their centromeres more close together really occurred, they would anyhow be very difficult to analyse. Formations which might be such quadrivalents were sometimes noticed, but they were interpreted as interlockings of two bivalents.

In the same manner as in *Allium fistulosum*, the localisation of the chiasmata was not absolute, non-localised chiasmata were observed also in *Allium Porrum*, although very seldom. Such chiasmata could give rise to quadrivalents of a new type. They were built up of two bivalents in which two cross-arms, one from each bivalent, were joined to a subterminal chiasma (Fig. 2 e). This chiasma did not influence the orientation of the quadrivalent on the spindle. The frequency of non-localised chiasmata was lower than in *Allium fistulosum*, where some forms had up to 1,7 non-localised chiasmata per cell. In one slide of *Allium Porrum* 3 such chiasmata were counted in 250 cells, and their average occurrence probably does not exceed 1 % of the cells.

Anaphase I and the second division take place normally. The pollen grains contain 16 chromosomes, so the few quadrivalents of meiosis evidently give rise to but slight disturbances.

II. DISCUSSION.

In the present paper a record is given of meiosis of an autotetraploid *Allium* species with almost absolute localisation of the chiasmata. DARLINGTON (1939) regards this proximal type of localisation only as a special case. Its type depends on the location of the original segment of pairing at zygotene. In other cases, for instance, tetraploid species of *Tradescantia*, chiasmata are formed only at the chromosome ends, due to the fact that the pairing starts distally.

Another independent genetic variable determining the meiosis type is the time limit of effective pairing. The time limit determines the degree of localisation. If the time of effective pairing is unlimited, the chromosomes will be paired along their whole length and their chiasmata will be distributed at random. The reason why the ends of the chromosomes of the proximal type of localisation do not form any chiasmata is, still according to DARLINGTON, not due to a precocious reproduction of the chromosomes in these parts, as HUSKINS and SMITH (1934) suggested, but is instead due to the time limit, which cuts off the effective pairing before it has reached the whole chromosome length.

The shorter this time limit, the more absolute is the localisation of chiasmata. In such asynaptic forms as *Allium amplectens* (LEVAN, 1940) the time limit must be nil. Since, however, a clear morphological pachytene pairing is seen both in localised and asynaptic *Allium* species;

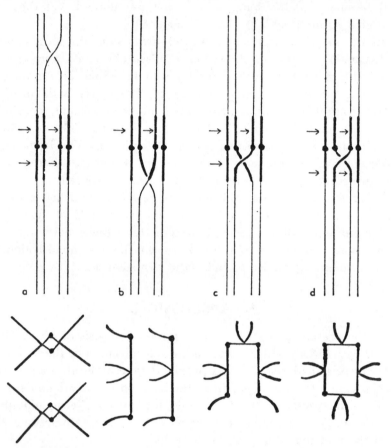

Fig. 3. Scheme of the possible cases of partner exchange in the pachytene quadrivalents and the result at metaphase I. Chiasmata are formed at the arrows. For further explanation see the text.

a distinction must be made between the apparent, visible pairing and the effective pairing, i. e. the pairing which gives rise to chiasmata.

It is clear that the original segment of pairing in such a species as *Allium Porrum* cannot be localised to a single exact point of the four homologous chromosomes. In that case quadrivalents could hardly be formed. The frequent occurrence of quadrivalents at pachytene indicates, in my opinion, that the pairing starts in different places even in chromosomes with absolute localisation of chiasmata. A quadrivalent can be formed only if the pairing starts in at least two places within different pairs of the 4 homologous chromosomes. These starting points are possibly located close to the centromeres in *Allium Porrum*, otherwise the conclusion must be drawn that the zone of effective pairing is not always the zone of the earliest pairing.

However this may be, it is certain that quadrivalents are formed in a greater number at pachytene than survive until metaphase. In later stages chiasmata are found only on both sides of the centromeres. If we suppose, in accordance with DARLINGTON, that chiasmata arise at the same place as they are found at metaphase I the following possibilities are valid within each four-group of homologous chromosomes (see Fig. 3, where each pachytene chromosome is represented by a thickly drawn central portion corresponding to the zone of effective pairing, and thinner end-portions, where chiasmata are formed only in exceptional cases). If the exchange of partners can occur anywhere within different parts of the chromosomes, it is evident that the commonest case will be Fig. 3 *a*, where the exchange occurs outside the central portion. This leads to the formation of 2 typical, localised bivalents, which in earlier stages are often joined by a torsion pairing, but which are free at metaphase I. If the exchange of partners takes place close enough to the central zone (Fig. 3 *b*), it is probable that the exchange will interfere with the pairing of the central zone on that side of the centromere, so that chiasmata can be formed only on the other side of the centromere. The result will be 2 bivalents with one chiasma each. This is probably the cause of the relatively high frequency of this kind of bivalents in *Allium Porrum*, as compared with diploid species with localised chiasmata. If in the former case the pairing is inhibited only in one pair of threads there will originate a localised chain quadrivalent with 3 chiasmata (Fig. 3 *c*). Finally, if the exchange of partners occurs exactly at the centromeres, so that chiasmata can be formed on both sides of the exchange, a localised ring quadrivalent will result. It is quite comprehensible that this last case must be very rare, since it depends for its realisation on the fulfilment of quite special conditions.

SUMMARY.

Meiosis is examined in *Allium Porrum*, an autotetraploid species with complete localisation of chiasmata. Quadrivalents are formed rather frequently at zygotene, but most of them disappear before metaphase I is reached. Chiasmata are formed only in the neighbourhood of the centromeres. A new type of metaphase quadrivalents is described: chains and rings with localised chiasmata.

Svalöf, January 20th, 1940.

———

LITERATURE CITED.

1. DARLINGTON, C. D. 1937. Recent advances in cytology. 2nd Ed. — London.
2. — 1939. The evolution of genetic systems. — Cambridge.
3. HUSKINS, C. L. and SMITH, S. G. 1934. Chromosome division and pairing in *Fritillaria Meleagris*: The mechanism of meiosis. — Journ. of Gen. 28: 397—406.
4. LEVAN, A. 1933. Cytological studies in *Allium*, IV. *Allium fistulosum*. — Svensk Bot. Tidskr. 27: 211—232.
5. — 1940. The cytology of *Allium amplectens* and the occurrence in nature of its asynapsis. — Hereditas XXVI: 353—394.

Part VII

QUANTITATIVE METHODS FOR PREDICTING AUTOPLOID MEIOTIC CONFIGURATIONS

Editors' Comments
on Papers 28 and 29

28 **JACKSON and CASEY**
Cytogenetic Analyses of Autopolyploids: Models and Methods for Triploids to Octoploids

29 **JACKSON and HAUBER**
Autotriploid and Autotetraploid Cytogenetic Analyses: Correction Coefficients for Proposed Binomial Models

The two papers of Part VII address the problem of predicting meiotic configurations expected in autopolyploids. An earlier article by Driscoll and associates (1979) gave tables and equations for predicting meiotic configurations in wheat and wheat hybrids, and these methods were later utilized by Alonso and Kimber (1981) and Kimber and Alonso (1981) in deriving methods for analysis of polyploid hybrids to determine genome affinities. Driscoll and coworkers (1979) proposed that their method also applied for autopolyploids.

Paper 28 provides binomial equations for meiotic configurations expected in autotriploids through autooctoploids. The methods differ from those used by Driscoll and colleagues in two important aspects. First, chiasmata were allocated after synapsis had occurred. Second, the chiasmata were distributed in a nonrandom fashion when the chiasmata number was reduced to the number of theoretical bivalents or their equivalents. Although binomial distribution of chiasmata implies their complete absence in some bivalents or their equivalents, the paper stresses reservations that such a system exists in normal and natural populations.

Data from synthetic and natural autotetraploids soon showed that the reservations concerning random distribution of chiasmata were well-founded. *Haplopappus* autoploids did not fit either the completely random or the partly random model, and the method proposed by Driscoll gave even greater deviation from observed values. Jackson and Hauber (Paper 29) developed equations to allocate chiasmata in such a way that all potential bivalents or their equivalents had at least one chiasma. Only after this condition is met are the remaining chiasmata distributed randomly among the paired configurations according to their expected frequencies. This paper also offers a solution to another problem—the completely random distribution of chiasmata in a pachytene quadrivalent with only two chiasmata available. If the chiasmata are randomly distributed among the four quadrivalent arms, a trivalent and univalent is expected two-thirds of the time. However, such disparate autotetraploids as found in *Haplopappus* and *Triticum* do not have the predicted number of trivalent-univalent configurations. The lack of these configurations is apparently caused by the preferential localization of the two chiasmata in opposite arms of the pachytene quadrivalent so that at diakinesis only two chain bivalents result. This problem is corrected in the equations by theoretically converting expected trivalent-bivalent configurations to two chain bivalents. Data we have analyzed since Paper 29 was published indicate the equations are highly accurate in terms of giving good agreement between observed and expected meiotic configuration frequencies.

REFERENCES

Alonso, L. C., and G. Kimber, 1981, The Analysis of Meiosis in Hybrids. II. Triploid Hybrids, *Can. J. Genet. Cytol.* **23:**221–234.

Driscoll, C. J., L. M. Bielig, and N. L. Darvey, 1979, An Analysis of Frequencies of Chromosome Configurations in Wheat and Wheat Hybrids, *Genetics* **91:**755–767.

Kimber, G., and L. C. Alonso, 1981, The Analysis of Meiosis in Hybrids. III. Tetraploid Hybrids, *Can. J. Genet. Cytol.* **23:**235–254.

Copyright ©1982 by the Botanical Society of America

Reprinted from Am. J. Bot. **69**:487–501 (1982)

CYTOGENETIC ANALYSES OF AUTOPOLYPLOIDS: MODELS AND METHODS FOR TRIPLOIDS TO OCTOPLOIDS[1]

R. C. JACKSON AND JANE CASEY

Department of Biological Sciences, Texas Tech University, Lubbock, Texas 79409

ABSTRACT

Methods are presented for determining the frequencies and numbers of various meiotic configurations expected in autopolyploids. This allows one to test polyploids of unknown origin for agreement with expected meiotic configurations. Rejection of the autoploid hypothesis may indicate the presence of *Ph*-like genes or some type of alloploid. The models consider mean chiasma frequencies of 2, 3, and 4 per bivalent for triploids and tetraploids and 2 per bivalent for pentaploids, hexaploids, heptaploids, and octoploids. Literature data for a known autotriploid, autotetraploids, allotetraploids, and allopentaploids were tested against expectations of the models. There was generally good agreement between number of observed autoploid meiotic configuration and those expected in the models.

POLYPLOIDS are usually classified as auto-polyploids, segmental allopolyploids, or true allopolyploids (Stebbins, 1947, 1950, 1970). Criteria for such a classification depend on a considerable amount of information, but a major factor is the kind of chromosome pairing in a hybrid or an individual that is known or presumed to have given rise to a polyploid. This kind of information is generally lacking in studies of natural polyploids.

While it is recognized that among evolving natural populations all degrees of chromosomal differentiation may be found, the terms autopolyploid, segmental allopolyploid, and allopolyploid are semantically useful for comparative purposes even though the terms represent two extremes and the hiatus between them (Jackson, 1976). Also, one must recognize that structural changes or gene mutations affecting chromosome pairing may not affect the external phenotype (Jackson, 1971, 1976; Stebbins, 1971) so that the type of polyploid is independent of taxonomic rank. A complicating factor in polyploid classification is the presence in some populations and species of genes affecting homoeologous chromosome pairing. The best known of these is the dominant *Ph* gene in hexaploid and tetraploid wheats which prevents homoeologous pairing. Instead, only homologous pairing of identical chromosomes is the rule, and only bivalents are observed in normal plants (Okamoto, 1957; Sears and Okamoto, 1958; Riley and Chapman,

1958). When the chromosome arm or segment carrying this dominant gene is removed from a plant, intergenomal pairing occurs and multivalents are produced (Sears, 1976, 1977). *Ph*-like but not necessarily dominant genes have been reported for diploid wheats, and *Ph*-like effects have been suggested for other plant species (Avivi, 1976; Sears, 1976).

Evolutionists, plant breeders, and systematists working with natural polyploids generally have no quantitative methods for cytologically distinguishing between autoploids and alloploids. A major difficulty is determining the correct frequencies of meiotic configurations that are possible at diakinesis or metaphase I. Heretofore, there have been no definitive guidelines, but in general those polyploids with bivalents and no multivalents were considered alloploids while those with multivalents were treated as autoploids or segmental alloploids. We now know that some segmental alloploids may behave as strict alloploids due to *Ph*-like genes, but there has been no testable method to distinguish between autopolyploids and segmental allopolyploids in terms of multivalents.

It seemed a logical construct to us that a prerequisite to classifying polyploids was the development of a method to determine expected pairing frequencies in autopolyploids. It should then follow that behavior deviant from that expected in an autoploid would have to be attributed to *Ph*-like genes, structural changes, or to a more gradual genome divergence which would lead eventually to the extreme of no pairing in an F₁ diploid hybrid but bivalent formation in the derived, true alloploid due to what has been described as preferential pairing or differential affinity.

We have developed autopolyploid models

[1] Received for publication 12 December 1980; revision accepted 17 May 1981.

We thank Professor Gordon Kimber, G. Doyle, L. C. Alonso, and A. Epinasse for their helpful comments on the manuscript.

that predict expected frequencies of various meiotic configurations for triploids through octoploids. The models thus allow a quantitative analysis of the observed and expected meiotic configurations which is amenable to statistical testing to determine if the meiotic behavior fits that of an autoploid. In conjunction with other analyses, the methods will aid in detecting *Ph*-like genes that cause perturbations of random pairing among homologous or homoeologous chromosomes.

While discussing the importance of chiasma frequency in estimating meiotic configurations at a recent polyploidy symposium (Lewis, 1980), Dr. Gordon Kimber called our attention to a paper then in press (Driscoll, Bielig and Darvey, 1979) which proposed a shortcut method very similar to the one we were developing for determining meiotic configuration frequencies in polyploids. After later reading this paper, we believed the quantification method was resolved and subsequently modified one of our tables accordingly (Jackson and Casey, 1980a, table 3). However, continued work on our methods has convinced us that they are more accurate for true autoploids. In addition, our models go beyond the ploidy levels already covered, and for triploids and tetraploids we consider systems with more than two chiasmata per bivalent. More recently, Kimber, Alonso and Sallee (1981), Alonso and Kimber (1981), and Kimber and Alonso (1981) have developed new theoretical methods of quantifying genome relationships. However, these methods utilize the expected meiotic configurations derived by Driscoll et al. (1979) which we believe are incorrect for terms involving two chiasmata for tetraploids and pentaploids as discussed later. Nevertheless, with certain modifications they should be useful for studying genome relationships with a rigor heretofore lacking.

THEORY AND METHODS—It is elementary but important to emphasize that all initial assumptions in the models we propose begin with synapsis at the pachytene stage of meiosis. It is clear also that the meiotic pairing configurations observed at this stage may persist until diakinesis or metaphase I only if there are sufficient and properly placed chiasmata. Calculations of the numbers and placement of chiasmata among the various pachytene configurations are essential for determining the expected frequencies of the configurations.

The kinds and frequencies of pachytene configurations depend upon the number of synaptic sites among homologous arms. The greater the number of synaptic sites, the greater the probability of multivalents with a concomitant reduction in bivalents. The number of synaptic sites may be positively correlated with occurrence of chiasmata if they are effective pairing sites. However, in the models discussed, normal synapsis and meiotic configuration frequencies are assumed, but the number of chiasmata may vary from 0 to N, while numbers of synaptic sites are constant for all arms.

In this paper we examine the results of pachytene pairing followed by random or mostly random distribution of chiasmata and their subsequent effects on meiotic configurations at diakinesis. Throughout all models, we consider that the hypothetical organism has a basic chromosome number of $x = 1$. The chromosomes are all essentially equal in length and are metacentric. At zygotene, each arm may pair at random with any of its homologues. In the one chiasma per arm models, each arm is considered to pair throughout its length with only one other arm (Jackson and Casey, 1980a, b). Models with more than one potential chiasma per arm may have random pairing partner exchanges in multivalents. We assume that no chiasmata are lost by terminalization before diakinesis.

Probability of random distribution of 0 to N chiasmata is described by the binomial $(p + q)^n$ where n is the maximum number of chiasmata expected for a set of chromosomes in a particular model. Exponents of p represent number of chiasmata while q is lack of chiasmata. In the new terms derived from the binomial, capital P is always the chiasma coefficient. This is equal to the c value of Driscoll et al. (1979). The chiasma coefficient is derived by dividing the theoretically possible number of chiasmata per bivalent into the average number of chiasmata observed. Capital Q is $1 - P$ or lack of chiasmata. Exponents of P are power functions representing numbers of possible chiasmata while those of Q are possible lack of chiasmata.

In each binomial expansion, the separate terms are considered individually until all terms and coefficients have yielded their expected diakinesis configurations. To determine the various meiotic configuration frequencies at diakinesis, it is first necessary to consider each of the pachytene configurations and their frequencies. These pachytene frequencies are multiplied by each binomial coefficient to obtain the fraction applicable to a particular pachytene configuration. Then the distribution of the appropriate number of chiasmata for that configuration is determined. The fractions representing each of the derived configurations from a particular pachytene configuration are then multiplied by the pachytene coefficient to give the final diakinesis coefficient for a configuration. The complete results for each term are given so that they may be compared to other methods. The terms are later summed, reduced, and placed in tables. All coefficients have been rounded to four decimal places.

The general method used throughout the paper is as follows. Suppose we are dealing with a tetraploid in which the model calls for a maximum of four chiasmata for the four homologous chromosomes. This is equal to two chiasmata for each of the two bivalents or four per quadrivalent. The descriptive binomial is $(p + q)^4$ which expands to $p^4 + 4p^3q^1 + 6p^2q^2 + 4p^1q^3 + q^4$. Beginning at the left, we designate the terms as 1 through 5 for convenience. Term 1, p^4, has a coefficient of 1 and there are 4 chiasmata. As discussed later, the expected frequency of quadrivalents at pachytene for this model is 2/3 and the expected frequency of two bivalents is 1/3 (Table 1). Since the maximum number of chiasmata is available for each of the four synapsed arms, the pachytene configurations frequencies will persist until diakinesis. Because the coefficient of p^4 is 1, $2/3 \times 1 = 0.6667(P^4)$ for a circle quadrivalent. For the same binomial term, there should be $1/3 \times 1 = 0.3333$ for two circle bivalents, but because there are two bivalents the product is doubled to yield $0.6667(P^4)$ for circle bivalents. Term 2, $4(p^3q^1)$, describes only three chiasmata available. In a pachytene configuration there will be chiasmata formed in only three arms of the quadrivalent which on opening out at diplotene would yield a chain quadrivalent. Thus $2/3 \times 4 = 2.6667(P^3Q^1)$ describes the new term and chiasmata coefficient for chain quadrivalents. If there are two bivalents at pachytene and a limitation of one chiasma per arm, then one bivalent would have two chiasmata and the other could have only one. Thus $1/3 \times 4 = 1.3333(P^3Q^1)$ for one circle bivalent and $1.3333(P^3Q^1)$ for one chain bivalent. Other terms are discussed in the section on autotetraploids. However, it is clear that term 3 with two chiasmata could not yield quadrivalents at diakinesis because there would not be sufficient chiasmata to hold the pachytene quadrivalent together. Instead only trivalents and univalents or bivalents could be produced.

When chiasmata number equals or exceeds theoretical bivalent number in all ploidy levels of our partly random method, the chiasmata are partitioned among bivalents and multivalents so that every configuration has the equivalent of at least one chiasma per bivalent. Numbers in excess of one per bivalent are distributed randomly. If chiasmata are less than one per bivalent, they are distributed randomly among the pairing sites of the bivalents and multivalents. In our random method, chiasmata are distributed randomly among the pairing sites under all conditions.

In some organisms, the chromosomes are of very different sizes and may have significantly different numbers of chiasmata. Under such conditions, the chromosomes are divided into subsets. The basic number (x), chiasma frequencies, and P values are then determined for each subset. This method yields greater accuracy than considering one P value for all chromosomes and is essential under the conditions described.

The following symbols are used in the remainder of this paper: I = univalent; oII = circle bivalent with two chiasmata; cII = chain bivalent with one chiasma; oIV, oVI, oVIII refer to circle quadrivalent, hexavalent, and octavalent with 4, 6, and 8 chiasmata respectively; cIII, cIV, cV, cVI, cVII, cVIII refers to chain multivalents with 2, 3, 4, 5, 6, and 7 chiasmata respectively; cx before any bivalent indicates a complex configuration with more than the number of chiasmata needed for chain or circle formation. The use of cx before a multivalent indicates that it either has more chiasmata than are required for a circle or chain or that the chiasmata are usually interstitial and may range from 3 to 8 in multivalents of triploids and tetraploids. Σ in the tables refers to total or sum of a particular configuration. Primary multivalents refer to synaptic associations at pachytene, but these may not persist until diakinesis because of inadequate numbers of chiasmata. Definitive multivalents are those held together until diakinesis by adequate numbers of chiasmata, as defined by Oksala (1952).

AUTOTRIPLOID ($2n = 3x = 3$)—There are two kinds of pachytene configurations possible, III and II, I. Their frequencies depend on the number of possible synaptic sites. Synaptic sites are considered to vary from 2 to 4 per II or III and the number of chiasmata to range from 0–4, depending on the model used.

Model with 0–2 chiasmata per II or III—The possible permutations of pairing arm arrangements with one synaptic site per arm and no pairing partner exchange are given in Table 1. Each arm may pair randomly with any homologue, and the opposite chromosome arms pair independently of one another. For each arm, this gives three possible pairing combinations so that the permutations can be determined in a 3×3 matrix. The total possibilities

TABLE 1. *Ploidy levels and theoretical chromosomal configurations and frequencies at pachytene. Based on $x = 1$, random pairing of chromosome ends, and one effective pairing site per arm*

Ploidy level	Chromosomal configurations and frequencies
$3x$	III (2/3); II, I (1/3)
$4x$	IV (2/3); 2II (1/3)
$5x$	V (8/15); IV, I (2/15); III, II (4/15); 2II, I (1/15)
$6x$	VI (8/15); IV, II (6/15); 3II (1/15)
$7x$	VII (48/105); VI, I (8/105); V, II (24/105); IV, III (12/105); IV, II, I (6/105); III, 2II (6/105); 3II, I (1/105)
$8x$	VIII (48/105); VI, II (32/105); 2IV (12/105); IV, 2II (12/105); 4II (1/105)

TABLE 2. *Coefficients and terms for diakinesis configurations in autotriploids with 0–4 chiasmata per II or III*

Chiasmata per II or III	Configurations	Coefficients, terms, and x
2	Σ cIII	$0.6667(P^2) \cdot x$
	oII, I	$0.3333(P^2) \cdot x$
	cII, I	$2(P^1Q^1) \cdot x$
	Σ II	$0.3333P(1 + 5Q) \cdot x$
	I[a]	$3(Q^2) \cdot x$
	Σ I	$0.3333(1 + 4P^1Q^1 + 8Q^2) \cdot x$
3	cxIII	$0.8889(P^3) \cdot x$
	cIII	$2(P^2Q^1) \cdot x$
	Σ III	$0.8889P^2(1 + 1.25Q) \cdot x$
	cxII, I	$0.1111(P^3) \cdot x$
	oII, I	$(P^2Q^1) \cdot x$
	cII, I	$3(P^1Q^2) \cdot x$
	Σ II	$0.1111P(1 + 7P^1Q^1 + 26Q^2) \cdot x$
	I[a]	$3(Q^3) \cdot x$
	Σ I	$(1 - 0.8889P^3 - 2P^2Q^1 + 2Q^3) \cdot x$
4	cxIII	$0.963P^3(1 + 2.6922Q) \cdot x$
	cIII	$4(PQ)^2 \cdot x$
	Σ III	$0.963P^2(1 + 1.6922P^1Q^1 + 3.1537Q^2) \cdot x$
	cxII, I	$0.037P^3(1 + 11.0135Q) \cdot x$
	oII, I	$2(PQ)^2 \cdot x$
	cII, I	$4(P^1Q^3) \cdot x$
	Σ II	$0.037P^3(1 + 11.0135Q) + 2P^1Q^2 (1 + Q) \cdot x$
	I[a]	$3(Q^4) \cdot x$
	Σ I	$0.037P^3(1 + 11.0135Q) + 2Q^2 (1 + 0.5Q^2) \cdot x$

[a] This I class results from complete failure of chiasmata in a II or a III.

are nine of which six are a III and three are a II, I. Thus with $2n = 3x = 3$, one would expect at pachytene a III $\frac{2}{3}$ of the time and a II, I $\frac{1}{3}$ of the time (Table 1).

The random distribution of 0–2 chiasmata is $(p + q)^2$ which expands to $p^2 + 2p^1q^1 + q^2$. Term 1 describes a chiasma for each pairing arm so that pachytene frequencies will be the same as those at diakinesis to yield $2/3 \times 1 = 0.6667(P^2)$cIII, $1/3 \times 1 = 0.3333(P^2)$oII and $0.3333(P^2)$I. Term 2 has only one chiasma so pachytene trivalents cannot persist and at diakinesis there is a cII, I so that $2/3 + 1/3 \times 2 = 2(P^1Q^1)$cII and $2(P^1Q^1)$I. Term 3 describes no chiasma so the unpaired chromosomes at diakinesis are derived from a pachytene III or II, I to yield $2/3 + 1/3 \times 1 = 1$ set of three univalents $= 3(Q^2)$I. Collected and reduced terms for all meiotic configurations are given in the two chiasmata per II section of Table 2.

Model with 0–3 chiasmata per II or III—The number of pairing sites is three per II or III while chiasmata may vary from 0 to 3. The number of permutations of zygotene synaptic sites is nine of which 8/9 yields a pachytene III and 1/9 a II, I.

The descriptive binomial is $(p + q)^3$ for chiasmata distribution which expands to $p^3 + 3p^2q^1 + 3p^1q^2 + q^3$. Term 1 reflects the maximum number of three chiasmata so the diakinesis yield is the same as the configuration frequency at pachytene: $8/9 \times 1 = 0.8889(P^3)$cxIII, $1/9 \times 1 = 0.1111(P^3)$cxII, $0.1111(P^3)$I. Term 2 yields $2(P^2Q^1)$cIII, $1(P^2Q^1)$oII and $1(P^2Q^1)$I. Term 3 gives $3(P^1Q^2)$cII and $3(P^1Q^2)$I. Term 4 lacks chiasmata thus has only univalents and is described by $3(Q^3)$I. Collected and reduced coefficients and terms are given in Table 2.

Model with 0–4 chiasmata per II or III—The synaptic sites are four per II or III. Permutations of the four synaptic sites is $(3)^3$ of which 26/27 yields a III and 1/27 a II, I. The descriptive binomial for chiasmata distribution is $(p + q)^4$ which expands to $p^4 + 4p^3q^1 + 6p^2q^2 + 4p^1q^3 + q^4$. Term 1 yields at diakinesis the same meiotic configuration frequencies that occur at pachytene: $26/27 \times 1 = 0.963(P^4)$cxIII, $1/27 \times 1 = 0.037 (P^4)$cxII and $0.037(P^4)$I. Term 2: $3.5555(P^3Q^1)$cxIII, $0.4445(P^3Q^1)$cxII, and $0.4445(P^3Q^1)$I. Term 3: $4(P^2Q^2)$cIII, $2(P^2Q^2)$oII, and $2(P^2Q^2)$I. Term 4: $4(P^1Q^3)$cII, and $4(P^1Q^3)$I. Term 5 lacks chiasmata so the three chromosomes are univalents and give $3(P^4)$I. Collected and reduced terms are given in Table 2.

AUTOTETRAPLOID $(2n = 4x = 4)$—The number of synaptic sites depends on the model used. Chiasmata may vary from 0 to N in the various models presented. Permutations of pairing arm arrangements vary with the model and are given where appropriate

Model with 0–2 chiasmata per II—The number of pairing sites is two per II or four per IV. Permutations of synaptic sites are the same as the two chiasmata per II model of trivalents. At pachytene, there should be a primary IV $\frac{2}{3}$ of the time and 2II $\frac{1}{3}$ of the time (Table 1). The descriptive binomial for chiasmata distribution is $(p + q)^4$ which expands to $p^4 + 4p^3q^1 + 6p^2q^2 + 4p^1q^3 + q^4$. At diakinesis, Term 1, yields the same meiotic frequencies expected at pachytene; $2/3 \times 1 = 0.6667(P^4)$ oIV, $1/3 \times 1 = 0.3333 \times 2 = 0.6667(P^4)$oII. Term 2 gives $2/3 \times 4 = 2.6667(P^3Q^1)$cIV, $1/3 \times 4 = 1.3333(P^3Q^1)$oII, and $1.3333(P^3Q^1)$cII.

Term 3 is treated in two ways, randomly and partly randomly. Both methods have $2/3 \times 6 = 4$ as the fraction of the coefficient allocated to a primary IV and $1/3 \times 6 = 2$ allocated to two II's at pachytene. In both methods, the pachytene IV has four pairing arms but only two chiasmata. There are six ways to randomly distribute the chiasmata among the arms, and this results in $2/3 \times 4 = 2.6667$ cIII, and 2.6667 I; $1/3 \times 4 = 1.3333 \times 2 = 2.6667$ cII at diakinesis.

In the second part of the original coefficient, the two methods differ. Random treatment of the two pachytene II's with four pairing arms yields $1/3 \times 2 = 0.6667(P^2Q^2)$oII, $2/3 \times 2 = 1.3333$ I plus 2.6667 from the primary IV products above give $4(P^2Q^2)$I. The remainder of the two pachytene II coefficient gives $2/3 \times 2 = 1.3333 \times 2 = 2.6667$ cII's which are added to the 2.6667 cII from the pachytene IV breakdown to yield a total $5.3333(P^2Q^2)$cII.

The partly random method of handling the pachytene II coefficient of 2 has as its basic assumption the idea that as long as there is an average of one chiasma per II available, then each II will have a chiasma. This means that I's and oII's will not occur normally. The basic assumption is based on the fact that our examination of large numbers of meiocytes rarely turn up a significant number of univalents. Certainly chiasma failure does not occur regularly in diploids. To the contrary, selection has acted to achieve at least one chiasma per bivalent. Those in excess of one may vary randomly. With the partly random method, the frequency of a III, I is the same as the random method. However, only cII's are allowed for the pachytene II fraction so that the new combined terms and coefficients resulting from the third term are $2.6667(P^2Q^2)$cIII, $2.6667(P^2Q^2)$I, and $6.6667(P^2Q^2)$cII.

Term 4 has only one chiasma available. Thus $2/3 \times 4 = 2.6667$ for a primary pachytene IV which at diakinesis gives a II and two I's;

$1/3 \times 4 = 1.3333$ for two pachytene II's which breakdown to a II and two I's. These yield $2.6667 + 1.3333 = 4(P^1Q^3)$cII and $2.6667 + 1.3333 = 4 \times 2 = 8(P^1Q^3)$I.

Term 5 has no chiasmata and so has one set of four unpaired chromosomes to give $4(Q^4)$I.

The diakinesis coefficients and terms are collected and reduced in Table 3. The random and partly random terms are given separately.

Model with 0–3 chiasmata per II—There are three synaptic sites per II and six per IV with nine permutations of crossover sites, 8/9 IV and 1/9 two II's at pachytene. The descriptive binomial for chiasmata distribution is $(p + q)^6$ which expands to $p^6 + 6p^5q^1 + 15p^4q^2 + 20p^3q^3 + 15p^2q^4 + 6p^1q^5 + q^6$.

The diakinesis configurations and their derived coefficients are the following. Term 1, (P^6): 0.8889 cxIV, 0.2222 cxII. Term 2, (P^5Q^1): 5.3333 cxIV, 0.6667 cxII, 0.6667 oII. Term 3, (P^4Q^2): 12.6667 cxIV, 0.6667 cxII, 3.3333 oII, 0.6667 cII. Term 4, (P^3Q^3) random: 11.5555 cIV, 3.5555 cxIII, 0.2222 cxII, 4.6667 oII, 4.6667 cII, 4 I; partly random: 11.5555 cIV, 3.5555 cxIII, 4.8889 oII, 4.8889 cII, 3.5555 I. Term 5, (P^2Q^4) random: 8 cIII, 2 oII, 10 cII, 12 I; partly random: 8 cIII, 1.3333 oII, 11.3333 cII, 10.6667 I. Term 6, (P^1Q^5): 6 cII, 12 I. Term 7, (Q^6): 4 I. Collected and reduced coefficients and terms for the configurations are given in Table 3.

Model with 0–4 chiasmata per II—There are four synaptic sites per II and eight per IV. The number of synaptic permutations is 27, 26/27 for IV and 1/27 for two II. The descriptive binomial for chiasmata distribution among eight pairing sites is $(p + q)^8$ which expands to $p^8 + 8p^7q^1 + 28p^6q^2 + 56p^5q^3 + 70p^4q^4 + 56p^3q^5 + 28p^2q^6 + 8p^1q^7 + q^8$.

Diakinesis configurations and their derived coefficients are the following. Term 1, (P^8): 0.963 cxIV, 0.0741 cx II. Term 2, (P^7Q^1): 7.7037 cxIV, 0.5926 cxII. Term 3, (P^6Q^2): 26.6667 cxIV, 2.4074 cxII, 0.2593 oII. Term 4, (P^5Q^3): 52.3704 cxIV, 3.6296 cxII, 3.3333 oII, 0.2963 cII. Term 5, (P^4Q^4) random: 58.2222 cxIV, 3.8519 cxIII, 3.037 cxII, 9.7778 oII, 2.9629 cII, 4 I: partly random: 58.2222 cxIV, 3.8519 cxIII, 3.037 cxII, 9.778 oII, 3.037 cII, 3.8519 I. Term 6, (P^3Q^5) random: 30.3704 cIV, 14.0741 cIII, 0.8926 cxII, 10.6667 oII, 10.6667 cII, 15.8519 I; partly random: 30.3704 cIV, 14.0741 cIII, 0.5926 cxII, 10.963 oII, 10.963 cII, 15.2592 I. Term 7, (P^2Q^6) random: 16 III, 4 oII, 16 cII, 24 I; partly random: 16 III, 3.5555 oII, 16.8889 cII, 23.1111 I. Term 8, (P^1Q^7): 8 cII, 16 I. Term

TABLE 3. *Coefficients and terms for diakinesis configurations in autotetraploids with 0–4 chiasmata per II*

Chias-mata per II	Configuration	Coefficients, terms and x	
		Partly random	Random
2	oIV	$0.6667(P^4) \cdot x$	same as partly random
	cIV	$2.6667(P^3Q^1) \cdot x$	same as partly random
	Σ IV	$0.6667P^3(1 + 3Q) \cdot x$	same as partly random
	Σ cIII, I	$2.6667(PQ)^2 \cdot x$	same as partly random
	oII	$0.6667P^3(1 + Q) \cdot x$	$0.6667(P^2) \cdot x$
	cII	$1.3333PQ(1 + 3P^1Q^1 + 2Q^2) \cdot x$	$1.3333PQ(1 + 2PQ + 2Q^2) \cdot x$
	Σ II	$0.6667P(1 + P^2Q^1 + 7 P^1Q^2 + 5Q^3) \cdot x$	$0.6667P(1 + P^2Q^1 + 6P^1Q^2 + 5Q^3) \cdot x$
	I[a]	$4Q^3(1 + P) \cdot x$	$4Q^3(1 + P) \cdot x$
	Σ I	$4Q^2(1 - 0.3333P^2) \cdot x$	$4(Q^2) \cdot x$
3	Σ IV[b]	$0.8889P^3(1 + 3P^2Q^1 + 11.25P^1Q^2 + 12Q^3) \cdot x$	same as partly random
	cxIII, I	$3.5555(P^3Q^3) \cdot x$	same as partly random
	cIII, I	$8(P^2Q^4) \cdot x$	same as partly random
	Σ III, I	$3.5555P^2Q^3(1 + 1.2188Q) \cdot x$	same as partly random
	cxII	$0.2222P^4(P^2 + 3Q) \cdot x$	$0.2222(P^3) \cdot x$
	oII	$0.6667P^2Q^1(1 + 2P^2Q^1 + 4.3333P^1Q^2 + Q^3) \cdot x$	$0.6667P^2Q^1(1 + 2P^2Q^1 + 4P^1Q^2 + 2Q^3) \cdot x$
	cII	$0.6667P^1Q^2(1 + 4.3333P^2Q^1 + 14 PQ^2 + 8Q^3) \cdot x$	$0.6667PQ^2(1 + 4P^2Q^1 + 12P^1Q^2 + 8Q^3) \cdot x$
	Σ II	$0.2222P(1 + P^4Q^1 + 11P^3Q^2 + 34P^2Q^3 + 52P^1Q^4 + 26Q^5) \cdot x$	$0.2222P(1 + P^4Q^1 + 11P^3Q^2 + 32P^2Q^3 + 49P^1Q^4 + 26Q^5) \cdot x$
	I[a]	$4Q^3(1 + 2P) \cdot x$	$4Q^3(1 + 2P) \cdot x$
	Σ I	$3.5555Q^3(1 + 0.375P^1Q^2 + 0.125Q^3) \cdot x$	$4(Q^3) \cdot x$
4	Σ IV[b]	$0.963P^3(1 + 3P^4Q^1 + 17.6912P^3Q^2 + 44.3825P^2Q^3 + 55.4592P^1Q^4 + 30.5373Q^5) \cdot x$	same as partly random
	cxIII, I	$3.8519P^3Q^4(1 + 2.6538Q) \cdot x$	same as partly random
	cIII, I	$16(P^2Q^6) \cdot x$	same as partly random
	Σ III, I	$3.8519P^2Q^4(1 + 1.6538 P^1Q^1 + 3.1538Q^2) \cdot x$	same as partly random
	cxII	$0.0741P^6(1 + 3P^4Q^1 + 22.4885P^3Q^2 + 38.9825P^2Q^3 + 35.9851P^1Q^4 + 6.9973Q^5) \cdot x$	$0.0741P^6(1 + 3P^4Q^1 + 22.4885P^3Q^2 + 38.9825P^2Q^3 + 35.9851P^1Q^4 + 11.0459Q^5) \cdot x$
	oII	$0.2593P^2Q^2(1 + 8.855P^3Q^1 + 31.7083P^2Q^2 + 38.2792P^1Q^3 + 12.7119Q^4) \cdot x$	$0.2593P^2Q^2(1 + 8.855P^3Q^1 + 31.7083P^2Q^2 + 37.1364P^1Q^3 + 14.4261Q^4) \cdot x$
	cII	$0.2963P^1Q^3(1 + 6.2497P^3Q^1 + 31P^2Q^2 + 53P^1Q^3 + 26Q^4) \cdot x$	$0.2963P^1Q^3(1 + 6P^3Q^1 + 30P^2Q^2 + 50P^1Q^3 + 26Q^4) \cdot x$
	Σ II	$0.0741P^7(1 + 6.9973Q) + 8P^1Q^6(1 + 1.5555P) + 2.6667P^3Q^2(1 + 2.9444P^1Q^2 + 7.4445Q^3 - 0.2778P^2Q^1) \cdot x$	$0.074P^7(1 + 6.9973Q) + 2.6667P^3Q^2(1 + 2.9166P^1Q^2 + 7.3346Q^3 - 0.2778P^2Q^1) + 8P^1Q^6(1 + 1.5P) \cdot x$
	I[a]	$4Q^7(1 + 3P) \cdot x$	$4Q^7(1 + 3P) \cdot x$
	Σ I	$3.8519Q^4(1 - 0.0385P^3Q^1 + 0.1538P^1Q^3 + 0.0384Q^4) \cdot x$	$4Q^4(1 - 0.037P^3Q^1) \cdot x$

[a] This class represents I's derived from the parts of the binomial distribution with less than an average of 1 chiasma per II or complete lack of chiasmata. It is referred to in the text as the "paired" I class.
[b] IV's are not broken down into discrete classes because of their complexity.

9, (Q^8): 4 I. Collected and reduced terms and coefficients for the configurations are given in Table 3.

AUTOPENTAPLOID ($2n = 5x = 5$)—*Model with 0–2 chiasmata per II*—The maximum number of chromosomes in a pachytene association is five, but only four arms are available for pairing. The model is based on whole arm pairing with a maximum of one chiasma per arm or four in a V. Expected pachytene frequencies for the various configurations are 8/15 V, 2/15 IV, I, 4/15 III, II, and 1/15 2II, I

(Table 1). The pairing permutations for each arm are 15 to give a 15×15 matrix for meiotic configurations. However, this can be reduced to 15×1 for the same frequencies. The descriptive binomial for chiasmata distribution is $(p + q)^4$ which expands to $p^4 + 4p^3q^1 + 6p^2q^2 + 4p^1q^3 + q^4$.

Diakinesis configurations and their derived coefficients are as follows: Term 1, (P^4) 0.5333 cV, 0.1333 oIV, 0.2667 cIII, 0.4 oII, 0.2 I. Term 2, (P^3Q^1): 1.6 cIV, 1.6 cIII, 0.8 oII, 2.4 cII, 2.4 I. Term 3, (P^2Q^2) random: 2.4 cIII, 0.4 oII, 6.4 cII, 9.2 I; partly random: 2.1333 cIII, 7.3333

TABLE 4. *Coefficients and terms for diakinesis configurations in autopentaploids with 0–2 chiasmata per II*

Configuration	Coefficients, terms and x	
	Partly random	Random
Σ cV	$0.53333(P^4)\cdot x$	same as partly random
oIV	$0.1333(P^4)\cdot x$	same as partly random
cIV	$1.6(P^3Q^1)\cdot x$	same as partly random
Σ IV	$0.1333P^3(1 + 11Q)\cdot x$	same as partly random
Σ cIII	$0.2667P^2(1 + 4P^1Q^1 + 7Q^2)\cdot x$	$0.2667P^2(1 + 4P^1Q^1 + 8Q^2)\cdot x$
oII	$0.4P^3(1 + Q)\cdot x$	$0.4(P^2)\cdot x$
cII	$2.4PQ(1 + 1.2222P^1Q^1 + 0.6667Q^2)\cdot x$	$2.4PQ(1 + 0.667P^1Q^1 + 0.6667Q^2)\cdot x$
Σ II	$0.4P(1 + 5P^2Q^1 + 16.3333P^1Q^2 + 9Q^3)\cdot x$	$0.4P(1 + 5P^2Q^1 + 14P^1Q^2 + 9Q^3)\cdot x$
Σ I	$0.2(1 + 8P^3Q^1 + 34.6667P^2Q^2 + 56P^1Q^3 + 24Q^4)\cdot x$	$0.2(1 + 8P^3Q^1 + 40P^2Q^2 + 56P^1Q^3 + 24Q^4)\cdot x$

cII, 8.1333 I. Term 4, (P^1Q^3): 4 cII, 12 I. Term 5, (Q^4): 5 I. Collected and reduced terms for the configurations are given in Table 4.

AUTOHEXAPLOID $(2n = 6x = 6)$—*Model with 0–2 chiasmata per II*—The maximum number of synaptic sites is one for each set of pairing arms. The model is based on whole arm synapsis, no within-arm pairing partner exchange, and one chiasma per arm. Expected permutations of pairing arms arrangements are 15 so that at pachytene there are 8/15 VI, 6/15 IV, II, and 1/15 3II among the theoretical population of cells and chromosomes (Table 1). The descriptive binomial for chiasmata distribution is $(p + q)^6$ which expands to $p^6 + 6p^5q^1 + 15p^4q^2 + 20p^3q^3 + 15p^2q^4 + 6p^1q^5 + q^6$.

Diakinesis configurations and their derived coefficients are as follows: Term 1, (P^6): 0.53333 oVI, 0.4 oIV, 0.6 oII. Term 2, (P^5Q^1) 3.2 cVI, 0.8 oIV, 1.6 cIV, 2.4 oII, 1.2 cII. Term 3, (P^4Q^2) random: 3.2 V, 0.4 oIV, 6.4 cIV, 4.8 cIII, 3.6 oII, 9.6 cII, 6 I; partly random: 3.2 V, 6.6286 cIV, 4.9143 cIII, 3.5714 oII, 10.3429 cII, 4.9143 I. Term 4, (P^3Q^3) random: 4.8 cIV, 9.6 III, 2.4 oII, 21.6 cII, 24 I; partly random: 3.2 cIV, 11.7333 cIII, 26.9333 cII, 18.1333 I. Term 5, (P^2Q^4): 4.8 cIII, 0.6 oII, 19.2 cII, 36 I. Term 6, (P^1Q^5): 6 cII, 24 I. Term 7, (Q^6): 6 I. Collected and reduced terms and coefficients for the configurations are given in Table 5.

AUTOHEPTAPLOID $(2n = 7x = 7)$—*Model with 0–2 chiasmata per II*—There is one synaptic site and one chiasma per arm and no within-arm pairing partner exchange. The number of permutations of pairing arms is 105 and give expected pachytene configurations and frequencies of 48/105 VII, 8/105 VI, I, 24/105 V, II, 12/105 IV, III, 6/105 IV, II, I, 6/105 III, 2II, 3/105 3II, I (Table 1). The descriptive

binomial for chiasmata distribution is $(p + q)^6$ which expands to $p^6 + 6p^5q^1 + 15p^4q^2 + 20p^3q^3 + 15 p^2q^4 + 6p^1q^5 + q^6$.

Diakinesis configurations and their derived coefficients are the following. Term 1, (P^6): 0.4571 cVII, 0.0762 oVI, 0.2286 cV, 0.1714 oIV, 0.1714 cIII, 0.4284 oII, 0.1428 I. Term 2, (P^5Q^1): 1.3714 cVI, 1.3714 cV, 0.3429 oIV, 2.0571 cIV, 2.0571 cIII, 1.7143 oII, 2.5715 cII, 2.5713 I. Term 3, (P^4Q^2) random: 2.0571 cV, 0.1714 oIV, 5.4857 cIV, 7.8858 cIII, 2.5714 oII, 13.7142 cII, 15.8571 I; partly random: 1.8285 cV, 5.649 cIV, 8.0408 cIII, 2.551 oII, 14.555 cII, 14.9265 I. Term 4, (P^3Q^3) random: 3.4285 cIV, 10.2857 cIII, 1.7143 oII, 25.7142 cII, 40.5714 I; partly random: 2.2857 cIV, 11.1619 cIII, 30.8191 cII, 35.7334 I. Term 5, (P^2Q^4): 4.2856 cIII, 0.4286 oII, 20.5709 cII, 50.1428 I. Term 6, (P^1Q^5): 6 cII, 30 I. Term 7, (Q^6): 7 I. Collected and reduced terms and coefficients are given in Table 6.

AUTOOCTOPLOID $(2n = 8x = 8)$—*Model with 0–2 chiasmata per II*—There is one synaptic site and one chiasma per arm and no within-arm pairing partner exchange. The 105 permutations of pairing arm arrangements give expected pachytene configurations and frequencies of 48/105 VIII, 32/105 VI, II, 12/105 2IV, 12/105 IV, 2II, 1/105 4II (Table 1). The descriptive binomial for chiasmata distribution is $(p + q)^8$ which expands to $p^8 + 8p^7q^1 + 28p^6q^2 + 56p^5q^3 + 70p^4q^4 + 56p^3q^5 + 28p^2q^6 + 8p^1q^7 + q^8$.

Diakinesis configurations and their derived coefficients are the following. Term 1, (P^8): 0.4571 oVIII, 0.3048 oVI, 0.3429 oIV, 0.5714 oII. Term 2, (P^7Q^1): 3.6571 cVIII, 0.6095 oVI, 1.8286 cVI, 1.3714 oIV, 1.3714 cIV, 3.4286 oII, 1.1426 cII. Term 3, (P^6Q^2) random: 3.6571 VII, 7.3142 cVI, 5.4857 V, 2.0571 oIV, 10.9714 cIV, 6.8571 cIII, 8.5714 oII, 13.7142 cII, 8 I; partly

TABLE 5. *Coefficients and terms for diakinesis configurations in autohexaploids with 0–2 chiasmata per II*

Config-urations	Coefficients, terms, and x	
	Partly random	Random
oVI	$0.5333(P^6) \cdot x$	same as partly random
cVI	$3.2(P^5Q^1) \cdot x$	same as partly random
Σ VI	$0.5333P^5(1 + 5Q) \cdot x$	same as partly random
cV	$3.2(P^4Q^2) \cdot x$	same as partly random
oIV	$0.4P^5(1 + Q) \cdot x$	$0.4(P^4) \cdot x$
cIV	$1.6P^3Q^1(1 + 2.1429P^1Q^1 + Q^2) \cdot x$	$1.6P^3Q^1(1 + 2P^1Q^1 + 2Q^2) \cdot x$
Σ IV	$0.4P^3(1 + 3P^2Q^1 + 13.5715P^1Q^2 + 7Q^3) \cdot x$	$0.4P^3(1 + 3P^2Q^1 + 14P^1Q^2 + 11Q^3) \cdot x$
Σ cIII	$4.8P^2Q^2(1 + 0.0238P^2 + 0.4444P^1Q^1) \cdot x$	$4.8(P^2Q^2) \cdot x$
oII	$0.6P^4(1 + 2P^1Q^1 + 4.9523Q^2) + 0.6(P^2Q^4) \cdot x$	$0.6(P^2) \cdot x$
cII	$1.2P^1Q^1(1 + 4.6191P^3Q^1 + 16.4444P^2Q^2 + 12P^1Q^3 + 4Q^4) \cdot x$	$1.2P^1Q^1(1 + 4P^3Q^1 + 12P^2Q^2 + 12P^1Q^3 + 4Q^4) \cdot x$
Σ II	$0.6P(1 + P^4Q^1 + 13.1905P^3Q^2 + 34.8888P^2Q^3 + 28P^1Q^4 + 9Q^5) \cdot x$	$0.6P(1 + P^4Q^1 + 12P^3Q^2 + 30P^2Q^3 + 28P^1Q^4 + 9Q^5) \cdot x$
Σ I	$6Q^2(1 - 0.1809P^4 - 0.9778P^3Q^1) \cdot x$	$6(Q^2) \cdot x$

random: 3.6571 VII, 7.4497 cVI, 5.5534 cV, 1.8637 oIV, 11.1797 cIV, 6.96 cIII, 8.7202 oII, 14.2398 cII, 6.9502 I. Term 4, (P^5Q^3) random: 5.4853 cVI, 10.9714 cV, 1.3714 oIV, 24.6857 cIV, 26.9714 cIII, 11.2571 oII, 51.7714 cII, 48.1145 I; partly random: 3.6567 cVI, 11.4103 cV, 25.8723 cIV, 29.3248 cIII, 11.4021 oII, 56.5421 cII, 41.1922 I. Term 5, (P^4Q^4) random: 5.4857 cV, 0.3429 oIV, 21.9428 cIV, 41.1429 cIII, 8 oII, 91.6953 cII, 119.0858 I; partly random: 3.6571 cV, 17.3714 cIV, 50.7429 cIII, 111.7714 cII, 95.5429 I. Term 6, (P^3Q^5): 6.8571 cIV, 27.4286 cIII, 3.4286 oII, 85.7144 cII, 160 I. Term 7, (P^2Q^6): 6.8571 cIII, 0.5715 oII, 41.1428 cII, 120 I. Term 8 (P^1Q^7): 8 cII, 48 I. Term 9, (Q^8): 8 I. Collected and reduced terms and coefficients are given in Table 7.

ANALYSES—Although there are many published papers on autopolyploids, most of them lack the degree of quantification of meiotic configurations required by the various models presented here. Data needed are the following: 1) number of meiocytes analyzed; 2) mean and maximum chiasma frequency; 3) relative size classes of the chromosomes so that they may be partitioned into smaller subsets if necessary; 4) number and types of multivalents and bivalents, including numbers of chiasmata in each; 5) number of univalents; 6) all configurations for each cell so that like cell classes later may be collected for analysis.

Use of the expected meiotic frequencies tables (2–7) for euploid series—Data needed to

TABLE 6. *Coefficients and terms for diakinesis configurations in autoheptaploids with 0–2 chiasmata per II*

Config-uration	Coefficients, terms, and x	
	Partly random	Random
Σ VII	$0.4571(P^6) \cdot x$	same as partly random
oVI	$0.0762(P^6) \cdot x$	same as partly random
cVI	$1.3714(P^5Q^1) \cdot x$	same as partly random
Σ VI	$0.0762P^5(1 + 17Q) \cdot x$	same as partly random
cV	$0.2286P^4(1 + 4P^1Q^1 + 7Q^2) \cdot x$	$0.2286P^4(1 + 4P^1Q^1 + 8Q^2) \cdot x$
oIV	$0.1714P^5(1 + Q) \cdot x$	$0.1714(P^4) \cdot x$
cIV	$2.0571P^3Q^1(1 + 0.7461P^1Q^1 + 0.1111Q^2) \cdot x$	$2.0571P^3Q^1(1 + 0.6667P^1Q^1 + 0.6667Q^2) \cdot x$
Σ IV	$0.1714P^3(1 + 11P^2Q^1 + 29.958P^1Q^2 + 12.3355Q^3) \cdot x$	$0.1714P^3(1 + 11P^2Q^1 + 30.0052P^1Q^2 + 19.0029Q^3) \cdot x$
Σ cIII	$0.1714P^2(1 + 8.0017P^3Q^1 + 40.9125P^2Q^2 + 61.1219P^1Q^3 + 24.0035Q^4) \cdot x$	$0.1714P^2(1 + 8.0017P^3Q^1 + 40.0082P^2Q^2 + 56.0099P^1Q^3 + 24.0035Q^4) \cdot x$
oII	$0.4284P^4(1 + 2.0016P^1Q^1 + 4.9547Q^2) + 0.4286P^2Q^4 \cdot x$	$0.4284(P^2) \cdot x$
cII	$2.5715P^1Q^1(1 + 1.6601P^3Q^1 + 5.9849P^2Q^2 + 4P^1Q^3 + 1.3333Q^4) \cdot x$	$2.5715PQ(1 + 1.3333P^3Q^1 + 4P^2Q^2 + 4P^1Q^3 + 1.3333Q^4) \cdot x$
Σ II	$0.4284P(1 + 5.0042P^4Q^1 + 29.93P^3Q^2 + 61.94P^2Q^3 + 44.0184P^1Q^4 + 14.0056Q^5) \cdot x$	$0.4284P(1 + 5.0042P^4Q^1 + 28.0149P^3Q^2 + 54.0254P^2Q^3 + 44.0184P^1Q^4 + 13.0056Q^5) \cdot x$
Σ I	$(0.1428P^6 + 2.5713P^5Q^1 + 14.9265P^4Q^2 + 35.7334P^3Q^3 + 50.1428P^2Q^4 + 30P^1Q^5 + 7Q^6) \cdot x$	$(0.1428P^6 + 2.5713P^5Q^1 + 15.8571P^4Q^2 + 40.5714P^3Q^3 + 50.1428P^2Q^4 + 30P^1Q^5 + 7Q^6) \cdot x$

TABLE 7. *Coefficients and terms for diakinesis configurations in autooctoploids with 0–4 chiasmata per II*

Config-uration	Coefficients, terms, and x	
	Partly random	Random
oVIII	$0.4571(P^8) \cdot x$	same as partly random
cVIII	$3.6571(P^7Q^1) \cdot x$	same as partly random
Σ VIII	$0.4571P^7(1 + 7Q) \cdot x$	same as partly random
Σ cVII	$3.6571(P^6Q^2) \cdot x$	same as partly random
oVI	$0.3048P^7(1 + Q) \cdot x$	same as partly random
cVI	$1.8286P^5Q^1(1 + 2.074P^1Q^1 + Q^2) \cdot x$	$1.8286P^5Q^1(1 + 2P^1Q^1 + 2Q^2) \cdot x$
Σ VI	$0.3048P^5(1 + 5P^2Q^1 + 21.4413P^1Q^2 + 10.997Q^3) \cdot x$	$0.3048P^5(1 + 5P^2Q^1 + 20.9967P^1Q^2 + 16.9964Q^3) \cdot x$
Σ cV	$5.5534P^4Q^2(1 + 0.0546P^1Q^1 - 0.3415Q^2) \cdot x$	$5.4857(P^4Q^2) \cdot x$
oIV	$0.3429P^6(1 + 2P^1Q^1 + 4.4351Q^2) \cdot x$	$0.3429(P^4) \cdot x$
cIV	$1.3714P^3Q^1(1 + 4.152P^3Q^1 + 12.8656P^2Q^2 + 8.6669P^1Q^3 + 4Q^4) \cdot x$	$1.3714P^3Q^1(1 + 4P^3Q^1 + 12.0004P^2Q^2 + 10.1393P^1Q^3 + 4Q^4) \cdot x$
Σ IV	$0.3429P^3(1 + 2.9988P^4Q^1 + 28.0385P^3Q^2 + 65.4514P^2Q^3 + 45.6602P^1Q^4 + 18.9974Q^5) \cdot x$	$0.3429P^3(1 + 2.9988P^4Q^1 + 27.995P^3Q^2 + 65.991P^2Q^3 + 59.9918P^1Q^4 + 18.9974Q^5) \cdot x$
Σ cIII	$6.96P^2Q^2(1 + 0.2133P^3Q^1 + 1.2906P^2Q^2 - 0.0591P^1Q^3 - 0.0148Q^4) \cdot x$	$6.8571P^2Q^2(1 - 0.0666P^3Q^1) \cdot x$
oII	$0.5714P^5(1 + 3P^2Q^1 + 12.2611P^1Q^2 + 18.9547Q^3) + 0.5714P^2Q^3(1 + 5P) \cdot x$	$0.5714P^2(1 + 0.0007P^4Q^2 + 0.2991P^3Q^3 + 0.9993P^2Q^4) \cdot x$
cII	$1.1426P^1Q^1(1 + 6.4626P^5Q^1 + 34.4855P^4Q^2 + 77.822P^3Q^3 + 60.017P^2Q^4 + 30.0018P^1Q^5 + 6Q^6) \cdot x$	$1.1428P^1Q^1(1 + 6.0005P^5Q^1 + 30.3022P^4Q^2 + 60.2374P^3Q^3 + 60.0038P^2Q^4 + 30.0017P^1Q^5 + 6Q^6) \cdot x$
Σ II	$0.5714P^1(1 + P^6Q^1 + 19.186P^5Q^2 + 83.91P^4Q^3 + 160.61P^3Q^4 + 135P^2Q^5 + 66P^1Q^6 + 13Q^7) \cdot x$	$0.5714P^1(1 + P^6Q^1 + 18.0017P^5Q^2 + 75.3054P^4Q^3 + 139.4755P^3Q^4 + 135P^2Q^5 + 66P^1Q^6 + 13Q^7) \cdot x$
Σ I	$6.9602P^1Q^2(1 + 0.9182P^4Q^1 + 3.727P^3Q^2 + 12.9878P^2Q^3 + 12.2409P^1Q^4 + 5.8963Q^5) + 8(Q^8) \cdot x$	$8P^1Q^2(1 + 1.0143P^4Q^1 + 4.8857P^3Q^2 + 10P^2Q^3 + 10P^1Q^4 + 5Q^5) + 8(Q^8) \cdot x$

use these tables are the P values (described earlier), maximum and mean chiasma frequency per II, and basic chromosome number (x). Driscoll et al. (1979) have shown how aneuploids may be analyzed, and our tables may be used in the same way. Univalents in the tables are included with the III's with which they occur in the triploids and tetraploids. The I class in tetraploids represents I's caused by chiasma failure in pachytene II's and IV's or only 1 chiasma in a IV and are called "paired" I's in the text.

Detailed procedure for the first autotetraploid example in Table 9 is as follows. The P value is 0.9825, $x = 7$, and there are 45 cells in the sample. Consider that the maximum

TABLE 8. *Analysis of observed and expected numbers of meiotic configurations in an autotriploid of* Collinsia tinctoria $(x = 7)$. *Mean II or III chiasma frequency = 1.8, P = 0.9, sample size = 12 meiocytes. Meiotic data from Dhillon and Garber (1961)*

	Configurations				Proba-bility[a]
	III	II, I	I	χ^2	
Observed	42	39	6		
Expected	45.36	37.80	2.52	0.29	> 0.5

[a] Expected class sizes of less than 5 were not included in χ^2 calculations.

number of chiasmata per II is two so that the terms and coefficients from this part of Table 3 are used for the analysis. The expected number of definitive IV's is $0.6667 \times (0.9825)^4 + 2.6667 \times (0.9825^3 \times 0.0175) = 0.6655 \times 7 = 4.6585$ (mean) $\times 45 = 209.63$. The III, I number is $2.6667 (0.9825^2 \times 0.0175^2) = 0.00079 \times 7 = 0.0055$ (mean) $\times 45 = 0.25$. The number of oII's is $0.6667 \times (0.9825)^4 + 1.3333 \times (0.9825^3 \times 0.0175) = 0.6434 \times 7 = 4.5036$ (mean) $\times 45 = 202.66$. Chain II number is $1.333 \times (0.9825^3 \times 0.0175) + 6.6667 (0.9825^2 \times 0.0175^2) + 4 \times (0.9825 \times 0.0175^3) = 0.0241 \times 7 = 0.16885$ (mean) $\times 45 = 7.60$. Univalent "pairs" resulting from chiasma failure in II's and IV's at pachytene and one chiasma in IV's are calculated as $8 \times (0.9825 \times 0.0175^3) + 4 \times 0.0175^4 = 0.00004 \times 7 = 0.0003$ (mean) $\times 45 = 0.01$. Calculations of the "paired" I class in this way makes it less likely that observed data will be rejected because of the incalculable variable of chiasma failure inherent in a particular genotype, such as chromosome specific asynapsis or desynapsis, not included in the binomial distribution model for crossovers. We doubt that the binomial probability for paired I's is valid for many species; most I's observed probably result from too few chiasmata in a multivalent as discussed elsewhere.

TABLE 9. *Analysis of observed and (expected) numbers of meiotic configurations in autotetraploids (2n = 4x = 28) with a maximum of two chiasmata per II using the partly random method of Table 3*

| Species | No. cells | P value | Numbers of observed and (expected)[d] configurations | | | | | χ^2 probability |
			IV	III, I	oII	cII	I	
Hordeum vulgare[a] (OAC 21)	45	0.9825	209 (209.63)	1 (0.25)	206 (202.66)	4 (7.6)	0 (0.01)	> 0.3
Triticum longissimum[b] (7214)	100	0.9261	453 (452.86)	1 (8.74)	402 (398.08)	90 (77.69)	0 (2.18)	< 0.05
Triticum longissimum[b] (7011)	53	0.9164	180 (238.1)	1 (5.81)	301 (206.27)	77 (47.14)	5 (1.66)	< 0.001
Oenothera affinis[c] (Santa Fe I)	15	0.7714	52 (54.17)	9 (8.71)	39 (39.48)	41 (40.33)	12 (8.89)	> 0.7

[a] Kasha and Sadasiviah (1971).
[b] Avivi (1976).
[c] Laws (1967).
[d] Classes with expected numbers less than 5 are not included in χ^2 calculations.

Testing the models—As mentioned earlier, literature data are lacking for some of the models. Observed data have been used when possible even though the configurations have not been classified as completely as desired. Where means and not total numbers of configurations were given by authors, cell sample size was multiplied by the given means and the answer rounded to the nearest whole number. We have used our partly random methods for all tests because we believe it is more accurate than the random method and more nearly approximates natural genetic systems.

Triploids—Data suitable for testing only the two chiasmata per II model was found in the literature. Although the number of meiocytes sampled was small, observed and expected meiotic configurations were at acceptable levels (Table 8).

Autotetraploids—These are the models which we expect to be most used because of the more frequent occurrence of this ploidy level in nature and in cultivated plants.

Several autotetraploid examples were analyzed. The first type involves plants that have essentially equal-armed chromosomes of a uniform size and two chiasmata per II (Table 9). The partly random method in Table 3 was used for the analyses of expected frequencies.

In Table 9, expected numbers of meiotic configurations closely approximate the observed numbers for *Hordeum vulgare*.

The two examples for *Triticum longissimum* are of particular interest because they deviate from expected. Number 7214 is a line which in crosses with hexaploid wheat brings about intermediate homoeologous pairing, presum-

ably by counteracting the effects of genes suppressing such pairing (Avivi, 1976; Mello-Sampayo, 1971). The autotetraploid of this strain behaves as expected (Table 9) for IV's and oII's, but there is a deficiency of III's and an excess of cII's. This may be caused by two chiasmata IV's having the chiasmata preferentially occur in opposite and not adjacent arms. At diakinesis such a IV would break down to 2 cII's with no III's and I's. Number 7011 contains a suppressor of homoeologous pairing such that in crosses with cultivated wheat low levels of pairing are observed (Avivi, 1976). However, its autotetraploid produces meiotic configuration frequencies which are significantly different from that expected. Multivalents are decreased while II's are increased.

The last example, *Oenothera affinis* (Table 9), as with many members of the genus, has preadapted chromosomes that fit the model of two terminal chiasmata per II. One configuration given in the original data was removed from the calculation because the number of chromosomes analyzed would have been greater than possible. The fit of the remaining observed data with expected meiotic configurations is very good ($P = >0.7$) considering the small sample size of 15 cells.

It would appear from data published on autoploids of other species that *Ph*-like genes may be present in some populations such as various subspecies and races of *Dactylis glomerata* (McCollum, 1958) and in certain inbred strains of rye (Hazarika and Rees, 1967). Additional possibilities are cited by Avivi (1976) and Sears (1976).

One example of an autotetraploid with three chiasmata per II is given in Table 10. The data

TABLE 10. *Analysis of observed and (expected) numbers of meiotic configurations of large (L), medium (M), and small (S) chromosomes with different chiasma frequencies in autotetraploid cells of* Schistocerca paranensis. *Raw data are from 25 cells analyzed by John and Henderson (1962). Maximum II chiasmata are 4 in (L), 3 and 2 in (M), and 2 in (S)*[a]

Chromosomes type	x	P	Number of observed and (expected) configurations				χ^2	Probability
			IV	III, I	II	I		
L	3	0.7500	70 (69.66)	0 (0.96)	8 (8.69)	0 (0.16)	0.056	> 0.95
M	$5^{(2)}_{(3)}$	0.7240 0.9271	92 (87.47)	1 (3.55)	64 (67.27)	0 (1.37)	0.396	> 0.5
S	3	0.5703	13 (14.15)	4 (8.00)	115 (105.68)	6 (0)	2.92	> 0.2

[a] Expected classes with numbers less than 5 are not included in χ^2 calculations.

are only for the 20 medium-sized chromosomes (M) of *Schistocerca paranensis,* a South American locust, carefully analyzed by John and Henderson (1962). Only IV's and II's had large enough sample sizes for a χ^2 test, and these gave a probability >0.5.

The 12 large chromosomes (L) in the tetraploid tissue of *Schistocerca paranensis* had a maximum of four chiasmata per II. The observed and expected configurations for the four chiasmata model gave very good agreement ($P > 0.9$) for IV's and II's which were the only classes large enough to test (Table 10).

The example of *Schistocerca* points to the necessity of analyzing different size classes separately. None of the models gives an acceptable level of agreement for autoploidy when all chromosomes are considered together. However, when treated accurately as subsets, the data fit the models.

The careful and detailed analysis of tetraploid cells of *Schistocerca* by John and Henderson (1962) made the analysis of the small chromosomes (S) possible even though chiasmata distribution was not random. A total of 25 cells were scored for meiotic configurations, but detailed chiasma analysis of the configurations could be made only for 16 of these. Data were given cell by cell which was necessary for the analysis in this case as it was with the M chromosomes. There are a total of 12 S chromosomes ($x = 3$) in the tetraploid cells, but examination of the 25 cells scored showed that apparently only eight of these could result in multivalent formation at diplotene or diakinesis. Thus, for calculating multivalent frequencies, a change to $x = 2$ was necessary for the S chromosomes capable of forming two chiasmata. Furthermore, the relatively low P value should have produced a considerable number of I's, but only three "paired" univalents were observed. This latter indicated that chiasmata were not occurring at random because there was usually one chiasma

per II. The only chiasmata varying at random were those in excess of one per II with $x = 2$. The regular P value was used to calculate the multivalent numbers based on $x = 2$ even though it included the chiasmata of the two II's which apparently had a maximum of one highly localized chiasma per II because for all practical purposes they could not be separated. We further reasoned that because "paired" I's should not occur normally, the expected II number could be calculated as the total chromosome population sampled minus the number involved in expected multivalents as

Expected II =
$$\frac{2n(N) - 0.6667P^2(P^2 + 4Q) \cdot x \cdot 4 \cdot N}{2},$$

where $2n$ = somatic chromosome number, N = number of cells sampled, and 4 is the chromosome number in the IV and III, I configurations. The results of such an analysis are given in Table 10 and show an acceptable probability (>0.2) for the modified system (Table 10). This method is applicable only when P = 0.5 or higher and the chiasmata are limited or localized so that single chiasmata II's normally do not yield I's.

It is quite possible that many other autotetraploid organisms have modified chiasma distribution systems in the sense that I's are extremely rare, and some II's have only one highly localized chiasma or there is only one effective pairing site so that multivalent formation after pachytene is impossible or quite rare.

Allotetraploids—It has been pointed out by Driscoll et al. (1979) that the techniques used to predict pairing in autotetraploids is applicable to allotetraploids if all the genomes are equally distant. If this is true, agreement with the autoploid model should indicate the genomes are equally distant. Table 11 presents an analysis of observed and expected meiotic

TABLE 11. *Analysis of observed and expected number of meiotic configurations in an allotetraploid* Triticum aestivum *(Chinese Spring)* × Secale cereale *(Imperial Rye) 2n = 4x = 28.[a] 200 cells scored by Driscoll et al. (1979). P = 0.03*

Configurations	IV	III, I	oII	cII	I	χ^2	Probability[d]
Observed	0.000	3.0	0.00	182	5,226		
Expected (J&C)[b]	0.000	3.16	0.05	161.28	5,264	2.80	> 0.05
Expected (D)[c]	0.000	4.0	1.07	157.44	5,266	3.97	< 0.05

[a] The hybrid contains 3 genomes from wheat (ABD) and 1 from rye.
[b] Partly random method of Jackson & Casey (1980b).
[c] Method of Driscoll et al. (1979).
[d] Expected numbers less than 5 were not used in χ^2 calculations.

configurations in an allotetraploid hybrid of *Triticum aestivum* $(2n = 6x = 42)$ with *Secale cereale* $(2n = 2x = 14)$ using our partly random method. The F_1 hybrid has four genomes, ABD from wheat and R from rye. The analysis of the hybrid using our partly random method (Table 11) gives agreement of observed and expected numbers with a χ^2 probability greater than 0.05 as does our random method, while the method of Driscoll et al. (1976) would reject the possibility of the genomes being equally divergent if 0.05 is the level of significance desired.

Allopentaploids—No autopentaploid data were available, but Driscoll et al. (1979) gave data on various hybrids of strains of *Triticum aestivum* × *T. variabilis*. The observed meiotic configurations differed significantly from expected with a χ^2 test using both our partly random method and the method proposed by Driscoll et al. (1979). However, Driscoll (1980) has given data on crosses of the same species except *Ph* mutants of *T. aestivum* were used in the crosses. If these mutants truly remove

the effect of the *Ph* gene, and if the genomes of the species crossed are equally distinct from one another, then the meiotic configurations observed should fit our autopentaploid partly random model, assuming the model is correct. Table 12 gives observed and expected numbers of configurations using both our model and the Driscoll et al. (1979) model. With our method, both *ph* 1a and *ph* 2 mutant hybrids give acceptable agreement, indicating that the five genomes possibly are equally divergent. Significantly different results from expected are obtained with the other model for the *ph* 1a mutant hybrid. The hybrid with the high pairing mutant *ph* 1b gene differed significantly from expected, but if one tests only for expected multivalents, II's, and I's, the probability is >0.05. It appears that the *ph* 1b gene caused an increase in III's and II's and a higher than expected proportion of oII's. The excess could have been due to slight preferential premeiotic association of the genomes by sets of two and three genomes.

We do not agree with the conclusion of Driscoll (1980) that there is good agreement be-

TABLE 12. *Analysis of observed and expected number of meiotic configurations in allopentaploid hybrids derived from crossing three Ph mutants of* Triticum aestivum *(2n = 6x = 42)* × T. variabilis *(2n = 4x = 28)[a]*

F_1 type with	No. cells	P value		Meiotic configurations							χ^2 probability
				V	oIV	cIV	III	oII	cII	I	
ph 1a	100	0.15	obs.	0	0	1	21	3	356	2,715	
			(J&C)[b]	0.19	0.05	3.2	30.39	1.75	350.75	2,698	> 0.20
			(DBD)[c]	0.19	0.05	3.2	33	6.98	318	2,740	< 0.01
ph 1b	40	0.65	obs.	17	3	111	55	284			
			(J&C)	26.65	6.66	43.06	87.32	41.52	207.87	307.1	< 0.01
			(DBD)	26.65	6.66	43.06	96.8	48	176	329.6	< 0.001
ph 2	50	0.21	obs.	0	0	0	23	7	232	1,203	
			(J&C)	0.36	0.09	4.62	24.83	2.32	225.6	1,201.12	> 0.70
			(DBD)	0.36	0.09	4.62	30.5	6.75	203.5	1,219.5	> 0.10

[a] Mean meiotic configuration and mutant sources are from Driscoll (1980).
[b] Expected numbers according to our partly random method.
[c] Expected numbers according to the method of Driscoll et al. (1979).
[d] Expected numbers of less than 5 were not included in χ^2 calculations.

tween his expected configuration frequencies and those observed in the allopentaploid hybrids with or without the *Ph* mutants except with *ph* 2. He gave only mean configuration frequencies, but these are subject to superficial misinterpretation until actual numbers are analyzed. We also believe that his conclusion that there is reasonable agreement of observed and expected configurations in autotetraploid *Triticum monococcum* is without support because a test of his data gives a χ^2 of 117.35, df = 5, $P = 0.001$. This is due probably to the fact that diploid *T. monococcum* has more than two chiasmata in some II's, and there is a strong possibility that this is true also in the autotetraploid. If some of the chromosomes in the tetraploid have more than two chiasmata per II, one would expect the higher multivalent and lower II and I frequencies observed. Unfortunately, the analysis was done with a two chiasmata per II model.

In view of the discrepancy between his observed and expected meiotic frequencies in both the pentaploids and the tetraploid, we do not believe that Driscoll (1980) can use these data to exclude Feldman's spatial model of chromosome pairing.

DISCUSSION—In most polyploids tested by the models we propose, there is acceptable to good agreement between observed and expected numbers of meiotic configurations. Those examples with unacceptable deviation from the model predictions can be explained by genes known to affect one kind of association more than others. The best example of this is in the low pairing strain of *Triticum longissimum*. Other similar examples are known from the literature citations given earlier.

The two chiasmata per bivalent model for triploids has meiotic configuration coefficients identical to those of Driscoll et al. (1979). However, our tetraploid model differs significantly from theirs in both our random and partly random version. Analysis of their table 6 of expected frequencies of meiotic configurations for four homologous chromosomes show equivalence of coefficients derived from 4, 3, 1, and 0 chiasmata probabilities. The differences occur with $6(p^2q^2)$ or two chiasmata results. The binomial coefficient indicates there are only six ways to randomly partition two chiasmata. We first determined that $2/3 \times 6$ was the fraction allocated to IV's and $1/3 \times 6$ gave the fraction for two II's at pachytene. The fraction of 6 allocated for IV's is 4/6, but only two chiasmata are available for four pairing arms. Random distribution of the two chias-

TABLE 13. *Comparison of III, II, and I frequencies resulting from the distribution of two chiasmata among 4 and 5 homologous chromosomes with a maximum of one chiasma per arm. The first column (DBD) are results from Driscoll et al. (1979) and the last two columns are from methods proposed herein*

Ploidy level	Configuration	DBD	Random	Partly random
4x	III	3.4286	2.6667	2.6667
	oII	0.8571	0.6667	0.
	cII	3.4286	5.3333	6.6667
	I	5.1429	4.	2.6667
5x	III	2.7692	2.4	2.1333
	oII	0.4615	0.4	0.
	cII	5.5385	6.4	7.7333
	I	9.6923	9.2	8.1333

mata among the four arms can be done in only six ways. Of these, 4/6 lead to a III, I while 2/6 gave two cII's. Thus $4/6 \times 4 = 2.6667$ III, 2.6667 I, and $2/6 \times 4 = 1.3333 \times 2 = 2.6667$ cII's. The second fraction of the coefficient contains two pachytene II's. If two chiasmata are partitioned at random among the four pairing arms, there are only six ways to do it. This yields $2/6$ oII $\times 2 = 0.667$, $8/6$ cII's $\times 2 = 2.6667$, $4/6 \times 2 = 1.3333$ I's. If the second fraction of pachytene II's is treated such that each of the two II's has one of the two chiasmata, the yield is four cII's. This latter method is more conservative, but results so far lead us to believe it is the normal system encountered.

As far as we can determine, Driscoll et al. (1979) partitioned the two chiasmata in seven ways and not the six ways required by random distribution. Their results were 4/7 III, 1/7 oII, 4/7 cII, and 6/7 I. When these fractions are multiplied by the coefficient 6 for the third term of the binomial, the derived coefficients for the meiotic configurations are those in Table 13. Apparently, they did not first partition the original binomial coefficient between pachytene IV's and two II's. We always assume that synaptic fractions must first be allocated and then the distribution of chiasmata considered. However, even if they did not first partition into pachytene configurations, their results are still in error because they used seven ways to allocate the two chiasmata.

As can be seen from Table 13, there is considerable difference between our methods and those of Driscoll et al. (1979) in the coefficients for meiotic configurations derived from the two chiasmata term of the binomial for both tetraploids and pentaploids. How this affects acceptance or rejection of observed data will depend on the average chiasma frequency. The effect probably will be greatest at P values centering around 0.5 and of little significance with high P values because the configurations most affected would be expected in very low frequencies.

It remains to be determined how well our pentaploid, hexaploid, heptaploid, and octoploid models conform to actual observations. However, the logic used in these models was the same as with lower ploidy levels, and they are presented for testing to those with available data.

The proposed models are powerful methods for analyzing natural or synthetic polyploids. However, they cannot solve all problems and will identify only autoploids that have genetic systems on which the models are based. The two chiasmata per bivalent models appear to fit a number of grasses and at least some species of Onagraceae. Undoubtedly other equivalent groups will be found. However, even in the Gramineae, the karyotypes of some taxa have considerable variation in size and centromere position. Furthermore, there is significant variation in mean chiasma frequency among members of a single complement. In *Zea mays* (Darlington, 1934) the average chiasmata number per II may vary from less than 2 to 3.65. In cases of this sort, it will be necessary to divide the chromosomes into subsets and then calculate expected frequencies of each subset as with *Schistocerca* because calculations based on cell means to determine II means will give incorrect results. We believe that many species will have to be analyzed by subsets of chromosomes.

The autoploids analyzed have relatively high chiasma frequencies, but many species do not share this characteristic. Some species probably have evolved genetic systems in which chiasmata are not distributed at random in the sense that I's are not expected as long as there is an average of one chiasma per II. Because of this, models based on completely random chiasmata distribution will have to be modified. In autotetraploids, for example, methods of estimating IV's and oII's are accurate as we have given them, but "paired" I's and III, I's may not occur as expected. Instead, one might expect the I and III, I part of the binomial distribution to be allocated mostly to cII's. If this is true, cII's can be estimated as the sum of chromosomes remaining from the other configurations, divided by 2.

In situations where *Ph*-like genes are suspected, colchicine treatment of premeiotic cells may disrupt the gene directed tendency or certainty of II formation (Driscoll, Darvey and Barber, 1967). If multivalents are then produced by colchicine treatment when they were not present normally, or if there is an increase over normally observed numbers, there is reason to believe *Ph*-like genetic effects are present and that one is not dealing with a true allopolyploid or segmental allopolyploid as the case may be. Further, Kimber and Alonso (1981) have shown that even true alloploid behavior may be exhibited by segmental alloploids due to the low probability of certain pairing arrangements. Thus sample size becomes very important.

Certainly, all problems dealing with classification of polyploids are not solved. Each natural polyploid should be analyzed as carefully as possible using the guidelines proposed. Various models can then be used as needed, but we suspect that many natural examples may not clearly fit one or another model. A combination of models, one for each subset of chromosomes, may be necessary. Moreover, some chromosomes, such as the small subset in *Schistocerca*, may not have random distribution of chiasmata, but models can be constructed that explain their behavior. Each example will need a knowledgeable and thoughtful analysis.

LITERATURE CITED

ALONSO, L. C., AND G. KIMBER. 1981. The analysis of meiosis in hybrids. II. Triploid hybrids. Can. J. Genet. Cytol. 23: 221–234.

AVIVI, L. 1976. The effect of genes controlling different degrees of homoeologous pairing on quadrivalent frequency in induced autotetraploid lines of *Triticum longissimum*. Can. J. Genet. Cytol. 18: 357–364.

DARLINGTON, C. D. 1934. The origin and behavior of chiasmata. VII. *Zea mays*. Z. Indukt. Abstamm. Verebungsl. 67: 96–114.

DHILLON, T. S., AND D. E. GARBER. 1961. The genus *Collinsia*. XIV. A cytogenetic study of four induced autoploids. Indian J. Genet. Plant Breed. 21: 206–211.

DRISCOLL, C. J. 1980. Mathematical comparison of homologous and homoeologous chromosome configurations and the mode of action of the genes regulating pairing in wheat. Genetics 92: 947–951.

———, L. M. BIELIG, AND N. L. DARVEY. 1979. An analysis of frequencies of chromosome configurations in wheat and wheat hybrids. Genetics 91: 755–767.

———, N. L. DARVEY, AND H. N. BARBER. 1967. Effect of colchicine on meiosis of hexaploid wheat. Nature 216: 687–688.

HAZARIKA, M. H., AND H. REES. 1967. Genotypic control of chromosome behavior in rye. X. Chromosome pairing and fertility in autotetraploids. Heredity 22: 317–32.

JACKSON, R. C. 1971. The karyotype in systematics. Ann. Rev. Ecol. Syst. 2: 327–68.

———. 1976. Evolution and systematic significance of polyploidy. Ann. Rev. Ecol. Syst. 7: 209–234.

———, AND J. CASEY. 1980a. Cytogenetics of polyploids. *In* W. Lewis [ed.], Polyploidy: biological relevance, p. 17–44. Plenum Press, Corp., New York.

———, AND J. CASEY. 1980b. Quantitative cytogenetic

analysis of polyploids. Bot. Soc. Amer. Misc. Ser. Publ. 158: 55.

JOHN, B., AND S. A. HENDERSON. 1962. Asynapsis and polyploidy in *Schistocerca paranensis*. Chromosoma 13: 111–147.

KASHA, K. J., AND R. S. SADASIVIAH. 1971. Genome relationships between *Hordeum vulgare* L. and *H. bulbosum* L. Chromosoma 35: 264–287.

KIMBER, G., ALONSO, L. C., AND P. J. SALLEE. 1981. The analysis of meiosis in hybrids. I. Aneuploid hybrids. Can. J. Genet. Cytol. 23: 209–219.

———, AND L. C. ALONSO. 1981. The analysis of meiosis in hybrids. III. Tetraploid hybrids. Can. J. Genet. Cytol. 23: 235–254.

LAWS, H. M. 1967. Cytology of induced polyploids in *Oenothera*, subgenus *Raimannia*. Cytologia 32: 125–141.

LEWIS, W. H. (ED.) 1980. Polyploidy: biological relevance. Plenum Press, New York.

MELLO-SAMPAYO, T. 1971. Promotion of homoeologous pairing in hybrids of *Triticum aestivum* × *Aegilops longissima*. Genet. Iber. 23: 1–9.

McCOLLUM, G. 1958. Comparative studies of chromosome pairing in natural and induced tetraploid *Dactylis*. Chromosoma 9: 571–605.

OKAMOTO, M. 1957. Asynaptic effect of chromosome V. Wheat Inf. Serv. Kyoto Univ. 5: 6.

OKSALA, T. 1952. Chiasma formation and chiasma interference in the Odonata. Hereditas 38: 449.

RILEY, R., AND V. CHAPMAN. 1958. Genetic control of the cytologically diploid behavior of hexaploid wheat. Nature 182: 713–15.

SEARS, E. R. 1976. Genetic control of chromosome pairing in wheat. Ann. Rev. Genet. 10: 31–51.

———. 1977. An induced mutant with homoeologous pairing in common wheat. Can. J. Genet. Cytol. 19: 585–593.

SEARS, E. R., AND M. OKAMOTO. 1958. Intergenomic chromosome relationships in hexaploid wheat. Proc. Int. Congr. Genet. X,2: 258.

STEBBINS, G. L. 1947. Types of polyploids: their classification and significance. Adv. Genet. 1: 403–29.

———. 1950. Variation and evolution in plants. Columbia University Press, New York.

———. 1970. Variation and evolution in plants. *In* Essays in evolution and genetics in honor of Theodosius Dobzhansky, p. 173–208. Appleton-Century-Croft, New York.

———. 1971. Chromosomal evolution in higher plants. Addison-Wesley, Reading, Mass.

Reprinted from *Am. J. Bot.* **69**:644–646 (1982)

AUTOTRIPLOID AND AUTOTETRAPLOID CYTOGENETIC ANALYSES: CORRECTION COEFFICIENTS FOR PROPOSED BINOMIAL MODELS[1]

R. C. JACKSON AND DONALD P. HAUBER

Department of Biological Sciences, Texas Tech University, Lubbock, Texas 79409

ABSTRACT

Correction coefficients have been derived for two-chiasmata models of autotriploids and autotetraploids to correct meiotic configuration frequencies obtained from the binomial method when the P values are ≥ 0.5. In addition, equations are derived for the direct calculation of meiotic configuration frequencies in autotriploids of the two-chiasmata model when P values are ≥ 0.5.

IN A PRECEDING ARTICLE in this issue (Jackson and Casey, 1982), methods and models were proposed which allow one to quantitatively determine the expected meiotic frequencies in autotriploids through autooctoploids. Chiasmata were allocated among the pachytene configurations according to a binomial distribution such that some cells had the maximum number while those at the other end of the distribution received none. It was realized that chiasmata distributions of this sort, although mathematically accurate and acceptable in a model, might not entirely fit natural genetic systems when the frequency was between 1 and about 1.8 per II. A mathematical solution was not apparent then, but one became more urgent when we recently observed a considerable disparity between observed and expected I's in synthetic and natural autoploids of *Haplopappus*. The disparity involved "paired" I's: those observed when homologous I's occur in 3's in autotriploids and in sets of 2 or 4 in autotetraploids. This caused us to doubt that natural genetic systems would permit more than one chiasma in some II's and none in others.

Recent examination of the literature and unpublished data on chiasma frequency in normal diploids revealed that chiasmata are not distributed randomly. Therefore, there is no a priori reason to expect them to be randomly distributed in autopolyploids.

Analysis of chiasmata distribution in normal diploids with a maximum of two per bivalent gave insight as to how they may be allocated in autotriploids and autotetraploids with P values 0.5 or greater. Solutions to the problems for two-chiasmata models for autotriploids and autotetraploids are presented here, but resolutions to the problems for three- and four-chiasmata models and for higher autopolyploids are not yet complete.

Terms and symbols are those used by Jackson and Casey (1982).

AUTOTRIPLOID WITH 1 TO 2 CHIASMATA PER II OR III AND P VALUES OF 0.5 OR GREATER—Since there is a maximum of two chiasmata per II or III, autotriploids that fit this model can be treated as normal diploids with like numbers of chiasmata because there are only two pairing arms in any configuration. Two chiasmata in a diploid give a oII at diakinesis. Two chiasmata in an autotriploid yield either a cIII or a oII. In normal diploids the expected frequency of oII's per cell is $x(1 - 2Q)$ when the maximum chiasmata per II is 2. Therefore, the expected frequency of oII equivalents (cIII's and oII's) in an autotriploid is also $x(1 - 2Q)$.

Because the pachytene frequency of III's is 0.6667, the expected diakinesis frequency per cell of cIII's is:

$$cIII = x(1 - 2Q)0.6667. \qquad \text{Eq. 1}$$

The pachytene frequency of a II, I is 0.3333. Therefore, the diakinesis frequency of a oII, I per cell is:

$$oII, I = x(1 - 2Q)0.3333. \qquad \text{Eq. 2}$$

The frequency of a cII, I per cell at diakinesis can be calculated directly as:

$$cII, I = x(2Q). \qquad \text{Eq. 3}$$

It should be emphasized that equations 1, 2, and 3 should be used in place of those in the two-chiasmata model of Jackson and Casey

[1] Received for publication 26 September 1981; revision accepted 24 November 1981.

We thank the Texas Tech University Graduate School for a Summer Research Assistantship to D. P. Hauber and Dr. Brian Murray for his helpful comments.

TABLE 1. *Equations for number of diakinesis configurations per cell in autotriploids with 1 to 2 chiasmata per II or III and P values of 0.5 or greater*

Configuration	Equation
cIII	$x(1 - 2Q)0.6667$
oII, I	$x(1 - 2Q)0.3333$
cII, I	$x(2Q)$

TABLE 2. *Coefficients, terms, correction coefficient (Cf) and x for diakinesis configurations per cell in autotetraploids with 1 to 2 chiasmata per II and P values of 0.5 or greater*

Configurations	Coefficients, terms, and x
oIV	$0.6667(P^4)Cf \cdot x$
cIV	$2.6667(P^3Q^1)Cf \cdot x$
oII	$0.6667P^3(1 + Q)Cf \cdot x$
cII	$\dfrac{4x - \Sigma^a\ ch\ oIV + cIV + oII}{2}$

[a] Σ ch = sum of chromosomes.

(1982) only when the P value is 0.5 or greater. The univalent error in table 2 of Jackson and Casey (1982) is insignificant at very high P values but becomes excessive as the 0.5 level is approached. Equations 1, 2, and 3 (Table 1) eliminate that error for the two-chiasmata model. When I's occur in observed data in addition to those of oII, I and cII, I configurations, they should be combined as a separate class for Chi-square tests.

For calculations based on P values of 0.5 or greater from table 2 of Jackson and Casey (1982), a correction coefficient can be applied which, as equations 1 and 2, take appropriate numbers of chiasmata from cIII's and oII's and allocate them to chromosomes that normally would be I's, converting them to cII, I configurations. The correction coefficient (Cf) is the oII equivalent frequency divided by that obtained from random distribution of chiasmata (P^2) so that:

$$Cf = \frac{1 - 2Q}{P^2}. \qquad \text{Eq. 4}$$

The Cf is multiplied by the expected numbers of cIII and oII, I in the two-chiasmata model (Jackson and Casey, 1982). The cII, I frequency per cell is obtained from equation 3.

AUTOTETRAPLOID WITH 1 TO 2 CHIASMATA PER II AND P VALUES OF 0.5 OR GREATER— The correction coefficient (Cf) is derived by first considering that the autotetraploid behaves as a normal diploid but with twice as many oII equivalents as:

$$2x(1 - 2Q). \qquad \text{Eq. 5}$$

The expected number of oII equivalents is then divided by those derived from summing the actual number and their equivalents from oIV and cIV configurations according to their binomial expectation: $0.67P^4(2) + 2.67P^3Q^1(1) + 0.67P^4(1) + 1.33P^3Q^1(1)$. These binomially derived oII's yield on multiplying by x the following:

$$oII\ equivalents = 2P^3(1 + Q)x. \quad \text{Eq. 6}$$

Equation 5 divided by 6 on further reduction gives the correction coefficient for autotetraploids as follows:

$$Cf = \frac{1 - 2Q}{P^3(1 + Q)}. \qquad \text{Eq. 7}$$

The Cf is then used as in Table 2. For Chi-square analysis, the product from each configuration is multiplied by the number of cells studied.

Note that Table 2 does not list III, I configurations. In some species this configuration may occur as predicted while in others it is absent, or rare, sporadic and unpredictable. The reason for its unpredictability may be the preferential occurrence of the two chiasmata in opposite arms of the pachytene cross configuration of quadrivalents (Jackson and Casey, 1982). The method of analysis in Table 2 assumes no III, I configurations so when testing data in which they occur, their values are doubled and added to the observed cII class because the opposite arm chiasmata distribution in quadrivalents leads to two cII's.

Jackson and Casey (1982) give data for an autotetraploid of *Triticum longissimum* (7214). According to the binomial method, the data did not fit an autopolyploid model because the expected III, I class was too large and the cII class was too small. By using the methods in Table 2 and adding the III, I and cII classes together, the results are compared in Table 3. In this known autopolyploid, the original method gave a fit to the model of $P < 0.05$ while the modified method gave a $P > 0.70$. In this case, it is clear that most of the deviation involved the III, I and cII configurations. The efficacy and accuracy of the Cf can be seen more readily when the P value is low and considerable I's are expected because the method is designed to compensate for both lack of I's and III, I's. We contend that in normal diploids and autopolyploids, I's do not normally occur as a result of chiasmata failure in paired pachy-

TABLE 3. *Comparison of observed meiotic configurations in an autotetraploid of* Triticum longissimum *(Tl) with P = 0.9261 and a natural tetraploid of the* Haplopappus spinulosus *group (Hs)[a] with P = 0.6823 using the binomial method (calc.) and the binomial plus the correction coefficient (exp.).*

Species	No. cells	oIV	cIV	III, I	oII	cII	I	P
Tl obs.	100	−453−		1	402	90	0	
calc.		−452.86−		8.74	398.08	77.69	2.18	< 0.05
exp.		−452.45−		0	397.72	97.38	0	> 0.70
Hs obs.	24	5	13	0	12	48	0	
calc.		6.93	12.92	6.01	13.39	25.69	10.36	< 0.001
exp.		6.04	11.25	0	11.67	49.75	0	> 0.90

[a] Only sets of four of chromosomes 1 and 2 of the 16 total were used in the analysis because the other eight rarely have two chiasmata per II.

tene II's. If they do occur, their numbers are unpredictable and sporadic. Their cause may be due to environmental effects or to chromosome specific desynapsis which can not be incorporated easily into a model. Table 3 gives data for a natural tetraploid of *Haplopappus spinulosus* in which the binomial method is compared to the modified method using the Cf. The two methods give significantly different results, and the natural tetraploid fits the autoploid model when the Cf method is utilized. The cytological behavior of the natural polyploid is very similar to synthetic tetraploids of this group.

The modifications of the original two-chiasmata models for autotriploids and autotetraploids given here very closely approximate natural genetic systems which distribute the chiasmata so that univalents do not normally occur when P values are 0.5 or greater. The modified methods give a better fit for known autoploids than those proposed earlier, and we recommend appropriate use of the modified methods in the testing of natural triploids and tetraploids.

LITERATURE CITED

JACKSON, R. C., AND J. CASEY. 1982. Cytogenetic analyses of autopolyploids: models and methods for triploids to octoploids. Amer. J. Bot. 69: 489–503.

EPILOGUE

It has become increasingly apparent to us, especially since 1980, that many of the current ideas concerning polyploidy are either erroneous or need to be interpreted in a different fashion. However, factors involved in the formation and evaluation of meiotic configurations in normal autopolyploids could not be more critically examined until their expected frequencies could be determined quantitatively.

Stebbins's classification (Paper 10) was based on the concept that chromosomes will pair preferentially with their most identical homologues. Homologous chromosomes were presumed to have changed, as a result of mutations, in such a way that they were unable to pair as well as complete homologues or had totally lost the ability to synapse. This loss of homology has been ascribed to various structural changes, such as inversions, translocations, duplications, and deletions, as well as to changes at the base level. However, several lines of evidence that have been accumulating suggest that the structural differences are not of major importance in causing a change in homology as applied to polyploidy. Probably the first significant discovery in this respect was the presence of a gene in bread wheat that causes this genomic allohexaploid to behave as a segmental alloploid (Papers 19, 20, and 21). Later, this mutant was localized and described as the dominant *Ph* (pairing homoeologous) gene on chromosome $5B^L$ (Wall et al., 1971). Since the discovery of the *Ph* gene, there have been claims for similarly acting mutants in a number of other genera, but such claims should be viewed with some skepticism until necessary quantitative data, such as indicated in Papers 28 and 29, can be obtained.

The second significant finding was that chromosomes in bread wheat are associated in mitotic cells in such a way that homologues are closer than homoeologues (Paper 22), and later work (Paper 23) showed that this association can be perturbed by the action of colchicine. Finally, a significant difference in quadrivalent frequency is found in autotetraploids from two different genotypes of *Triticum longissimum* (Paper 24).

The development of quantitative methods to predict frequencies of the various meiotic configurations in autopolyploids (Papers 28 and 29) allows one to test quantitatively for normal autoploid behavior. These methods have been used for the analysis of a number of known autotetraploids, and most fit the pairing models very well. The few examples that did not give a fit to the expected meiotic configurations are perhaps the most interesting because there is a high probability that they have mutant genes affecting the placement of the genomes on the nuclear membrane prior to zygotene.

Jackson (1982) has proposed that the kinds of pairing described for both segmental and genomic alloploids are caused by mutations that displace a genome from its normal prezygotene position on the nuclear membrane. This means that genomic alloploids theoretically could be derived from individuals of the same populations. Thus taxonomic rank per se has no bearing on the kind of polyploid derived from a diploid organism. However, there may be a higher probability that nuclear membrane attachment site mutations would characterize different populations, and Wains (1976) has suggested that what he called diploidizing genes may lead to speciation events.

If the above hypothesis is true (Jackson, 1982), then one should expect to find variations in pairing among hybrids and their derived polyploids from the same species. This variation in pairing that occurs can be seen from data presented in Paper 24. Also, it is interesting to note that genetic textbooks that refer to the classical allotetraploid origin of Raphanobrassica show only univalents in the F_1 hybrid but only bivalents in the derived genomic allotetraploid. However, different strains or varieties of *Raphanus sativa* and *Brassica oleracea* give different results when crossed, yielding up to nine bivalents in some microsporocytes. Derived tetraploids may then show varying numbers of quadrivalents. We should expect similar results in most other organisms.

From data available, it is clear that zygotene chromosomes are attached to the nuclear membrane. We infer that prezygotene chromosomes are also attached, as suggested earlier by others (see Avivi and Feldman, 1980, for review). Whatever the mechanism of attachment, it can be disrupted by the action of colchicine or vinblastine sulfate so that so-called homoeologous or "nonhomologous" chromosomes may synapse and form chiasmata. We may now deduce that lack of synapsis in allodiploid hybrids is not due to lack of homology of the two genomes. Rather, the chromosomes are not close enough together for synapsis because their attachment sites on the nuclear membrane are too far apart.

It follows from the above conclusions that pairing relationships in F_1 hybrids and their derived polyploids are not a measure of homology in the true sense. Chromosomes of two genomes could be completely homologous but unable to synapse due to mutation affecting nuclear membrane attachment sites (Jackson, 1982). The term *holoeologous* has traditionally been used to describe residual homoloy of originally completely homologous chromosomes. Homoeogy is generally identified in meiotic analysis by a decrease in pairing om that observed in parentals of F_1 diploid hybrids, or by the presenc of varying numbers of univalents and deviations from the expecte frequencies of various configurations in polyploids. It is now cle that what is meant generally by the term *homoeologous* is not wh s being observed. The term should be discontinued or used only wi the understanding that one is referring to displacement of genom on the nuclear membrane from their original positions.

Loally, we can no longer use the terms *differential affinity* or *prefereal pairing* as classically done in reference to polyploids. The two termean that the more-homologous chromosomes will pair in prefere to synapsis with less-homologous ones. However, in polyploidsreference is involved; chromosomes of the same type that are cl will synapse. If other chromosomes of the same kind are moreant, they are less likely to enter into the same pairing conftion. In the most extreme situations, if two completely homus chromosomes are too far apart when synapsis is initiated, no g can occur, and univalents will be present at pachytene andstages of the first meiotic division. Theoretically, this kind of situ could lead to a misinterpretation of genome relationships in theving way. Suppose that genome A^0 gives rise to genome A^1 by numembrane attachment site (NAS) mutation such that A^1 has bved far enough from the original site that univalents are p about half the time in meiocytes of A^0A^1 hybrids. Finally, a mutation in a subpopulation of A^1 that yields A^2 moves th ent site back toward the ancestral A^0 position. In hybrids A^2 composition, univalents may be lacking or of very cy. If we assume that some morphological difference lated after the origin of A^1 and A^2, then the diploid ty een described as separate taxa. Cytogenetic anal would then lead one to conclude on classic g A^2A^2 is more closely related to the ancestral A more distantly related to the other two taxa ave happened is unknown. However, this hyp l encourage us to use as many additional for phyletic relationships.

inally, we wish to emphasize the possibl ne of the papers and hypotheses discu luce intergenomal pairing and cros

1967 (Paper 23), but to our knowledge insufficient use has been made of this valuable tool. Perhaps it has been thought that this alkaloid was effective only in bread wheat and that this organism with its *Ph* gene was unique. However, colchicine can be utilized for two purposes. First, it can be used to determine if an organism is an ancient polyploid. The production of multivalents at pachytene and diakinesis stages following premeiotic colchicine treatment constitute good evidence for such a polyploid condition. The genus *Helianthus* is believed to be of polyploid origin, and Jackson and Murray (1983) have produced multivalents by premeiotic application of colchicine in so-called diploid species of this genus. We predict similar results for other organisms. Second, use of colchicine can have important consequences in plant breeding, as can be seen from the data on *Helianthus*. This is because genomic alloploids *(AABB)*, which usually do not have recombination between the *A* and *B* genomes, can now be induced to produce such intergenomic recombinants. This ability to produce intergenomal recombinants will aid the plant breeder to produce new genotypic combinations for further selection work.

The presence of nuclear membrane attachment site (NAS) mutations, as deduced from data presented earlier, means a large number of such mutants probably exist in widespread species and in c... by virtue of their isolation from other forms, either spatially ... hoice of growers. It seems likely that such mutations will ... ore easily in the same generation of their origin in sel... ility barriers may decrease the survival probability of su... tcrossers.

...he NAS mutants may be recognized by their asyna... heterozygotes and should fit the random chiasm... r diploids (Jackson, 1982) in that all chrom... of sometimes being univalents. In the... univalents may be present. The plant b that ...duce NAS mutants into commercial that ...suitable inbred lines, and then do ten this ...the F1 hybrids. NAS mutants t situati... ...otentially the most valuable possi... ...oretically should have all ...

Thus, the heterosis pr... ould be preserved i mic ...be highly fertile a olc ...idy per se. Othe ver ...ions to move... umbers of ...

6

F_1 heterozygotes, and polyploids derived from them would have varying numbers of multivalents, depending on the degree of genome displacement. However, any displacement should result in fewer multivalents than expected under normal conditions, and nuclear gene-controlled heterosis should persist longer than in normally behaving autoploids.

We believe that a new era in polyploid research is just beginning.

REFERENCES

Avivi, L., and M. Feldman, 1980, Arrangement of Chromosomes in the Interphase Nucleus of Plants, *Hum. Genet.* **55:**281-295.

Jackson, R. C., 1982, Polyploidy and Diploidy: New Perspectives on Chromosome Pairing and Its Evolutionary Implications, *Am. J. Bot.* **69:**1512-1523.

Jackson, R. C., and B. Murray, 1983, Colchicine Induced Quadrivalent Formation in *Helianthus:* Evidence for Ancient Polyploidy, *Theor. Appl. Genet.* **64:**219-222.

Wains, J. G., 1976, A Model for the Origin of Diploidizing Mechanisms in Polypoid Species, *Am. Nat.* **110:**415-430.

Wall, A. M., R. Riley, and M. D. Gale, 1971, The Position of a Locus on Chromosome 5B of *Triticum aestivum* Affecting Homoeologous Meiotic Pairing, *Genet. Res. Camb.* **18:**329-339.

AUTHOR CITATION INDEX

371

373

SUBJECT INDEX

About the Editors

R. C. JACKSON is a professor in the Department of Biological Sciences of Texas Tech University. He received the A.B. and A.M. degrees from Indiana University in 1952 and 1953, respectively, following a three-year enlistment in the U. S. Air Force. His A.M. thesis work was directed by Professor Charles B. Heiser, Jr., who first interested him in biosystematics and mechanisms of plant speciation. His doctoral work was completed at Purdue University in 1955 under the guidance of Dr. Arthur T. Guard. Jackson was first a member of the biology department faculty at the University of New Mexico and two years later he moved to the Department of Botany at the University of Kansas, where he taught graduate courses in cytogenetics and biosystematics for twelve years. From 1971-1978 he was chairman of the Department of Biological Sciences at Texas Tech University. At his present position he has taught courses in general genetics and cytogenetics. His research includes biosystematics and genetics of *Machaeranthera* and *Haplopappus* and the development of mathematical models for quantitative cytogenetic analyses of meiotic configurations in diploids and polyploids.

Dr. Jackson's publications include reviews of the role of karyotype analysis in systematics, cytogenetics of polyploids, and the systematic and evolutionary significance of polyploidy. Other articles concern cytogenetic and systematic relationships of plant species.

DONALD P. HAUBER is a part-time instructor and a doctoral candidate in the Department of Biological Sciences at Texas Tech University. Throughout his graduate and undergraduate studies he has done extensive research in cytogenetics. He received the B.S. and M.A. degrees from the University of Kansas in 1978 and 1980, respectively, in cell biology and botany, studying under Dr. W. L. Bloom. His thesis topic concerned translocation heterozygosity in *Clarkia*. Under the direction of Dr. R. C. Jackson at Texas Tech University he has concentrated his research efforts on the study of both synthetic and natural polyploidy in a variety of plant species, focusing on quantitative meiotic analysis, which allows distinction of autoploids from alloploids. His interests also include the agronomic potential of colchicine in polyploid crops.